Cell Motility

From molecules to organisms

Cell Motility

From molecules to organisms

Edited by

Anne Ridley
Ludwig Institute for Cancer Research, London, UK

Michelle Peckham
University of Leeds, UK

Peter Clark
Imperial College London, UK

John Wiley & Sons, Ltd

Telephone (+44) 1243 779777

E-mail (for orders and customer service enquiries): cs-books@wiley.co.uk
Visit our Home Page on www.wileyeurope.com or www.wiley.com

Other Wiley Editorial Offices

John Wiley & Sons Inc., 111 River Street, Hoboken, NJ 07030, USA

Jossey-Bass, 989 Market Street, San Francisco, CA 94103-1741, USA

Wiley-VCH Verlag GmbH, Boschstr. 12, D-69469 Weinheim, Germany

John Wiley & Sons Australia Ltd, 33 Park Road, Milton, Queensland 4064, Australia

John Wiley & Sons (Asia) Pte Ltd, 2 Clementi Loop #02-01, Jin Xing Distripark, Singapore 129809

John Wiley & Sons Canada Ltd, 22 Worcester Road, Etobicoke, Ontario, Canada M9W 1L1

Wiley also publishes its books in a variety of electronic formats. Some content that appears in print may not be available in electronic books.

British Library Cataloguing in Publication Data

A catalogue record for this book is available from the British Library

ISBN 0 470 84872 3

Typeset by Dobbie Typesetting Ltd, Tavistock, Devon
Printed and bound in Great Britain by Antony Rowe Ltd., Chippenham, Wiltshire
This book is printed on acid-free paper responsibly manufactured from sustainable forestry in which at least two trees are planted for each one used for paper production.

*This book is dedicated to Joan Heaysman,
longtime collaborator of Michael Abercrombie,
to mark her reaching her seventieth year.*

Contents

Preface

The study of cell motility encompasses a wide range of approaches and techniques. This book provides a series of reviews by experts on different aspects of cell motility, from those studying molecules *in vitro* to those studying whole organisms. The reviews were commissioned from speakers at the 5th Abercrombie Symposium on Cell Motility, held in Oxford, UK in September 2003. These symposia are held every five years to commemorate the work of Michael Abercrombie, who was one of the pioneers in studying cell behaviour. Many of the concepts on how cells move that we now take for granted were established through his careful analysis. He made numerous timelapse films of moving cells cultured *in vitro*, and established that they extended lamellipodia and that when they met each other normal cells stopped moving (contact inhibition), rather than crawling over each other. The Abercrombie Symposia have become a forum for presenting the latest results in Cell Motility research.

Several articles in this volume report the enormous progress that has been made in the last few years in establishing at a molecular level how cells extend lamellipodia. The biochemical basis for the actions of actin-regulatory proteins such as the Arp2/3 complex, cofilin, profilin and capping proteins has been intensively investigated, and is discussed by Tom Pollard. A major discovery of the last five years is that the WASp-related proteins are central players in signal transduction from the plasma membrane to the Arp2/3 complex, and the regulation and action of WASp proteins is the topic of articles by the groups of Laura Machesky, Robert Insall and Tadaomi Takenawa.

Severing of actin filaments as well as *de novo* nucleation is important for altering cell morphology, as described in the chapters by Daniel Louvard and John Condeelis. A new player in the actin dynamics field is dynamin, a GTPase first characterized for its role in vesicle fission; the involvement of dynamin in cell motility is introduced by Orth and colleagues.

Lamellipodium extension is required for cell migration, but the cell body needs to move to follow the extension. Several events are critical for this. First, cell adhesion to its surroundings is important for the cell to exert a traction

Cell Motility. Edited by Anne Ridley, Michelle Peckham and Peter Clark.
 ISBN 0 470 84872 3

force. Meg Titus discusses the contribution of talin, which binds to transmembrane integrin receptors, and myosins with sequence homology to talin, in cell adhesion. Second, loss of cell adhesion by cell detachment selectively at the rear of the cell is essential for productive locomotion; Anna Huttenlocher reviews the role of the protease calpain in this process. Third, actin interaction with myosin is important for generating contractile forces and movement of actin filaments inside cells, and Soldati and Kistler discuss how class I myosins contribute to these processes.

Following the movement of and interaction between proteins within living cells is essential for understanding how they contribute to cell motility. Mark Holt and colleagues describe different microscopy techniques for tracing molecules in living cells. Delivery of new membrane components to the plasma membrane is often essential for initiating and/or maintaining membrane protrusion, for example during cell migration and phagocytosis. Pierre Chavier's group describe how the small GTPase ARF6 contributes to this process.

As well as the actin cytoskeleton, microtubules play a crucial role in cell migration in many cell types. This has been known for many years, but it is only recently that the molecular basis for the contribution and regulation of microtubules has been revealed. The Rho GTPases that are well known to regulate actin polymerization turn out to be central to microtubule dynamics as well, as reviewed here by Wittmann and Waterman-Storer. Recently it has become clear that microtubules are important for regulating the turnover of integrin-mediated adhesions to the substratum, as illustrated in the review by Alexander Bershadsky and colleagues.

One of the model systems that has been most informative for carrying out a genetic analysis of proteins important for cell migration is *Dictyostelium*, a slime mould that produces cAMP to attract other *Dictyostelium* cells under starvation conditions. The chapters by Rick Firtel's and Robert Insall's groups describe how *Dictyostelium* has been used to investigate how cells polarize and migrate towards a source of chemoattractant. This work has firmly established the crucial role of the generation of membrane phosphoinositides in cell polarization.

In vivo most cells do not move alone – they interact with other cells. The mechanisms whereby cells recognize and respond to other cells vary depending on the two types of cells involved. Epithelial cells meeting other epithelial cells form stable cell–cell adhesions, and Jason Ehrlich and colleagues describe how these adhesions form and mature. On the other hand, neuronal cells can be attracted or repulsed, depending on the stimulus. Kate Nobes' group describe how the transmembrane receptors ephrins and Ephs signal to the cytoskeleton to induce cell retraction, leading to loss of contact rather than stabilization of contact.

The last five years has seen an explosion of research monitoring the migration of cells in living organisms. For Michael Abercrombie this was not possible, but it is now because of technical advances in microscopy and in genetic manipulation of cells. Three chapters demonstrate the power of tracking cells *in vivo* both in the context of normal development (groups of Ray Keller and Rick Horwitz) and in cancer cell migration (Condeelis' laboratory). Genetic manipulation in mice has been crucial for identifying proteins important for migration of cell populations during development, and this approach is described by Carmen Birchmeier.

We hope that this book will provide an overview of the field of cell motility research in the early 21st century and will serve as a reference for both novices and experts. We thank all the authors for contributing to this book, and Lene Harbott and Kate Nobes for providing the cover picture.

Anne Ridley

Michelle Peckham

Peter Clark

List of Contributors

Hannalore Asmussen Department of Cell Biology, University of Virginia Health Sciences Center, Charlottesville, Virginia 22908, USA

Rafika Athman Laboratoire de morphogénèse et signalisation cellulaires, Institut Curie UMR 144, 26 rue d'Ulm, 75248 Paris cedex 05, France

Jonathan M. Backer Molecular Pharmacology, Albert Einstein College of Medicine, 1300 Morris Park Avenue, Bronx, NY 10461, USA

Christoph Ballestrem Department of Molecular Cell Biology, The Weizmann Institute of Science, Rehovot 76100, Israel

Alexander Bershadsky Department of Molecular Cell Biology, The Weizmann Institute of Science, Rehovot 76100, Israel

Carmen Birchmeier Max-Delbrück-Centrum for Molecular Medicine, Robert-Rössle-Str. 10, 13125 Berlin-Buch, Germany

Simone Blagg School of Biosciences, Birmingham University, Birmingham B15 2TT, UK

Philippe Chavrier Institut Curie-Section Recherche/CNRS UMR144, Membrane and Cytoskeleton Dynamics Laboratory, F-75248 Paris, France

Emma Colucci-Guyon Institut Curie-Section Recherche/CNRS UMR144, Membrane and Cytoskeleton Dynamics Laboratory, F-75248 Paris, France

John Condeelis Anatomy and Structural Biology, Albert Einstein College of Medicine, 1300 Morris Park Avenue, Bronx, NY 10461, USA

Lance Davidson Department of Cell Biology, University of Virginia, Charlottesville, VA 22904, USA

Ian M. Dobbie Genome Stability Laboratory, Department of Biochemistry, National University of Ireland, Galway, University Road, Galway, Ireland

Karen Donais Department of Cell Biology, University of Virginia Health Sciences Center, Charlottesville, Virginia 22908, USA

Cell Motility. Edited by Anne Ridley, Michelle Peckham and Peter Clark.
© 2004 John Wiley & Sons, Ltd. ISBN 0 470 84872 3

Thierry Dubois Institut Curie-Section Recherche/CNRS UMR144, Membrane and Cytoskeleton Dynamics Laboratory, F-75248 Paris, France

Graham A. Dunn The Randall Centre, New Hunt's House, Guy's Campus, King's College London, London SE1 1UL, UK

Jason S. Ehrlich Department of Molecular and Cellular Physiology, Beckman Center for Molecular and Genetic Medicine, Stanford University School of Medicine, Stanford, CA 94305-5435, USA

Richard A. Firtel Center for Molecular Genetics, University of California, San Diego, 9500 Gilman Drive, La Jolla, CA 92093-0634, USA

Satoru Funamoto Department of Neuropathology, Faculty of Medicine, University of Tokyo, 7-3-1 Hongo, Bunkyo-ku, Tokyo 113-0033, Japan

Noah W. Gray 1702 Guggenheim, Mayo Clinic and Foundation, Rochester, MN 55905, USA

Marc D. H. Hansen Department of Molecular and Cellular Physiology, Beckman Center for Molecular and Genetic Medicine, Stanford University School of Medicine, Stanford, CA 94305-5435, USA

Lene K. Harbott Centre for Cell and Molecular Dynamics, Department of Anatomy and Developmental Biology, University College London, Gower Street, London WC1E 6BT, UK

Mark R. Holt The Randall Centre, New Hunt's House, Guy's Campus, King's College London, London SE1 1UL, UK

Alan F. Horwitz Department of Cell Biology, University of Virginia Health Sciences Center, Charlottesville, VA 22908, USA

Anna Huttenlocher Department of Pharmacology, 3780 MSC, 1300 University Ave., Madison, WI 53706, USA

Robert Insall School of Biosciences, Birmingham University, Birmingham B15 2TT, UK

Ray Keller Department of Biology, University of Virginia, Charlottesville, VA 22904, USA

Claudia Kistler Max-Planck-Institute for Medical Research, Heidelberg, Germany

Daniel Louvard Laboratoire de morphogénèse et signalisation cellulaires, Institut Curie UMR 144, 26 rue d'Ulm, 75248 Paris cedex 05, France

Laura M. Machesky School of Biosciences, University of Birmingham, Birmingham B15 2TT, UK

Natalia Magid Department of Molecular Cell Biology, The Weizmann Institute of Science, Rehovot 76100, Israel

Daniel J. Marston Centre for Cell and Molecular Dynamics, Department of Anatomy and Developmental Biology, University College London, Gower Street, London WC1E 6BT, UK

Mark A. McNiven 1702 Guggenheim, Mayo Clinic and Foundation, Rochester, MN 55905, USA

James Monypenny Light Microscopy, Cancer Research UK, London Research Institute, Lincoln's Inn Fields Laboratories, London WC2A 3PX, UK

Shin-ichi Murase Department of Cell Biology, University of Virginia Health Sciences Center, Charlottesville, VA 22908, USA

W. James Nelson Department of Molecular and Cellular Physiology, Beckman Center for Molecular and Genetic Medicine, Stanford University School of Medicine, Stanford, CA 94305-5435, USA

Florence Niedergang Institut Curie-Section Recherche/CNRS UMR144, Membrane and Cytoskeleton Dynamics Laboratory, F-75248 Paris, France

Catherine D. Nobes Centre for Cell and Molecular Dynamics, Department of Anatomy and Developmental Biology, University College London, Gower Street, London WC1E 6BT, UK

James D. Orth 1702 Guggenheim, Mayo Clinic and Foundation, Rochester, MN 55905, USA

Thomas D. Pollard Department of Molecular, Cellular and Developmental Biology, Yale University, New Haven, CT, USA

Magali Prigent Institut Curie-Section Recherche/CNRS UMR144, Membrane and Cytoskeleton Dynamics Laboratory, F-75248 Paris, France

Sylvie Robine Laboratoire de morphogénèse et signalisation cellulaires, Institut Curie UMR 144, 26 rue d'Ulm, 75248 Paris cedex 05, France

Jeffrey Segall Anatomy and Structural Biology, Albert Einstein College of Medicine, 1300 Morris Park Avenue, Bronx, NY 10461, USA

Stewart J. Sharp School of Biosciences, University of Birmingham, Birmingham B15 2TT

Michael Shtutman Department of Molecular Cell Biology, The Weizmann Institute of Science, Rehovot 76100, Israel

Thierry Soldati Department of Biological Sciences, Sir Alexander Fleming Building, Imperial College London, South Kensington Campus, London SW7 2AZ, UK

Xiaoyan Song Anatomy and Structural Biology, Albert Einstein College of Medicine, 1300 Morris Park Avenue, Bronx, NY 10461, USA

Daniel Y. H. Soong The Randall Centre, New Hunt's House, Guy's Campus, King's College London, London SE1 1UL, UK

Craig F. Stovold School of Biosciences, University of Birmingham, Birmingham B15 2TT, UK

Shiro Suetsugu Department of Biochemistry, Institute of Medical Science, University of Tokyo, Shirokanedai, Minato-ku, Tokyo 108-8639, Japan

Tadaomi Takenawa Department of Biochemistry, Institute of Medical Science, University of Tokyo, Shirokanedai, Minato-ku, Tokyo 108-8639, Japan

Heather M. Thompson 1702 Guggenheim, Mayo Clinic and Foundation, Rochester, MN 55905, USA

Margaret A. Titus Department of Genetics, Cell Biology and Development, University of Minnesota, 6-160 Jackson Hall, 321 Church Street SE, Minneapolis, MN 55455, USA

Clare M. Waterman-Storer Department of Cell Biology, The Scripps Research Institute, 10550 North Torrey Pines Road, La Jolla, CA 92037, USA

Donna J. Webb Department of Cell Biology, University of Virginia Health Sciences Center, Charlottesville, Virginia 22908, USA

Torsten Wittmann Department of Cell Biology, The Scripps Research Institute, 10550 North Torrey Pines Road, La Jolla, CA 92037, USA

Jeffrey Wyckoff Anatomy and Structural Biology, Albert Einstein College of Medicine, 1300 Morris Park Avenue, Bronx, NY 10461, USA

Daniel Zicha Light Microscopy, Cancer Research UK, London Research Institute, Lincoln's Inn Fields Laboratories, London WC2A 3PX, UK

Julia Zonis Department of Molecular Cell Biology, The Weizmann Institute of Science, Rehovot 76100, Israel

1

Molecular Mechanisms Regulating Actin Filament Dynamics at the Leading Edge of Motile Cells

Thomas D. Pollard

We have analysed the molecular basis of pseudopod extension using biochemical and biophysical approaches. We and others have determined the atomic structures of the key proteins including actin, profilin, Arp2/3 complex, capping protein and ADF/cofilin as well as the rate and equilibrium constants for their interactions. Arp2/3 complex interacts with actin monomers and filaments to generate new filament branches. A pool of actin bound to profilin provides subunits to elongate the ends of the branches and to push forward the plasma membrane. Capping protein terminates branch elongation. ADF/cofilin and profilin promote the disassembly of older actin filaments and the recycling of actin subunits to a pool ready for elongation of new filaments.

Students of cellular motility, marvelled for years about how cells advance their leading edges by spreading lamellae as they migrate over substrates and through the extracellular matrix. After the discovery of actin in non-muscle cells (Hatano and Oosawa, 1966), electron microscopy (Abercrombie *et al.*, 1971; Small *et al.*, 1978; Svitkina *et al.*, 1997) and later fluorescence microscopy (Lazarides and Weber, 1974) revealed that actin filaments are

---

Cell Motility. Edited by Anne Ridley, Michelle Peckham and Peter Clark.
 ISBN 0 470 84872 3

concentrated in the leading lamella. Seminal experiments with photobleaching of fluorescent actin (Wang, 1985) and photoactivation of caged-fluorescent actin (Theriot and Mitchison, 1991) demonstrated that these actin filaments assemble near the leading edge and turn over on a time scale of tens of seconds deeper in the cytoplasm. More recent work with fluorescent speckle microscopy (Watanabe and Mitchison, 2002) suggests that assembly and disassembly are actually distributed processes: assembly is strongest at the leading edge but occurs deeper as well; and filaments turn over in a broad zone behind the leading edge. A key point is that external signals from chemotactic attractants and repellents guide assembly temporally and spatially. On a time scale of seconds cells can re-orientate toward a new source of attractant or turn away from repellents (Gerisch, 1982; Bourne and Weiner, 2002).

Deciphering the molecular basis of this localized assembly and disassembly of actin filaments might have proven impossible, given that it requires coordinating the activities of millions of protein molecules. However, the system has proven to be remarkably amenable to reductionist analysis coupled with insightful microscopy. All of the major components have been identified and their cellular concentrations measured in selected systems. A crystal structure of each component is available. Rate and equilibrium constants for most of the reactions are known. *In vitro* motility assays based on the propulsive comet tails of bacteria allow reconstitution of the whole assembly and disassembly process (Loisel *et al.*, 1999). Many of the reactions of the purified proteins have also been visualized in real time by fluorescence microscopy (Maciver *et al.*, 1991; Amann and Pollard, 2001; Ichetovkin *et al.*, 2002; Fujiwara *et al.*, 2002). Experiments in genetically tractable organisms allow tests for physiological functions, and mathematical models (Mogilner and Edelstein-Keshet, 2002) of the whole cycle of assembly and disassembly now guide mechanistic experiments.

We now have a first generation, quantitative model for the process (Figure 1.1) called the dendritic nucleation, array treadmilling hypothesis (Mullins *et al.*, 1998a; Svitkina and Borisy, 1999; Pollard *et al.*, 2000). This model is an over-simplification, since it considers only five protein components and ATP, but these molecules are sufficient to reconstitute motility of bacterial comet tails (Loisel *et al.*, 1999) and thus are likely to be the core components that operate a more elaborate system in cells. I will explain our current understanding of this machine and indicate some points that need clarification or possible revision.

Inventory of components

In vitro reconstitution of bacterial motility from purified proteins (Loisel *et al.*, 1999) showed that the essential components of the system are actin, Arp2/3

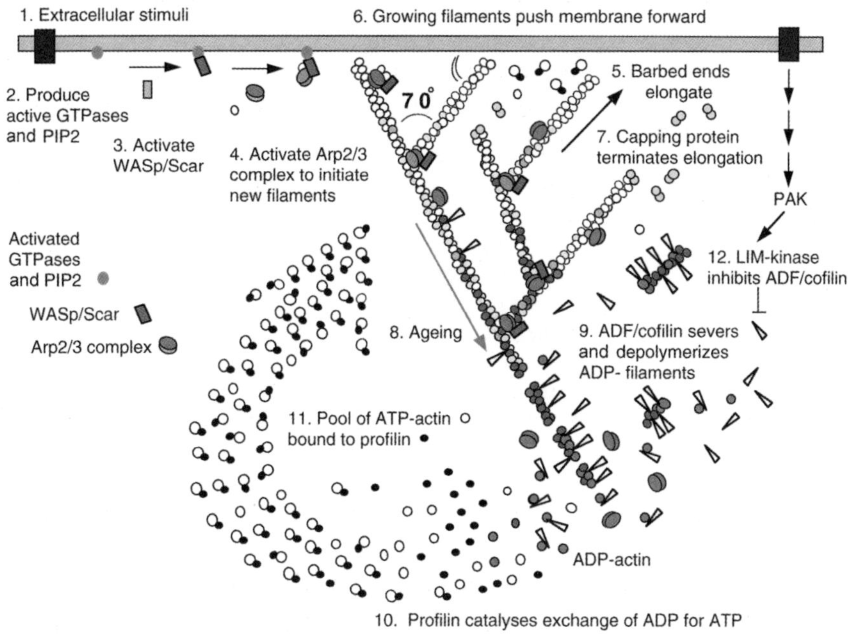

Figure 1.1 Dendritic nucleation, array-treadmilling model for the leading edge of motile cells. (1) Chemoattractants activate plasma membrane receptors. (2) Signalling from the receptors generates activated Rho-family GTPases and PIP_2, which (3) activate WASp/Scar proteins on the inner surface of the plasma membrane. (4) WASp/Scar assembles a complex consisting of the Arp2/3 complex and an actin monomer on the side of an actin filament, initiating a branch. (5) Actin-profilin elongates new barbed ends, which (6) push the membrane forward until they are (7) capped. (8) Hydrolysis of ATP bound to actin subunits and dissociation of the γ-phosphate makes the filaments targets for (9) ADF/cofilin, which promotes severing and depolymerization of the filaments. (10) Profilin promotes exchange of ADP for ATP, (11) refilling the cytoplasmic pool of subunits. (12) Parallel signalling through PAK and LIM-kinase tends to stabilize filaments through phosphorylation, which inhibits ADF/cofilin. Not shown here are the events which lead to remodelling of the branched network into long, unbranched filaments. (Modified from Pollard *et al.*, 2000). Reprinted with permission from the *Annual Review of Biophysics and Biomolecular Structure*, Volume 29 © 2000 by Annual Reviews

complex, a barbed end capping protein and ADF/cofilin (Table 1.1). Profilin increases the rate of movement. An actin filament crosslinking protein, α-actinin, also promotes movement. These are all ancient proteins with genes that arose before the earliest branches of the eukaryotic lineage. Their active sites are conserved to such a remarkable degree that components from any eukaryote interact with their partners from any other eukaryote. Thus the entire system is ancient and highly conserved. It is noteworthy that genetic experiments have established that the assembly and motility of actin filament

Table 1.1 Inventory of proteins required for dendritic nucleation and actin filament turnover

Protein	Required for biochemical reconstitution of *Listeria* motility	Essential for viability and assembly of actin filament structures in fungi
Actin	Yes	Yes
Arp2/3 complex	Yes	Yes
ADF/cofilin	Yes	Yes
Capping protein	Yes	No
Profilin	Enhances	Yes
α-actinin	Enhances	No
VASP	Enhances	No

patches are defective in yeast cells lacking actin, Arp2/3 complex, ADF/cofilin or profilin.

Actin is the subunit for the filaments. Arp2/3 complex, an assembly consisting of actin related proteins Arp2 and Arp3 with five other protein subunits, initiates new filaments growing in the barbed direction as branches on the side of existing filaments (Mullins *et al.*, 1998a). ADF/cofilin promotes the dissociation of the γ-phosphate from polymerized actin that has hydrolysed its bound ATP (Blanchoin and Pollard, 1999) and hastens the disassembly of ADP-actin filaments by severing and other effects (Bamburg *et al.*, 1999). Heterodimeric capping protein blocks the barbed end of actin filaments (Isenberg *et al.*, 1980). Neither budding nor fission yeast absolutely requires capping protein, but deletion of either gene encoding fission yeast capping protein is synthetically lethal with another barbed-end capping protein, so this activity is essential (D. Kovar, J. Kuhn and T.D. Pollard, submitted). In animals, gelsolin and its relatives also block barbed ends and may supplement capping protein, although cells lacking gelsolin still move (Witke *et al.*, 1995). Profilin catalyses the exchange of nucleotide bound to actin monomers and, with capping protein, maintains a pool of unpolymerized actin far above the 'critical concentration' for polymerization at either end of actin filaments. Both proteins are required: profilin blocks addition of actin to pointed ends (Pollard and Cooper, 1984) and capping protein blocks the barbed ends. The actin monomer sequestering protein, thymosin-β4, is an example of a protein likely to participate in actin dynamics in animal cells (Safer and Nachmias, 1994), but which is not required for *in vitro* reconstitution of actin polymerization (or even found in the genomes of lower eukaryotes).

The Arp2/3 complex consists of seven subunits with structures that have been conserved since the early branches of the eukaryotic lineage (Figure 1.2;

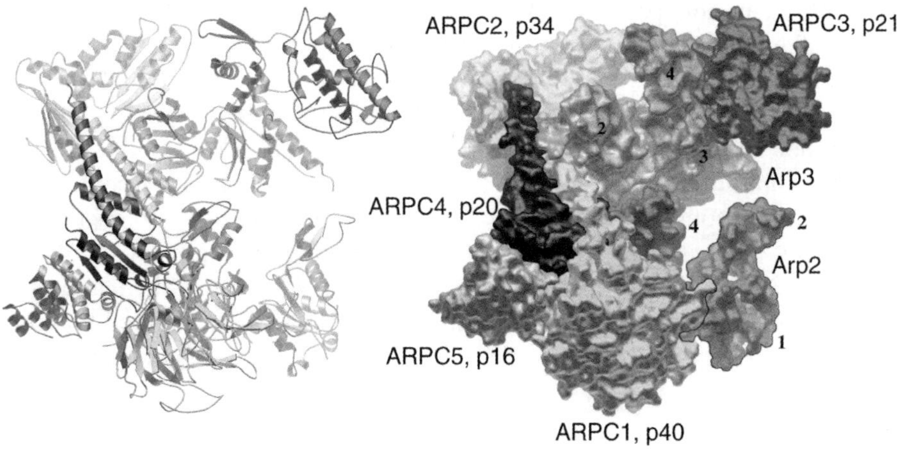

Figure 1.2 Crystal structure of inactive bovine Arp2/3 complex, illustrating the arrangement of the seven subunits in ribbon and space filling models. Separation of the Arps is postulated to account for the inactivity of the complex. Nucleation-promoting factors are postulated to stabilize a more compact conformation with the Arps arranged like adjacent subunits in an actin filament. (Modified from Robinson *et al.*, 2001)

Robinson *et al.*, 2001; C. Beltzner and T.D. Pollard, in press, 2004). The actin related proteins Arp2 and Arp3 are folded exactly like actin, but with amino acid substitutions and longer surface loops that participate in interactions with the other subunits. ARPC1 (also called p40) is a β-propeller protein with seven blades similar to a trimeric G-protein β subunit. A novel loop inserted between blades 6 and 7 is postulated to interact with actin filaments at branches. A dimer of two similar subunits (ARPC2 and ARPC4) holds the complex together through extensive interactions of most of the other subunits. ARPC3 and ARPC5 are α-helical subunits on the periphery. The conformation in the crystal is thought to be inactive, since physical separation of the two Arps prevents them from initiating a new actin filament.

The ground state of the system

In the absence of positive stimuli, physiological concentrations of these essential proteins will assemble a static gel. Roughly half of the actin will assemble into filaments and the remainder will be bound to profilin (and thymosin-β4 in vertebrates). Even pure actin filaments are quite stable under physiological conditions in ATP. Owing to a small difference in the critical concentrations for elongation at the two ends (Pollard, 1986), actin subunits flux slowly onto the barbed end and off the pointed end, but the rate is less than 0.1 subunit per second. Capping protein is expected to reduce treadmilling. The combination of barbed end caps and a high concentration of profilin allows cells to maintain a high concentration of unpolymerized ATP-actin ready for elongation of barbed ends when they appear.

But new barbed ends rarely appear without a positive stimulus. First, profilin inhibits the initiation of new actin filaments by spontaneous nucleation. Secondly, the Arp2/3 complex is inactive without nucleation-promoting factors. Thirdly, nucleation-promoting factors such as WASp (Wiskott–Aldrich Syndrome protein) are strongly auto-inhibited and thus inactive in the absence of positive signals. Thus the system is poised far from equilibrium ready to grow new actin filaments in response to positive stimuli.

Signalling pathways

No signalling pathway from chemoattractants or repellents to actin is fully defined, but it is clear that multiple receptor types participate. Some are from familiar families: receptor tyrosine kinases (such as the EGF receptor), seven helix receptors coupled to trimeric G-proteins (such as *Dictyostelium* cAMP receptors and human leukocyte receptor for f-Met-Leu-Phe), and integrins coupled to cytoplasmic tyrosine kinases. Other receptors with yet-to-be-defined transduction mechanisms also direct actin assembly. Examples are DCC, the receptor for the growth cone attractant netrin, and Robo, the receptor for the growth cone repellent Slit (Stein and Tessier-Lavigne, 2001). Endogenous activation mechanisms independent of external stimuli are also likely to exist.

Each family of receptors has its own downstream transduction hardware, so the opportunities for complexity are immense, but several of these pathways lead to a modest number of 'nucleation-promoting factors' that activate the Arp2/3 complex (reviewed by Higgs and Pollard, 2001). It is important to note that alternative pathways are likely to exist: *de novo* formation of new ends by formins (Sagot *et al.*, 2002; Pruyne *et al.*, 2002) or other proteins, or multiplication of ends of existing filaments by uncapping (Glogauer *et al.*, 2000) or severing (Zebda *et al.*, 2000).

The first nucleation-promoting factor for the Arp2/3 complex to be identified was ActA from *Listeria monocytogenes* (Welch *et al.*, 1998). This transmembrane protein suffices for the bacterium to usurp the cytoplasmic actin system to assemble a comet tail. The first eukaryotic proteins shown to activate the Arp2/3 complex were the WASp/Scar family (Machesky and Insall, 1998; Machesky *et al.*, 1999; Rohatgi *et al.*, 1999; Yarar *et al.*, 1999; Winter *et al.*, 1999; Egile *et al.*, 1999). WASp is the product of the gene mutated in the X-linked immunodeficiency and bleeding disorder called Wiskott–Aldrich syndrome. Newly recognized nucleation-promoting factors include cortactin (Weaver *et al.*, 2001), fungal myosin-I (Evangelista *et al.*, 2000; Lechler *et al.*, 2000; Lee *et al.*, 2000), fungal Abp1p (Goode *et al.*, 2001) and fungal Pan1p (Duncan *et al.*, 2001).

Nucleation-promoting factors activate the Arp2/3 complex, generally employing a sequence of acidic residues with a key tryptophan to bind the Arp2/3 complex. Most also bind one or two actin monomers which presumably become the first subunit(s) in the new filament. The activating part of WASp/Scar proteins is located near their C-terminus, consisting of a 'V' motif (for verprolin homology), a 'C' motif (for central or connecting) and an 'A' motif (for acidic). The VC region binds the actin monomer and the CA region binds Arp2/3 complex. In isolation, the VCA domain constitutively activates the Arp2/3 complex. The corresponding functional regions are located in the middle of the sequence of ActA.

Some nucleation-promoting factors are constitutively active such as ActA, but the WASp/Scar family are tightly regulated and responsive to activation by signalling pathways (Figure 1.3). WASp and N-WASp are auto-inhibited by virtue of an intramolecular interaction of the C motif

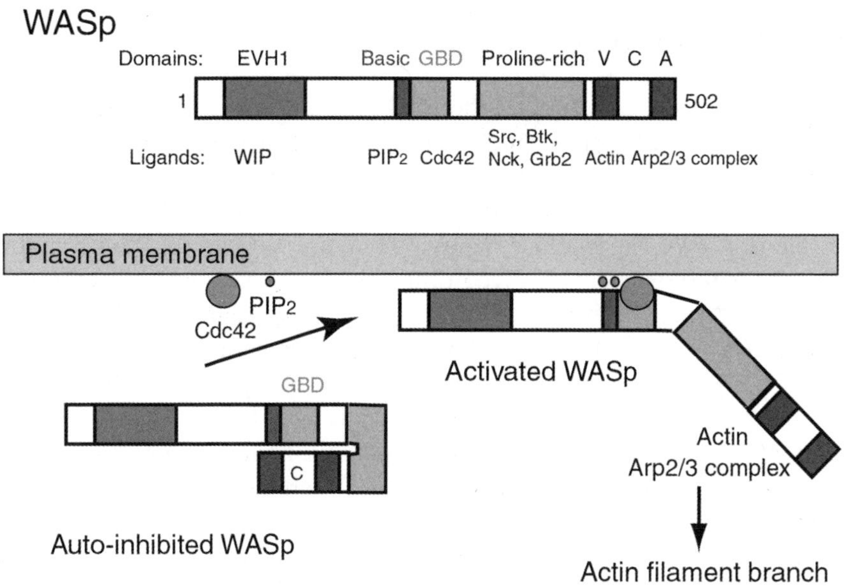

Figure 1.3 WASp domains and activation. Intramolecular binding of the C region to the GTPase binding domain (GBD) strongly auto-inhibits the nucleation promoting activity of the VCA domains at the C-terminus. Binding of the GBD to Cdc42 and the basic region to PIP_2 release the VCA domains, so that the VC region can bind an actin monomer and the CA region can bind Arp2/3 complex. Binding of this ternary complex to the side of an actin filament complete the activation process and initiates the formation of a new filament. (Redrawn from Higgs and Pollard, 2001)

with the GTPase binding domain (GBD) located in the middle of the polypeptide. This interaction prevents the VCA region from activating the Arp2/3 complex. The Rho family GTPase Cdc42 and the membrane lipid phosphatidylinositol(4,5)bisphosphate (PIP_2) act synergistically to overcome the auto-inhibition of WASp and N-WASp by freeing VCA to interact with the Arp2/3 complex. Other parts of WASp may contribute to activation, since the full-length protein activated by Cdc42 and PIP_2 is 100 times more active than VCA alone. Alternatively, the SH3 proteins Grb2 and Nck can activate N-WASp together with PIP_2 (Carlier *et al.*, 2000; Rohatgi *et al.*, 2001). The ability of multiple signalling molecules to activate WASp allows the protein to be a coincidence detector for signals flowing from diverse receptors (Prehoda *et al.*, 2000).

Scar (also known as WAVE) lacks a GBD and is not regulated by auto-inhibition. Instead, a complex of four other proteins interferes with the ability of Scar to activate Arp2/3 complex (Eden *et al.*, 2002). The GTPase Rac1 and the adapter protein Nck overcome this regulatory complex and allow Scar to activate the Arp2/3 complex (see also chapter by Blagg and Insall).

Activation of the Arp2/3 complex

Based on the crystal structure (Robinson *et al.*, 2001) and our analysis of the activation mechanism (Marchand *et al.*, 2001), we proposed that nucleation-promoting factors and actin filaments activate the Arp2/3 complex by stabilizing a conformation with the Arps juxtaposed like two subunits in an actin filament. We expect that the Arp2/3 complex visits this conformation rarely even without activators. Accordingly, high (micromolar) concentrations of purified Arp2/3 complex generate new actin filaments, but with a stoichiometry of only 0.001 new barbed ends per complex (Mullins *et al.*, 1998a), reflecting the very low fraction of active complex. Therefore a small fraction of the complex must be in the active conformation at any time. Confirmation that nucleation-promoting factors stabilize an active conformation will require crystal structures or other biophysical probes of activated complexes.

Kinetic and thermodynamic analysis of the activation mechanism (Marchand *et al.*, 2001) showed that the VCA domains from WASp and Scar bind actin and the Arp2/3 complex with submicromolar affinity and that both reactions are rapidly reversible on a subsecond time scale. This means that when WASp or Scar are activated, their VCA domains will rapidly bind the micromolar concentrations of actin and Arp2/3 complex diffusing in the cytoplasm. Given that VCA reacts faster and has higher affinity for actin than the Arp2/3 complex, we propose that when freed from inhibition VCA first binds actin and then the Arp2/3 complex. Although the Arp2/3 complex has only micromolar affinity for the sides of actin filaments, the presence of filaments increases the affinity of the Arp2/3 complex for the VCA domain. Thus VCA binding will increase the affinity of the Arp2/3 complex for actin filaments. This thermodynamic coupling suggests that VCA and filaments favour the same active conformation of the Arp2/3 complex.

Although we had evidence that the Arp2/3 complex binds to and forms branches on the sides of existing filaments (Mullins *et al.*, 1998a; Blanchoin *et al.*, 2000a), Pantaloni *et al.* (2000) suggested that the branches actually form at the barbed end of growing filaments, with one of the Arps incorporated into the mother filament and one into the daughter filament. Direct observation of branching by total internal reflection microscopy (Amann and Pollard, 2001b; Fujiwara *et al.*, 2002) and confocal microscopy (Ichetovkin *et al.*, 2002) confirms that branches form on the sides of mother filaments. These real-time assays also confirmed the observation of static samples (Amann and Pollard, 2001a) that branching is favoured on newly polymerized filaments. The mechanism of this bias toward new filaments is not established, but is likely to be related to nucleotide hydrolysis and/or a structural change in the mother filament as it ages.

We do not yet know how the nucleation-promoting factors are organized in cells, but most of their activators of WASp and N-WASp are associated with membranes: PIP_2 is part of the lipid bilayer; Cdc42 is tethered to the bilayer by a prenyl group; and Grb2 associates with active receptor tyrosine kinases. Thus the active fraction of WASp (and perhaps Scar) proteins is expected to be associated with the plasma membrane, making the inner surface of the plasma membrane a favoured site for activating the Arp2/3 complex. In fact, when PIP_2 and active Cdc42 are incorporated into small lipid vesicles, a huge cloud of branched actin filaments grows from their surface (Higgs and Pollard, 2000). A bias toward branching from newly formed filaments would also favour nucleation near the plasma membrane where growing filaments interact with the membrane.

A key unresolved point is why branches are so much more stable than the low affinity binding of the Arp2/3 complex to the sides of actin filaments (Mullins *et al.*, 1998b). The branch half-life *in vitro* is about 6 min (Blanchoin *et al.*, 2000b) and can be prolonged by inhibiting the release of phosphate from ADP-P_i subunits in the branch with phalloidin or using the stably bound phosphate analogue, BeF_3. On the other hand, ADF/cofilin promotes both the dissociation of the γ-phosphate and dissociation of branches. Thus, phosphate release from subunits in the branch seems to trigger dissociation of the pointed end of the daughter filament from Arp2/3 complex. Insight into this matter is likely to require a high-resolution model of branches from cryo-electron microscopy.

Growth of the branched actin filament network

Once a branch is initiated, its free barbed end will grow at a rate [10 $\mu M^{-1} s^{-1}$ (actin-profilin)] limited by diffusion of actin-profilin to the leading edge. Mogilner estimates that this reaction creates an actin-profilin sink at the leading edge, with a concentration about half that deeper in the cytoplasm (Mogilner and Edelstein-Keshet, 2002). Even so, elongation is expected to be fast, in the order of 100 subunits (0.25 μm) per second, enough to account for the observed expansion of the leading edge in the fastest cells. Theoretical calculations (Mogilner and Edelstein-Keshet, 2002; Mogilner and Oster, 1996; Carlsson, 2002) show that the concentration of growing filaments is sufficient to produce a force to push forward the membrane.

Electron microscopy (Svitkina *et al.*, 1997) indicates that the branches are short, on the order of 0.5 μm or less, so elongation must be terminated after a few seconds. This capping rate is consistent with the rate of capping by heterodimeric capping protein estimated from its concentration and rate constant for binding barbed ends (Schafer *et al.*, 1996).

Remarkably, most filaments in these branched networks have their barbed ends orientated toward the front of the cell. Maly and Borisy (2001) proposed a Darwinian model to explain this bias. Their model is based on the assumption that interactions of forward pointing filaments with the inside of the plasma membrane inhibits capping, whereas filaments pointing away from the membrane are capped rapidly and irreversibly. This assumption is plausible, since PIP_2 dissociates capping protein from filament ends (Schafer *et al.*, 1996). The remarkable flatness of a typical leading lamella means that growth forward is strongly favoured relative to growth in the dorsal direction. No one has proposed a mechanism to account for this bias.

Assembly of the branched network appears to be self-organizing – an inert plastic bead coated with a nucleation-promoting factor suffices to induce a force-producing comet tail in a cell extract (Cameron *et al.*, 1999). If the expanding network at the leading edge is anchored to the substrate via transmembrane attachments, growth of the filaments pushes the plasma membrane forward. If the actin network is not anchored to the substrate, its expansion at the leading edge results in the whole network sliding as a unit toward the cell centre. Some cells exhibit a mixture of forward motion of the cell and rear-ward motion of the network. The molecular clutch presumably consists of links between the network and cell adhesion molecules, but they still need to be identified.

Filament ageing, remodelling and disassembly

The zone of short, branched filaments at the leading edge is narrow, less than 1 μm wide. Actin filaments further from the front are long and unbranched

(Svitkina *et al.*, 1997). This means that a cell extending its leading edge at 0.2 μm per second must remodel the branched network in 5 s or less. Given this brief lifetime, the branched zone will be seen only in lamellae that are actively expanding at the time of preparation for microscopy. Choosing keratocytes, cells that move rapidly at a constant rate, may have contributed to the success of Svitkina and Borisy (Svitkina *et al.*, 1997) in preserving branches that were missing in earlier studies.

Remodelling must involve two steps: dissociation of branches; and the conversion of short filaments into long filaments. How does this happen? ATP hydrolysis and phosphate dissociation destabilize branches and are also likely to be the timer for disassembly. The rate constant for ATP hydrolysis is $0.3\,s^{-1}$ (Blanchoin and Pollard, 2002) and nothing has yet been found that influences this reaction rate. Thus newly assembled ATP-actin subunits hydrolyse their bound ATP with a half-time of about 2 s. Phosphate dissociation is much slower, with a half-time of 350 s (Carlier, 1987; Blanchoin and Pollard, 1999), far too slow to account for debranching in cells. However, ADF/cofilin strongly accelerates phosphate dissociation from ADP-P_i actin filaments (Blanchoin and Pollard, 1999) to rates that keep pace with hydrolysis. Rate of phosphate dissociation depends on the concentration of active ADF/cofilin and involves a very low-affinity transient interaction of ADP/cofilin with the filament. Phosphorylation of ADF/cofilin by LIM kinase downstream of PAK (p21-activated kinase) (Edwards *et al.*, 1999), blocks this and other interactions of ADF/cofilin with actin (Blanchoin *et al.*, 2000c) and is expected to slow phosphate dissociation and to stabilize branches.

Short filaments might be converted to long filaments by subunit addition to the filament ends or by end-to-end annealing, a very favourable reaction (Andrianantroandro *et al.*, 2001). Capping barbed ends is expected to prevent both of these reactions, so the cell must have a mechanism to avoid it. VASP appears to inhibit capping near the leading edge (Bear *et al.*, 2002), so it may also promote annealing in the presence of capping protein.

Actin filaments in the lamella turn over on a time scale of tens of seconds. Both the pioneering photoactivation observations (Theriot and Mitchison, 1991) and recent speckle microscopy experiments (Watanabe and Mitchison, 2002) show that depolymerization occurs broadly behind the leading edge. Pure actin filaments are stable for days, treadmilling at less than 0.1 subunits per second at steady state and capping will stabilize their dynamic barbed ends. So, how do filaments turn over rapidly in cells? ADF/cofilin is the prime candidate to drive filament disassembly in cells, since ADF/cofilin and profilin increase the turnover of filaments *in vitro* (Carlier *et al.*, 1997; Rosenblatt *et al.*, 1997). The higher affinity of ADF/cofilins for ADP-actin monomers than ADP-actin filaments provides the thermodynamic basis for their ability to depolymerize filaments (Blanchoin and Pollard, 1999) but does not reveal the pathway of disassembly. One factor is the ability of ADF/cofilin to sever filaments, creating ends for subunit dissociation. ADF/cofilin may also promote subunit dissociation from these ends. Further work is required to firm up both the mechanism and the kinetics of disassembly.

Some cellular actin filaments, including those in stress fibres and filopodia turn over slowly. Binding of tropomyosin may contribute to their stability, since it protects filaments from severing by ADF/cofilin (Maciver *et al.*, 1991) and also inhibits branching by the Arp2/3 complex (Blanchoin *et al.*, 2000b).

Recycling ADP-actin subunits

ADF/cofilin binds ADP-actin tighter than ATP-actin and also inhibits exchange of the bound ADP, so it might trap ADP-actin when it dissociates from filaments. However, both profilin and ADF/cofilin are in rapid equilibria with ADP-actin (Perelroizen *et al.*, 1994; Vinson *et al.*, 1998; Blanchoin and Pollard, 1998). Bound profilin promotes rapid dissociation of ADP and the high cytoplasmic concentration of ATP relative to ADP results in nucleotide-free actin binding ATP (Vinson *et al.*, 1998). ATP-actin binds profilin much better than ADF/cofilin, refilling the pool of ATP-actin (Rosenblatt *et al.*, 1995) ready for elongation of barbed ends.

Reaction to a chemoattractant

Chemoattractants activate parallel signalling pathways employing Rho-family GTPases, which promote actin polymerization locally by at least two mechanisms: creation of new barbed ends as branches by activating Arp2/3 complex; and inhibition of ADF/cofilin, which tends to stabilize existing filaments. Alternatively, or in addition, some activators may transiently activate ADF/cofilin by dephosphorylation, promoting severing and the growth of barbed ends which are favourable for branching nucleation by the Arp2/3 complex (Zebda *et al.*, 2000; Ichetovkin *et al.*, 2002).

Reaction to the withdrawal of a chemoattractant

The whole system runs down automatically in the absence of a positive signal for assembly. Lacking activators, nucleation-promoting factors will return to their inhibited states. The rate of decay will be determined by the rate of GTP hydrolysis by the Rho-family GTPases. It is not known if any of the nucleation-promoting factors are GTPase activators (GAPs), like effectors of trimeric G-proteins, but there is a large family of GAPs for the Rho-family GTPases. Without active nucleation-promoting factors creation of new filaments stops rapidly, since the Arp2/3 complex activated previously is consumed by branching nucleation. Networks in unstimulated parts of cells are predicted to disassemble in tens of seconds, perhaps hurried along by active ADF/cofilin relieved of inhibition by the PAK/LIM kinase pathway.

Acknowledgements

This work was supported by NIH Research Grant GM26338. The author thanks the members of this laboratory who contributed to this work.

References

Abercrombie, M., Heaysman, J. E. M. and Pegrum, S. M., 1971. The locomotion of fibroblasts in culture: IV. Electron microscopy of the leading lamella. *Exp. Cell Res.* **67**: 359–367.

Amann, K. J. and Pollard, T. D., 2001a. Arp2/3 complex nucleates filaments that branch from the side of pre-existing actin filaments. *Nature Cell Biol.* **3**: 306–310.

Amann, K. J. and Pollard, T. D., 2001b. Direct real-time observation of actin filament branching mediated by Arp2/3 complex using total internal reflection microscopy. *Proc. Natl. Acad. Sci. USA* **98**: 15009–15013.

Andrianantroandro, E., Blanchoin, L., Sept, D., McCammon, J. A. and Pollard, T. D., 2001. Kinetic mechanism of end to end annealing of actin filaments. *J. Molec. Biol.* **312**: 721–730.

Bamburg, J. R., McGough, A. and Ono, S., 1999. Putting a new twist on actin: ADF/cofilins modulate actin dynamics. *Trends Cell Biol.* **9**: 364–370.

Bear, J. E., Svitkina, T. M., Krause, M., Schafer, D. A., *et al.*, 2002. Antagonism between Ena/VASP proteins and actin filament capping regulates fibroblast motility. *Cell* **109**: 509–521.

Beltzner, C. C. and Pollard, T. D., 2004. Identification of functionally important residues of Arp2/3 complex by analysis of homology models from diverse species. *J. Mol. Biol.*, in press.

Blanchoin, L. and Pollard, T. D., 1998. Interaction of actin monomers with *Acanthamoeba* actophorin (ADF/cofilin) and profilin. *J. Biol. Chem.* **273**: 25106–25111.

Blanchoin, L. and Pollard, T. D., 1999. Mechanism of interaction of *Acanthamoeba* actophorin (ADF/cofilin) with actin filaments. *J. Biol. Chem.* **274**: 15538–15546.

Blanchoin, L. and Pollard, T. D., 2002. Hydrolysis of bound ATP by polymerized actin depends on the bound divalent cation but not profilin. *Biochemistry* **41**: 597–602.

Blanchoin, L., Amann, K. J., Higgs, H. N., Marchand, J. B., *et al.*, 2000a. Direct observation of dendritic actin filament networks nucleated by Arp2/3 complex and WASp/Scar proteins. *Nature* **404**: 1007–1011.

Blanchoin, L., Pollard, T. D. and Mullins, R. D., 2000b. Interaction of ADF/cofilin, Arp2/3 complex, capping protein and profilin in remodeling of branched actin filament networks. *Curr. Biol.* **10**: 1273–1282.

Blanchoin, L., Robinson, R. C., Choe, S. and Pollard, T.D., 2000c. Phosphorylation of *Acanthamoeba* Actophorin (ADF/cofilin) blocks interaction with actin without a change in atomic structure. *J. Mol. Biol.* **295**: 203–211.

Bourne, H. R. and Weiner, O., 2002. A chemical compass. *Nature* **419**: 21.

Cameron, L. A., Footer, M. J., van Oudenaarden, A. and Theriot, J. A., 1999. Motility of ActA protein-coated microspheres driven by actin polymerization. *Proc. Natl. Acad. Sci. USA* **96**: 4908–4913.

Carlier, M. F., 1987. Measurement of Pi dissociation from actin-filaments following ATP hydrolysis using a linked enzyme assay. *Biochem. Biophys. Res. Comm.* **143**: 1069–1075.

Carlier, M. F., Laurent, V., Santolini, J., Melki, R., *et al.*, 1997. Actin depolymerizing factor (ADF/cofilin) enhances the rate of filament turnover: implication in actin-based motility. *J. Cell Biol.* **136**: 1307–1322.

Carlier, M. F., Nioche, P., Broutin-L'Hermite, I., Boujemaa, R., *et al.*, 2000. Grb2 links signalling to actin assembly by enhancing interaction of neural Wiskott–Aldrich Syndrome protein (N-WASp) with actin-related-protein (Arp2/3) complex. *J. Biol. Chem.* **275**: 21946–21952.

Carlsson, A. E., 2002. Growth of branched actin networks against obstacles. *Biophys. J.* **81**: 1907–1923.

Duncan, M. C., Cope, M. J., Goode, B. L., Wendland, B. and Drubin, D. G., 2001. Yeast Eps15-like endocytic protein, Pan1p, activates the Arp2/3 complex. *Nat. Cell Biol.* **3**: 687–690.

Eden, S., Rohatgi, R., Podtelejnikov, A. V., Mann M. and Kirschner, M. W., 2002. Mechanism of regulation of WAVE1-induced actin nucleation by Rac1 and Nck. *Nature* **418**: 790–793.

Edwards, D. C., Sanders, L. C., Bokoch, G. M. and Gill, G. N., 1999. Activation of LIM-kinase by Pak1 couples Rac/Cdc4242 GTPase signalling to actin cytoskeletal dynamics. *Nature Cell Biol.* **1**: 253–259.

Egile, C., Loisel, T. P., Laurent, V., Li, R., *et al.*, 1999. Activation of the CDC42 effector N-WASp by the *Shigella flexneri* IcsA protein promotes actin nucleation by Arp2/3 complex and bacterial actin-based motility. *J. Cell Biol.* **146**: 1319–1332.

Evangelista, M., Klebl, B. M., Tong, A. H., Webb, B. A., *et al.*, 2000. A role for myosin-I in actin assembly through interactions with Vrp1p, Bee1p, and the Arp2/3 complex. *J. Cell Biol.* **148**: 353–362.

Fujiwara, I., Suetsugu, S., Uemura, S., Takenawa, T. and Ishiwata, S., 2002. Visualization and force measurement of branching by Arp2/3 complex and N-WASp in actin filament. *Biochem. Biophys. Res. Commun.* **293**: 1550–1555.

Gerisch, G. 1982. Chemotaxis in *Dictyostelium*. *Annu. Rev. Physiol.* **44**: 535–552.

Glogauer, M., Hartwig, J. and Stossel, T., 2000. Two pathways through Cdc42 couple the N-formyl receptor to actin nucleation in permeabilized human neutrophils. *J. Cell Biol.* **150**: 785–796.

Goode, B. L., Rodal, A. A., Barnes, G. and Drubin, D. G., 2001. Activation of the Arp2/3 complex by the actin filament binding protein Abp1p. *J. Cell Biol.* **153**: 627–634.

Hatano, S. and Oosawa, F., 1966. Isolation and characterization of plasmodium actin. *Biochim. Biophys. Acta* **127**: 488–498.

Higgs, H. N. and Pollard, T. P., 2000. Wiskott–Aldrich Syndrome protein (WASp) activation of Arp2/3 complex: effects of phosphatidylinositol-4,5-bisphosphate and Cdc42. *J. Cell Biol.* **150**: 1311–1320.

Higgs, N. H. and Pollard, T. D., 2001. Regulation of actin filament network formation through Arp2/3 complex: activation by a diverse array of proteins. *Ann. Rev. Biochem.* **70**: 649–676.

Ichetovkin, I., Grant, W. and Condeelis, J., 2002. Cofilin produces newly polymerized actin filaments that are preferred for dendritic nucleation by the Arp2/3 complex. *Curr. Biol.* **12**: 79–84.

Isenberg, G. H., Aebi, U. and Pollard, T. D., 1980. An actin binding protein from *Acanthamoeba* regulates actin filament polymerization and interactions. *Nature* **288**: 455–459.

Lazarides, E. and K. Weber, 1974. Actin antibody: The specific visualization of actin filaments in non-muscle cells. *Proc. Natl. Acad. Sci. USA* **71**: 2268–2272.

Lechler, T., Shevchenko, A. and Li, R., 2000. Direct involvement of yeast type I myosins in Cdc42-dependent actin polymerization. *J. Cell Biol.* **148**: 363–373.

Lee, W. L., Bezanilla, M. and Pollard, T. D. 2000. Fission yeast myosin-I, Myo1p, stimulates actin assembly by Arp2/3 complex and shares functions with WASp. *J. Cell Biol.* **151**: 789–800.

Loisel, T. P., Boujemaa, R., Pantaloni, D. and Carlier, M. F., 1999. Reconstitution of actin-based motility of *Listeria* and *Shigella* using pure proteins. *Nature* **401**: 613–616.

Machesky, L. M. and Insall, R. H., 1998. Scar1 and the related Wiskott–Aldrich syndrome protein WASp regulate the actin cytoskeleton through the Arp2/3 complex. *Curr. Biol.* **8**: 1347–1356.

Machesky, L. M., Mullins, D. M., Higgs, H. N., Kaiser, D. A., *et al.*, 1999. Scar, a WASp-related protein, activates nucleation of actin filaments by the Arp2/3 complex. *Proc. Natl. Acad. Sci. USA* **96**: 3739–3744.

Maciver, S. K., Zot, H. G. and Pollard, T. D., 1991. Characterization of actin filament severing by actophorin from *Acanthamoeba castellanii. J. Cell Biol.* **115**: 1611–1620.

Maly, I. V. and Borisy, G. G., 2001. Self-organization of a propulsive actin network as an evolutionary process. *Proc. Natl. Acad. Sci. USA* **98**: 11324–11329.

Marchand, J. B., Kaiser, D. A., Pollard, T. D. and Higgs, H. N., 2001. Interaction of WASp/Scar proteins with actin and vertebrate Arp2/3 complex. *Nat. Cell Biol.* **3**: 76–82.

Mogilner, A. and Edelstein-Keshet, L., 2002. Regulation of actin dynamics in rapidly moving cells: a quantitative analysis. *Biophys. J.* **83**: 1237–1258.

Mogilner, A. and Oster, G., 1996. Cell motility driven by actin polymerization. *Biophys. J.* **71**: 3030–3045.

Mullins, R.D., Heuser, J. A. and Pollard, T. D., 1998a. The interaction of Arp2/3 complex with actin: nucleation, high-affinity pointed end capping, and formation of branching networks of filaments. *Proc. Natl. Acad. Sci. USA* **95**: 6181–6186.

Mullins, R. D., Kelleher, J. F., Xu, J. and Pollard, T. D., 1998b. Arp2/3 complex from *Acanthamoeba* binds profilin and crosslinks actin filaments. *Mol. Biol. Cell.* **9**: 841–852.

Pantaloni, D., Boujemaa, R., Didry, D., Gounon, P. and Carlier, M. F., 2000. The Arp2/3 complex branches filament barbed ends: functional antagonism with capping proteins. *Nat. Cell Biol.* **2**: 385–391.

Perelroizen, I., Marchand, J. B., Blanchoin, L., Didry, D. and Carlier, M. F., 1994. Interaction of profilin with G-actin and poly(L-proline). *Biochem.* **33**: 8472–8478.

Pollard, T. D., 1986. Rate constants for the reactions of ATP- and ADP-actin with the ends of actin filaments. *J. Cell Biol.* **103**: 2747–2754.

Pollard, T. D. and Cooper, J. A., 1984. Quantitative analysis of the effect of *Acanthamoeba* profilin on actin filament nucleation and elongation. *Biochemistry* **23**: 6631–6641.

Pollard, T. D., Blanchoin, L. and Mullins, R. D., 2000. Molecular mechanisms controlling actin filament dynamics in nonmuscle cells. *Annu. Rev. Biophys.* **29**: 545–576.

Prehoda, K. E., Scott, J. A., Mullins, R. D. and Lim, W. A., 2000. Integration of multiple signals through cooperative regulation of the N-WASp-Arp2/3 complex. *Science* **290**: 801–806.

Pruyne, D., Evangelista, M., Yang, C., Bi, E., *et al.*, 2002. Role of formins in actin assembly: nucleation and barbed end association. *Science* **297**: 612–615.

Robinson, R. C., Turbedsky, K., Kaiser, D., Higgs, H., *et al.*, 2001. Crystal structure of Arp2/3 complex. *Science* **294**: 1660–1661.

Rohatgi, R., Ma, L., Miki, H., Lopez, M., *et al.*, 1999. The interaction between N-WASp and the Arp2/3 complex links Cdc42-dependent signals to actin assembly. *Cell* **97**: 221–231.

Rohatgi, R., Nollau, P., Ho, H. Y., Kirschner, M. W. and Mayer, B. J., 2001. Nck and phosphatidylinositol 4,5-bisphosphate synergistically activate actin polymerization through the N-WASp-Arp2/3 pathway. *J. Biol. Chem.* **276**: 26448–26452.

Rosenblatt, J., Agnew, B. J., Abe, H., Bamburg, J. R. and Mitchison, T. J., 1997. *Xenopus* actin depolymerizing factor/cofilin (XAC) is responsible for the turnover of actin filaments in *Listeria monocytogenes* tails [see comments]. *J. Cell Biol.* **136**: 1323–1332.

Rosenblatt, J., Peluso, P. and Mitchison, T. J., 1995. The bulk of unpolymerized actin in *Xenopus* egg extracts is ATP-bound. *Mol. Biol. Cell* **6**: 227–236.

Safer, D. and Nachmias, V. T., 1994. Beta thymosins as actin binding peptides. *Bioessays* **16**: 473–479.

Sagot, I., Rodal, A. A., Moseley, J., Goode, B. L. and Pellman, D., 2002. An actin nulceation mechanism mediated by Bni 1 and Profilin. *Nat. Cell Biol.* **4**: 626–631.

Schafer, D. A., Jennings, P. B. and Cooper, J. A., 1996. Dynamics of capping protein and actin assembly in vitro: uncapping barbed ends by polyphosphoinositides. *J. Cell Biol.* **135**: 169–179.

Small, J. V., Isenberg, G. and Celis, J. E., 1978. Polarity of actin at the leading edge of cultured cells. *Nature* **272**: 638–639.

Stein, E. and Tessier-Lavigne, M., 2001. Hierachical organization of guidance receptors: silencing of netrin attraction by slit through a Robo/DCC receptor complex. *Science* **291**: 1928–1938.

Svitkina, T. M. and Borisy, G. G., 1999. Progress in protrusion: the tell-tale scar. *Trends Biochem. Sci.* **24**: 432–436.

Svitkina, T. M., Verkhovsky, A. B., McQuade, K. M. and Borisy, G. G., 1997. Analysis of the actin-myosin II system in fish epidermal keratocytes: mechanism of cell body translocation. *J. Cell Biol.* **139**: 397–415.

Theriot, J. A. and Mitchison, T. J., 1991. Actin microfilament dynamics in locomoting cells. *Nature* **352**: 126–131.

Vinson, V. K., De La Cruz, E. M., Higgs, H. N. and Pollard, T. D., 1998. Interactions of *Acanthamoeba* profilin with actin and nucleotides bound to actin. *Biochem.* **37**: 10871–10880.

Wang, Y., 1985. Exchange of actin subunits at the leading edge of living fibroblasts: Possible role of treadmilling. *J. Cell Biol.* **101**: 597–602.

Watanabe, N. and Mitchison, T. J., 2002. Single-molecule speckle analysis of actin filament turnover in lamellipodia. *Science* **295**: 1083–1086.

Weaver, A. M., Karginov, A. V., Kinley, A. W., Weed, S. A., Li, Y., Parsons, J. T. and Cooper, J. A., 2001. Cortactin promotes and stabilizes Arp2/3-induced actin filament network formation. *Curr. Biol.* **11**: 370–374.

Welch, M. D., Rosenblatt, J., Skoble, J., Portnoy, D. A. and Mitchison, T. J., 1998. Interaction of human Arp2/3 complex and the *Listeria monocytogenes* ActA protein in actin filament nucleation. *Science* **281**: 105–108.

Winter, D., Lechler, T. and Li, R., 1999. Activation of the yeast Arp2/3 complex by Bee1p, a WASp-family protein. *Curr. Biol.* **9**: 501–504.

Witke, W., Sharpe, A. H., Hartwig, J. H., Azuma, T., *et al.*, 1995. Hemostatic, inflammatory, and fibroblast responses are blunted in mice lacking gelsolin. *Cell* **81**: 41–51.

Yarar, D., To, W., Abo, A. and Welch, M. D., 1999. The Wiskott–Aldrich syndrome protein directs actin-based motility by stimulating actin nucleation with the Arp2/3 complex. *Curr. Biol.* **9**: 555–558.

Zebda, N., Bernard, O., Bailly, M., Welti, S., *et al.*, 2000. Phosphorylation of ADF/cofilin abolishes EGF-induced actin nucleation at the leading edge and subsequence lamellipod extension. *J. Cell Biol.* **151**: 1119–1128.

2

The Role of Talin and Myosin VII in Adhesion – A FERM Connection

Margaret A. Titus

Cell–surface and cell–cell contact is mediated by surface receptors that are intimately linked to the actin cytoskeleton. A number of actin binding and signalling proteins are important for establishing these adhesions and it has recently been shown that unconventional myosins also play a role. *Dictyostelium* cells lacking a class VII myosin adhere poorly to surfaces and are defective in calcium-dependent cell–cell adhesion, a phenotype strikingly similar to talin null cells. Both proteins bind actin and possess one or more FERM domains, a module found in a large number of adhesion proteins. The conserved role of talin throughout phylogeny and the potential shared function between myosin VII and the closely related class X myosins of mammals suggests that lessons learned about *Dictyostelium* adhesion will provide insight into the conserved function for these proteins.

Adhesion between a cell and the substrate is essential for efficient, directed cell migration and phagocytosis and cell–cell adhesion is critically important for morphogenesis during multicellular development. Engagement of cell surface receptors by extracellular ligands results in the stimulation of cell signalling pathways that ultimately activate the cytoskeleton (Geiger *et al.*, 2001). This, in turn, drives the extension of an actin-rich protrusion such as a lamellipodium or phagocytic cup in the direction of the contact or results in

---

Cell Motility. Edited by Anne Ridley, Michelle Peckham and Peter Clark.
 ISBN 0 470 84872 3

the movement of epithelial sheets. The sequence of molecular events that start with adhesion of a cell to a surface or another cell and ends with cytoskeletal activation remains to be defined fully. *Dictyostelium discoideum* is a lower eukaryote that has been exploited in studies of chemotaxis, phagocytosis and development due to its well characterized chemotactic and phagocytic behaviours, simplicity of its developmental cycle, the ability to inactivate genes either by design or randomly and the wide range of assays available for analysing these mutants. Many of its behaviours can be readily manipulated in the laboratory and it is possible to obtain sufficiently large quantities of cells to enable the isolation and characterization of molecules involved in a process of interest. The recent discovery of interesting new mutants with adhesion defects that exhibit altered phagocytic, motility or developmental behaviours has focused new attention on this system for the study of links between cell surface receptors and the cytoskeleton.

Dictyostelium are highly motile and make close contact with the substratum over a broad region during migration (Gingell *et al.*, 1982; Weber *et al.*, 1995). This contrasts with the slower moving cultured cells that are more familiar to cell biologists. Cells such as fibroblasts make smaller contacts with the substratum, termed focal contacts while the cell is actively moving and focal adhesions when the cell is stationary (Wehrle-Haller and Imhof, 2002). The type of broad contact that *Dictyostelium* amoebae make with a substratum is not unique to this cell type. Leukocytes, similarly highly motile cells, also engage wide regions of substrate during movement indicating that this mechanism of adhesion is perhaps better suited for relatively rapid migration (Friedl *et al.*, 2001). Studies in an easily manipulatable system such as *Dictyostelium* should, therefore, provide general insights into how fast-moving cells from a range of organisms make optimal contacts with the surface along which they must move.

Dictyostelium amoebae are professional phagocytes and subsist on bacteria in the wild. Depletion of this food supply initiates the developmental programme that is characterized by a halt in cell division and expression of genes required for chemotaxis (Loomis, 1982), such as the chemoattractant receptor cAR1 and adenylyl cyclase (Devreotes, 1994). Cells begin to secrete the chemoattractant cAMP and as one source of cAMP begins to predominate cells migrate towards it via a signal-relay system. In addition to chemotaxis genes, genes encoding cell adhesion molecules are also turned on in sequence (Coates and Harwood, 2001). The cells begin to adhere to each other in a head-to-tail fashion at the same time as a combination of calcium- and magnesium-dependent cell–cell contacts are made (Loomis, 1982; Coates and Harwood, 2001). Observation of aggregating cells reveals that the formation of these contacts is preceded by the extension and interdigitation of filopodia from one cell to another (Choi and Siu, 1987; Sesaki and Siu, 1996). Typically, a total of 1×10^5 cells stream together to form a mound and during the later stages of this process they adhere to each other in a Ca^{2+}-independent manner. Once the mound is formed, differentiation and morphogenesis begin, culminating in the formation of a mature fruiting body comprised of a stalk and a spore head. The reappearance of food results in the germination of the spores and re-emergence of amoebae.

The adhesion systems in *Dictyostelium* amoebae are, in some ways, specific to this organism as in their native environments, such as the leaf litter of the forest floor, they encounter a wide range of surfaces along which to move or bacteria to ingest. Therefore, a given *Dictyostelium* amoebae extracellular receptor may have a wide range of ligands in order to provide the cell with maximal adhesion. Similarly, the molecules known to play a role in cell–cell adhesion during streaming and morphogenesis are specific for *Dictyostelium* but they do share general features with adhesion molecules in higher eukaryotes (Cornillon *et al.*, 2000; Fey *et al.*, 2002). Furthermore, adherens junctions have been observed close to the top of the stalk in mature fruiting bodies (Grimson *et al.*, 2000). These junctions appear to have actin filaments emanating from either side and a β-catenin homologue, *aardvark*, is localized to these structures. The β-catenin null mutant exhibits altered junctions and the fruiting bodies lack mechanical stability (Grimson *et al.*, 2000). Additionally, a novel *Dictyostelium* protein, AmpA, that possesses both disintegrin and ornitin domains appears to act as an anti-adhesion molecule, modulating cell–cell contacts that must occur during morphogenesis and late development (Varney *et al.*, 2002). Thus, while *Dictyostelium* employs organism-specific adhesion and anti-adhesion molecules throughout its life cycle, the general themes of cell–cell adhesion/de-adhesion, signalling and junction formation are found indicating a general conservation of mechanism throughout phylogeny.

Dictyostelium cell surface receptors are linked to cytoskeletal proteins, as observed for higher eukaryotes. Support for this comes from the observation that mutant strains, which lack various actin-binding proteins, have reduced adhesion (Niewöhner *et al.*, 1997; Tuxworth *et al.*, 2001; Han *et al.*, 2002). The loss of adhesion not only alters the motility of these cells but in many cases (but not all) phagocytosis is also impaired. This reflects the fact that pseudopod and phagocytic cup extension share the same general mechanism and machinery – engagement of cell surface receptors stimulates directed actin polymerization along a surface. In fact, the two behaviours appear to be mutually exclusive as it has been observed that cells typically stop to eat, recruiting leading edge components to the phagocytic cup (Maniak *et al.*, 1995). Mutants in VASP, talin and myosin VII (DdM7) are each defective in adhesion during motility and the talin and DdM7 mutants additionally show reduced phagocytic activity (Niewöhner *et al.*, 1997; Tuxworth *et al.*, 2001; Han *et al.*, 2002). Interestingly, both talin and M7 are linked to adhesion in higher eukaryotes, thus their function appears to be conserved throughout phylogeny.

Progress has been made in defining the molecules involved in adhesion during the growth phase as well as the early (i.e., during aggregation or streaming) and late (i.e., morphogenesis) developmental phases of *Dictyostelium*. The adhesion mechanism(s) employed during the growth phase and early development that are characterized by high levels of single cell migration will be highlighted here.

Adhesion receptors in *Dictyostelium*

The original studies of adhesion in *Dictyostelium* focused on molecules required for cell–cell contact during early development. Isolated antibodies directed against *Dictyostelium* cells were observed to block either early calcium-dependent or later calcium-independent cell–cell adhesion (Beug *et al.*, 1973). The molecules responsible for both of these activities have been isolated and characterized. Calcium-dependent adhesion is mediated by a 24 000 mw molecule DdCAD-1/gp24 (Knecht *et al.*, 1987; Brar and Siu, 1993) so named because it binds Ca^{2+}, it has some homology to the first and second extracellular repeats of E-cadherin, and forms homophilic interactions (Brar and Siu, 1993; Wong *et al.*, 1996). Antibodies specific for this molecule block calcium-dependent cell–cell adhesion during early development (Knecht *et al.*, 1987; Brar and Siu, 1993) and the null mutant also shows reduced calcium-dependent adhesion (Wong *et al.*, 2002). The DdCAD-1 null forms normal streams but completes the streaming process early, forming mounds 1–2 h ahead of the control cells. The resulting mounds are larger than normal and often form multiple finger structures yet the mutants do complete the developmental cycle and form normal fruiting bodies (Wong *et al.*, 2002). These findings indicate that another molecule must participate in adhesion during the initial aggregation process. Perhaps the as yet uncharacterized Mg^{2+}-dependent adhesions (Fontana, 1993) can account for the ability of the mutant cells to stream normally. It should be noted that a second DdCAD gene has been found in the *Dictyostelium* genome and this could encode the protein responsible for this second class of divalent cation dependent adhesion molecule (Coates and Harwood, 2001). However, it is clear that DdCAD-1 is required for correct mound formation and normal morphogenesis.

DdCAD-1 neither has a signal sequence nor a transmembrane domain and all available data indicate that it can exist both as soluble in the cytoplasm as well as on the surface of cells (Sesaki and Siu, 1996; Wong *et al.*, 1996). It appears to be exported out to the cell surface via an unusual transport pathway through the contractile vacuole (Sesaki *et al.*, 1997). Nothing is known about the mechanism by which this protein is linked to the external plasma membrane.

The second major adhesion molecule is contact site A (csA), also referred to as gp80, an 80 000 mw protein expressed late in the streaming process (8 h post-starvation). Surprisingly, csA/gp80 null mutants do not exhibit any developmental defects under standard laboratory conditions (Harloff *et al.*, 1989). This may due to the precocious expression of gp150/lagC (Wang *et al.*, 2000), a third adhesion molecule normally expressed during early mound formation (10 h post-starvation). However, when the csA/gp80 null cells are developed on soil plates that more closely mimic the native environment of *Dictyostelium*, the mutants do not develop efficiently and, interestingly, appear to adhere more avidly to the substratum (Ponte *et al.*, 1998). csA/gp80 is a GPI-anchored protein and has been found in association with lipid raft-like complexes in developed cells (12 h post-starvation), along with cytoskeletal proteins such as ponticulin, a major link between the plasma membrane and actin cytoskeleton (Hitt *et al.*, 1994; Harris and Siu, 2002; Harris *et al.*, 2003). While csA/gp80 is required for recruiting ponticulin into the lipid rafts, there is no evidence for a direct association of these two proteins and the molecular details of their (likely indirect) interaction remain to be elucidated (Harris *et al.*, 2003).

The adhesion receptors for vegetative cells are less well-described. Screens for phagocytosis mutants have provided clues about their properties and have resulted in the identification of at least one. Careful phenotypic characterization of the early phagocytosis mutants revealed that *Dictyostelium* bind particles non-specifically through a combination of glucose and hydrophobic receptors (Vogel *et al.*, 1980). The genes mutated in these early studies have not been cloned. However, *Dictyostelium* plasma membranes possess a major 130 kD glycoprotein that appears to be altered in one of these adhesion mutants, HV29 (Chia, 1996). Interestingly, the original antibodies made against gp130 recognize a carbohydrate epitope and these block particle uptake, presumably due to inhibiting adhesion (Chia and Luna, 1989). The 130 kD molecule has not yet been cloned. A later phagocytosis screen resulted in the identification of Phg1, an integral membrane protein with nine transmembrane domains (Cornillon *et al.*, 2000). Phg1 homologues are found throughout phylogeny but their function in other cell types is not yet known. The *phg1* mutant is defective in the uptake of latex beads and *E. coli* but not *Klebsiella aerogenes* bacteria, indicating that this molecule plays a role in adhesion to a subset of surfaces (Cornillon *et al.*, 2000).

A deceptively simple and direct screen for molecules required for cell–substrate adhesion resulted in the identification of an adhesion receptor with attributes of receptors found higher eukaryotes. A population of cells mutagenized through the random insertion of a plasmid was plated and then the dish subjected to gentle shaking. The non-adherent cells were collected and these cells were subjected to the same screening process for several rounds. Cloning of one mutant locus identified in this collection, *sad*A (surface-adhesion defective), revealed that the gene encodes an integral membrane protein with nine transmembrane domains, similar to Phg1, and an extracellular region that contains three EGF repeats (Fey *et al.*, 2002). In spite of the gross topological similarity between Phg1 and SadA, these two molecules are quite distinct from each other. *sadA* mutants exhibit defects in phagocytosis and actin organization of growth-phase cells and also have a cytokinesis defect. Interestingly, the growth phase *sad*A null cells move at a faster rate than wild type cells. Aggregation-phase cells do not exhibit alterations in either adhesion or actin organization and motility is normal. Consistent with its role in adhesion, SadA is localized around the periphery of the cell and is not particularly concentrated either in phagocytic cups or at the leading edge of migrating cells. A number as yet uncharacterized mutants were obtained in this screen (Fey *et al.*, 2002), as one would expect, and continued analysis of this collection should provide additional, valuable information about the means by which *Dictyostelium* adhere to surfaces, the various ligands that these cells potentially bind to and how this receptor is associated with the cytoskeleton.

Links between the *Dictyostelium* cytoskeleton and adhesion

Studies in mammalian cells have established clearly that integral membrane receptors are linked to the underlying actin cytoskeleton. Three different *Dictyostelium* actin binding proteins, VASP (vasodilator-stimulated phosphoprotein), talinA, DdM7 have been shown to play a role in adhesion during growth phase and early aggregation. Studies of how these proteins are linked to extracellular receptors are in their infancy but the available data do provide some clues and directions for future research.

The Ena/VASP family of proteins regulates actin assembly and plays a role in cell motility (Bear *et al.*, 2000; Krause *et al.*, 2002). Experiments in fibroblasts have shown that VASP competes with capping protein for the ends of actin filaments and cells that do not localize this protein to the leading edge have shorter, more branched actin filaments in the lamellipodium and move more rapidly. Conversely, increasing the concentration of VASP in the lamellipodium results in longer filaments and slower moving cells (Bear *et al.*, 2002). *Dictyostelium* has a single VASP gene and GFP-VASP is concentrated at the leading edge of chemotactic cells, transiently enriched in filopodia and found at sites of cell–cell contact (Han *et al.*, 2002). VASP null cells exhibit alterations in actin distribution and extend few, if any filopodia. A quantitative, three-dimensional analysis of the motility of these null cells revealed that protrusions extended off the substratum do not make contact with the surface for unusually long periods of time, suggesting an adhesion defect. These cells do move slower than controls and tend to move with reduced persistence, changing direction frequently. Additional support for a defect in adhesion comes from an analysis of particle binding – the VASP null cells exhibit significantly reduced binding of 1 μm latex beads when compared with controls. This lower binding, surprisingly, does not result in a corresponding phagocytosis defect (Han *et al.*, 2002).

Integrins are major cell–surface receptors in higher eukaryotes and binding to their extracellular ligand results in the recruitment of a number of cytoplasmic proteins, including signalling and actin-binding proteins (Geiger *et al.*, 2001). One of these proteins is talin. Talin interacts with integrin cytoplasmic tails, the focal adhesion protein vinculin and actin. Thus, it is positioned to play a key role in linking external attachments to the actin cytoskeleton (Calderwood *et al.*, 2000). The N-terminus of talin harbours a FERM domain, a region of shared homology between band 4.1, ezrin, radixin and moesin, and the remaining portion of the molecule is predicted to be rod-like. The rod domain harbours a binding site for vinculin and actin (Hemmings *et al.*, 1996). The FERM domain from a number of ERM proteins has been crystallized and found to consist of three subdomains arranged in a clover-leaf like shape (Hamada *et al.*, 2000; Pearson *et al.*, 2000). The structure of the 'F3' FERM subdomain is remarkably similar to that of a phosphotyrosine binding (PTB) domain and it interacts with the NpxY motif of the cytoplasmic tail of β integrin (Calderwood *et al.*, 1999; García-Alvarez *et al.*, 2003). The FERM domain of talin also binds to a specific phosphoinositol phosphate kinase type 1γ (PIPK1γ) isoform that is targeted to focal adhesion (Di Paolo *et al.*, 2002; Ling *et al.*, 2002). Stimulation of PIPK1γ activity by adhesion results in increased PIP(4,5)P_2 production that, in turn, enhances the interaction of vinculin with actin and talin.

Dictyostelium expresses two distinct forms of talin – talinA and talinB (Kreitmeier *et al.*, 1995; Tsujioka *et al.*, 1999) (Figure 2.1). These two proteins are highly similar at their N- and C-termini, but are otherwise quite distinct. Most notably, TalB has a villin headpiece at its C-terminus (Tsujioka *et al.*, 1999). TalA was identified in a search for actin binding proteins present in a triple null mutant lacking three major actin binding proteins (gelation factor, α-actinin and severin) by F-actin affinity chromatography (Kreitmeier *et al.*, 1995). TalA is present in the cytosol, enriched at the leading edge of chemotactic cells and present in filopodia (hence its original name, filopodin). TalA has also recently been shown to be present in small dot-like structures on the bottom of the cell that are reminiscent of focal contacts (Hibi *et al.*, 2003). The *tal*A null mutant is defective in adhesion to the substratum as observed by interference reflection contrast microscopy (IRM). The loss of substrate binding results in decreased phagocytic activity and abnormal motility but does not appear to affect the formation of filopodia (Niewöhner *et al.*, 1997; Tuxworth *et al.*, 2001). The *tal*A null cells also exhibit a mild cytokinesis defect and decreased membrane bending modulus (Niewöhner *et al.*, 1997;

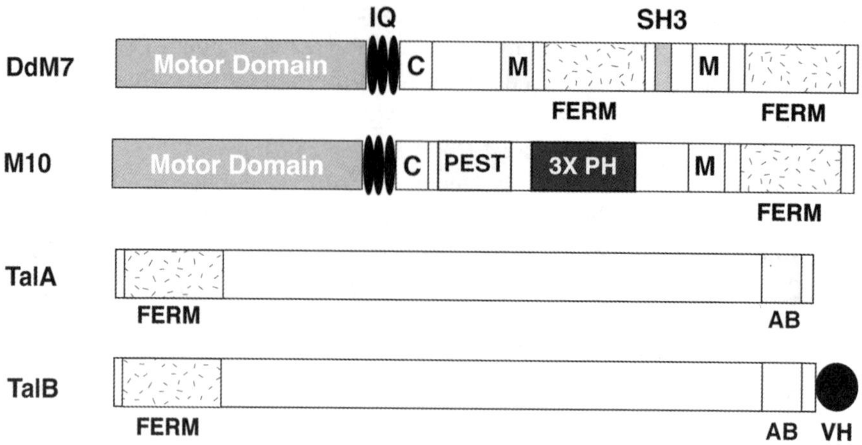

Figure 2.1 Schematic illustration of the FERM proteins M7, M10 and talin. Shown are box diagrams of each molecule, with the individual domains indicated. The motor domain of *Dictyostelium* M7 (DdM7) is shaded grey, the light chain-binding IQ motifs are dark ovals, the region of predicted coiled-coil is indicated by C, the MyTH4 domains by M, FERM domains by the stippled box and the SH3 domain by the lighter grey box. The heavy chain of M10 has a PEST sequence in the N-terminal region of the tail and the region spanning the 3 PH domains is indicated by the dark box (3X PH). TalinA (TalA) and TalinB (TalB) are also shown, each with their N-terminal FERM domains and C-terminal actin-biding regions (AB). TalB also possess a villin headpiece (VH) at its C-terminus

Simson *et al.*, 1998). Interestingly, Ca^{2+}-dependent cell–cell binding of aggregating cells is reduced significantly in the *tal*A null mutant (Niewöhner *et al.*, 1997).

Talins also play a role in multicellular development. *Dictyostelium* cells lacking TalB stream and form mounds normally but halt at this stage as they fail to form a tip that is associated with differentiation (Tsujioka *et al.*, 1999). However, cells in the *tal*B null mounds do undergo cell differentiation properly suggesting that the block in development is due to a failure in morphogenesis. These results indicate that the two talins have unique roles in *Dictyostelium* and continued analysis of TalB should reveal its role in adhesion in development. Deletion of talin from mouse embryonic stem cells results in the inability of undifferentiated cells to adhere to laminin and lower levels of $\beta 1$ integrin. Homozygous talin null mice are not viable and die at 8.5 to 9.5 days post-coitum, due to defects in cell migration at gastrulation (Monkley *et al.*, 2000). These results show that the role of talin has been generally conserved throughout evolution.

M7 is an actin-based motor protein essential for hearing in humans, mice and zebrafish and adhesion in *Dictyostelium* (Gibson *et al.*, 1995; Weil *et al.*, 1995; Titus, 1999; Ernest *et al.*, 2000; Tuxworth *et al.*, 2001). The M7a heavy chain comprises an N-terminal motor domain and a tail region that has a brief region of predicted coiled-coil followed by a tandem repeat of MyTH4 (myosin tail homology 4)/FERM domains separated by an SH3 (*src* homology 3) domain (Hasson, 1999) (Figure 2.1) Its expression is generally restricted to specialized cell types that extend actin-filled projections such as the retinal pigment epithelium of the eye, the sensory hair cells of the ear and the intestinal epithelium, to name a few (Hasson *et al.*, 1995). Mutations in M7a result in Usher syndrome type 1B in humans, one of a collection of Usher syndrome subtypes that are characterized by deafness and late onset retinitis pigmentosa (Petit, 2001). Mice and zebrafish lacking M7a are deaf but do not exhibit any overt visual defects. Detailed analyses of mouse M7a revealed that it is expressed in the inner and outer hair cells of the cochlea where it is both found in the cytoplasm as well as in the actin-rich stereocilia (SC) (Hasson *et al.*, 1997). M7a appears to be concentrated within the SC at the sites of lateral links between adjacent SC. These links are believed to be formed by at least two different types of cadherin, cadherin-23 and protocadherin-15 (Alagramam *et al.*, 2001; Di Palma *et al.*, 2001). Careful phenotypic analysis of the mouse M7a mutant, *shaker-1* (*sh1*), reveals that their cochlear hair cells do extend SC but instead of being tightly organized together in a bundle they are splayed apart (Self *et al.*, 1998). These findings suggested that M7a plays a role in anchoring receptors involved in forming the SC-SC links to the underlying polarized actin bundles.

Two independent lines of investigation provide support for this hypothesis. First, vezatin, a novel protein that binds to the C-terminal FERM domain of M7a is present at cell–cell junctions and it can recruit the tail of M7a to these regions of the cell (Küssel-Andermann *et al.*, 2000). Vezatin, in turn, is associated with the cadherin–catenin complex (Küssel-Andermann *et al.*, 2000). Secondly, a systematic effort to identify the genes mutated in the different subtypes of Usher syndrome type 1 that share the same gross phenotype has resulted in the identification of harmonin, a PDZ domain-containing protein that is mutant in Usher syndrome type 1C, as a M7a binding partner (Boëda *et al.*, 2002). The C-terminal FERM domain of M7a interacts with the N-terminal PDZ domain of harmonin. Interestingly, harmonin is also capable of binding actin via its C-terminus and the second N-terminal PDZ binding domain interacts with the cytoplasmic tail of cadherin-23. Mutations in cadherin-23 result in Usher syndrome type 1D in humans and the *waltzer* deafness mutation in mice (Bolz *et al.*, 2001; Di Palma *et al.*, 2001). As expected, the SC from the cochlear hairs cells of *waltzer* mice appear splayed apart in much the same manner as observed for those in the *sh1* mice. Together, these results provide strong support for the model in which M7a plays a key role in anchoring SC–SC links to the actin cytoskeleton and is essential for maintaining the integrity of the hair bundle in the sensory cells of the ear.

Dictyostelium express a single M7, DdM7 (Titus, 1999), a surprising finding when one considers the fact that these cells do not extend any stable, highly organized actin-based structures. DdM7 is restricted to the tips of filopodia, the leading edge of chemotactic cells and the phagocytic cup (Tuxworth *et al.*, 2001). Its localization is dynamic. For example, DdM7 is associated with the growing phagocytic cup, but once it is almost fully formed, DdM7 disappears. The DdM7 null mutant has a pronounced adhesion defect (Tuxworth *et al.*, 2001). These cells no longer make broad regions of close contact with the substratum and instead only the rear of the cell maintains a contact with a small surface area. The DdM7 null cells are capable of making pseudopodia but these often protrude off the surface and do not touch down for significant periods of time. These null cells also do not make filopodia. The DdM7 nulls bind particles more poorly and, as one would expect, have a significant phagocytosis defect regardless of the particle type (Titus, 1999; Tuxworth *et al.*, 2001). Aggregation-stage cells also exhibit a defect in Ca^{2+}-dependent cell–cell adhesion; however, the cells are capable of aggregating, undergoing morphogenesis and completing the developmental cycle. Analysis of two known adhesion receptors, Phg1 and gp130, in the DdM7 null mutant as well as localization studies indicates that this motor protein is not acting as a transport motor to supply adhesion receptors to the cell surface. The available data suggest instead that DdM7 acts at the plasma membrane to promote the binding of the cell to surfaces, possibly by cooperating in the formation of a high avidity complex of receptors linked to the underlying actin cytoskeleton (Tuxworth *et al.*, 2001).

It is informative to compare the DdM7 null mutant phenotype with that of the *sh1* mutant mouse. In both mutants, the observed phenotype is consistent with defects in adhesion. The DdM7 mutant does not bind a wide range of surfaces while the defect in the *sh1* mice is apparently highly specific – the link between neighbouring cochlear hair cell SC is not formed, resulting in a disordered and non-functional hair bundle. The SC–SC links are calcium-dependent and formed by cadherins and M7a is linked to one of these, cadherin-23, through an adaptor protein, harmonin. Aggregating DdM7 null mutant do not have Ca^{2+}-dependent cell–cell adhesion. This cell–cell interaction is mediated by DdCAD-1, a protein with some similarity to cadherins. The link between DdM7 and DdCAD-1 is not known at present however it is most certainly an indirect one, as is the case for M7a and cadherin 23 in mammals. Taken together, these observations suggest that the general function of M7a has been throughout evolution and that ongoing studies of DdM7 may provide additional insight into the function of M7a in higher eukaryotes.

A link between M7 and talin?

Comparison of the available *Dictyostelium* adhesion mutants reveals a strong similarity between the DdM7 and *tal*A null mutant phenotypes. These mutants exhibit reduced close contact surface area that results in aberrant motility, are highly defective in the internalization of a variety of particle types (i.e., latex beads, yeast, bacteria) and do not exhibit Ca^{2+}-dependent cell–cell adhesion during early aggregation (Niewöhner *et al.*, 1997; Tuxworth *et al.*, 2001). These observations suggest that TalA and DdM7 may interact with each other and preliminary co-immunoprecipitation experiments indicate that they do (Tuxworth, Stephens and Titus, unpublished).

It is interesting to note that both TalA and DdM7 have FERM domains (Figure 2.1), suggesting that their shared interaction or function may be dictated by the presence of this domain. Furthermore, a number of FERM domain proteins play a role in cell–substrate adhesion. Members of the ERM (ezrin, radixin, moesin) family of actin binding proteins are required for both the extension of actin-based structures such as microvilli, as well as for adhesion to surfaces (Takeuchi *et al.*, 1994; Paglini *et al.*, 1998; Bonilha *et al.*, 1999; Yonemura and Tsukita, 1999). Major components of the focal contact, such as FAK (focal adhesion kinase) possess a FERM domain (Schaller *et al.*, 1995). In the case of mammalian cells, direct links between FERM proteins and adhesion receptors have been established. The best example of this is the direct binding of the talin FERM domain to β integrin or the binding of the Na^+/H^+ exchanger NHE1 to ezrin (Calderwood *et al.*, 1999; Denker *et al.*, 2000). Additionally, a PDZ protein, EBP50, has been shown to bind to the FERM domain of ezrin and moesin, likely mediating the interaction of these proteins with the renal brush border Na^+/H^+ exchanger (Reczek *et al.*, 1997). One possibility is either that TalA and DdM7 bind to the same integral membrane receptor or transmembrane linker protein, providing a connection to DdCAD-1 and organizing the receptor(s) into a high avidity complex. Binding of just one of these proteins would be insufficient for promoting substrate binding. Alternatively, an adaptor protein, similar to EBP50, might first bind to both DdM7 and TalA and then this complex would bind to and organize receptors in the plane of the membrane.

An alternative explanation for the observed similarity of the DdM7 and *tal*A null mutant phenotypes is that TalA serves as a DdM7 receptor. This seems unlikely as the distribution of TalA appears to be broader than that of M7 (Kreitmeier *et al.*, 1995; Tuxworth *et al.*, 2001; Hibi *et al.*, 2003). While it is enriched at the leading edge of a chemotactic cell it is also found all around the periphery of the cell. Also, it is along the length of the filopodia in addition to being localized to the tip. This contrasts with that of M7, which is only found in dynamic regions of the cell, i.e., those areas of the cell undergoing extension, and appears to be largely at the tips of filopodia. Furthermore, while the phenotypes of the two null mutants are quite similar there are some notable differences. The *tal*A mutant extends filopodia while the DdM7 null does not, the phagocytic defect of the *tal*A mutant can be rescued partially by adhesive bacteria while the DdM7 null mutant phenotype can not (Niewöhner *et al.*, 1997; Titus, 1999; Tuxworth *et al.*, 2001). Another striking aspect of the dynamic localization of DdM7 is the gross similarity to that of the $PI(3,4)P_2$/$PI(3,4,5)P_3$ binding protein CRAC (cytosolic regulator of adenylyl cyclase), which in addition to being recruited to the sites of chemoattractant binding is also found in regions of dynamic actin polymerization such as phagocytic cups and macropinocytic crowns (Parent *et al.*, 1998), as is DdM7. It is possible that DdM7 is recruited to the plasma membrane through localized production of $PI(3,4)P_2$ and/or $PI(3,4,5)P_3$ that would occur immediately upon the interaction of a receptor and its substrate. Once DdM7 is recruited to the membrane it might then cooperatively interact with membrane-resident TalA to organize adhesion receptors into the necessary high avidity complex (Figure 2.2).

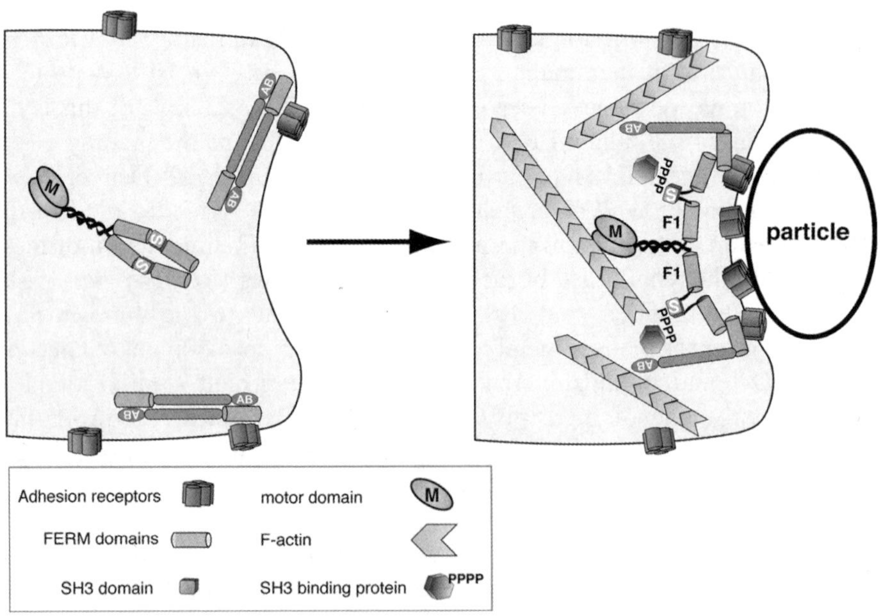

Figure 2.2 Speculative model of the shared contributions of *Dictyostelium* M7 and TalA to adhesion. Shown is a region of a *Dictyostelium* cell prior to (left) and after (right) binding a particle. Prior to engaging a particle or surface, cell surface adhesion receptors are distributed along the surface of the cell, perhaps in association with talin (dimer of elongated rods). A DdM7 dimer may be soluble in the cytoplasm. Initial contact with a particle may trigger release of PIP_2 (not depicted) and result in recruitment of DdM7 to the talin-receptor complex. Conformational changes in both DdM7 and TalA (only one TalA molecule per dimer is shown for clarity) results in the assembly of a high avidity adhesion receptor/TalA/DdM7 complex that then signals to the actin cytoskeleton, perhaps via exposure of the DdM7 SH3 domain that would, in turn, recruit proteins that stimulate localized actin polymerization at the plasma membrane and would drive an extension along the surface

The relationship of DdM7 to another FERM myosin, M10

A number of unconventional myosins possess the combined MyTH4/FERM module in their tail regions (Berg *et al.*, 2001). These include the class VII, X, XII and XV myosins. Inspection of phylogenetic trees of the myosin superfamily reveals that the class VII and X myosins are related to each other (Hodge and Cope, 2000; Berg *et al.*, 2001). Of interest is the fact that DdM7 comes off quite early after the class VII and X myosins diverge. Unlike mammalian M7, myosin X (M10) is expressed more widely as might be expected for a myosin with a shared function in higher and lower eukaryotes (Berg *et al.*, 2000). The M10 tail domain shares some features with that of the M7 tail. It has a stretch of predicted coiled-coil followed by three phospholipid binding PH domains and then a MyTH4/FERM module (Berg *et al.*, 2000) (Figure 2.1). While the presence of the three PH domains is unique to M10 it is worth noting that the FERM domain contains a PH fold (Pearson *et al.*, 2000), suggesting that there may be some shared or related function between these two modules. The PH domains of M10 have been shown to bind to the products of phosphatidylinositol 3-kinase (PI3K) (Isakoff *et al.*, 1998) and the FERM domain of ERM proteins has been shown to bind to PI(4,5)P_2 (Niggli *et al.*, 1995; Hirao *et al.*, 1996). Ongoing investigations into the membrane and potential phospholipid association of M7 and M10 may clarify the significance of the structural homology between the PH and FERM domains.

Further underscoring the potential shared functions of DdM7 and M10, the distribution of M10 is remarkably similar to that of DdM7. It is also found in dynamic, actin rich regions of the cell as well as in filopodia (Berg *et al.*, 2000; Berg and Cheney, 2002). M10 also appears to play a role in filopod formation as overexpression results in the production of increased numbers of filopodia (Berg and Cheney, 2002). In contrast to DdM7, M10 appears to participate in the movement of particles up and down the shaft of the filopodium and it has been proposed that motor-driven movement powers the transport of materials out to the tip of the filopod and the motor then is passively moved out of the filopodium via actin flow (Berg and Cheney, 2002). M10 is expressed in macrophages where it is localized to the phagocytic cup (Cox *et al.*, 2002). Treatment of these cells with the PI3K inhibitor wortmannin results in loss of this localization. Overexpression of the M10 tail region encompassing the three PH domains along with the MyTH4/FERM domain significantly inhibits phagocytosis in a size-dependent manner, with the internalization of 6 μm beads being inhibited while that of 2 μm beads is not. Microinjection of antibodies directed against the motor domain of M10 into macrophages also inhibits phagocytosis. The inhibitory effect of M10 tail expression does not appear to be due to a lack of adhesion but rather to a reduction in the ability of the cells to spread along a surface (Cox *et al.*, 2002) suggesting differences between the contributions of DdM7 and M10 to phagocytosis in these two cell types.

Conclusions

Studies of phagocytosis and adhesion in *Dictyostelium* have begun to uncover adhesion receptors and proteins that link them to the underlying actin cytoskeleton. Continued studies should result in the identification of links between them but the initial results indicate that while the substrates engaged by these receptors are specific to *Dictyostelium*, the same basic principles of operation are at work in this lower eukaryote. The binding of food particles, such as bacteria, stimulates the production of an actin-rich phagocytic cup. Loss of key cytoskeletal proteins such as talin and M7 disables this pathway and results in a severely reduced rate of phagocytosis. At least one receptor, Phg1, is conserved throughout evolution and another, SadA, contains motifs typical of mammalian adhesion receptors. Cell–cell contacts are essential for early development and morphogenesis and one class is Ca^{2+}-dependent and shares some features with the cadherins of higher eukaryotes. Finally, FERM proteins have been shown to play key roles in linking cell surface receptors to the actin cytoskeleton in *Dictysotelium* as has been observed in mammalian systems. This basic conservation of function throughout phylogeny suggests that studies of *Dictyosteium* adhesion mutants with the same phenotypes will expedite the process of characterizing the molecular details of these links between adhesion receptors and the underlying actin cytoskeleton.

Acknowledgements

Work in the Titus laboratory is supported by a grant from the National Institutes of Health as well as an Established Investigator Award from the American Heart Association. The author thanks Dr Richard Tuxworth, Greg Addicks, Stephen Stephens and Zach Ryan for their many contributions, past and present, to the ongoing analysis of DdM7.

References

Alagramam, K. N., Murcia, C. L., Kwon, H. Y., Pawlowski, K. S., *et al.*, 2001. The mouse Ames waltzer hearing-loss mutant is caused by mutation of *Pcdh15*, a novel protocadherin gene. *Nature Genetics* **27**: 99–102.

Bear, J. E., Loureiro, J. J., Libova, I., Fassler, R., *et al.*, 2000: Negative regulation of fibroblast motility by Ena/VASP proteins. *Cell* **101**: 717–728.

Bear, J. E., Svitkina, T. M., Krause, M., Schafer, D. A., *et al.*, 2002. Antagonism between Ena/VASP proteins and actin filament capping regulates fibroblast motility. *Cell* **109**: 509–521.

Berg, J. C., Powell, B. C. and Cheney, R. E., 2001. A millennial myosin census. *Molecular Biol. Cell* **12**: 780–794.

Berg, J. S. and Cheney, R. E., 2002. Myosin-X is an unconventional myosin that undergoes intrafilopodial motility. *Nature Cell Biol.* **4**: 246–250.

Berg, J. S., Derfler, B. H., Pennisi, C. M., Corey, D. P. and Cheney, R. E., 2000. Myosin-X, a novel myosin with pleckstrin homology domains, associates with regions of dynamic actin. *J. Cell Sci.* **113**: 3439–3451.

Beug, H., Katz, F. E. and Gerisch, G., 1973. Dynamics in antigenic membrane sites relating to cell aggregation in *Dictyostelium discoideum*. *J. Cell Biol.* **56**: 647–658.

Boëda, B., El-Amraoui, A., Bahloul, A., Goodyear, R., *et al.*, 2002. Myosin VIIa, harmonin and cadherin 23, three Usher I gene products that cooperate to shape the sensory hair cell bundle. *EMBO J.* **21**: 6689–6699.

Bolz, H., von Brederlow, B., Ramirez, A., Bryda, E.C., *et al.*, 2001. Mutation of *CDH23*, encoding a new member of the cadherin gene family, causes Usher syndrome type 1D. *Nature Genetics* **27**: 108–112.

Bonilha, V. L., Finnemann, S. C. and Rodriguez-Boulan, E., 1999. Ezrin promotes morphogenesis of apical microvilli and basal infoldings in retinal pigment epithelium. *J. Cell Biol.* **147**: 1533–1548.

Brar, S. K. and Siu, C. H., 1993. Characterization of the cell adhesion molecule gp24 in *Dictyostelium discoideum* – mediation of cell–cell adhesion via a Ca^{2+}-dependent mechanism. *J. Biol. Chem.* **268**: 24902–24909.

Calderwood, D. A., Zent, R., Grant, R., Rees, D. J., *et al.*, 1999. The talin head domain binds to integrin beta subunit cytoplasmic tails and regulates integrin activation. *J. Biol. Chem.* **274**: 28071–28074.

Calderwood, D. A., Shattil, S. J. and Ginsberg, M. H., 2000. Integrins and actin filaments: reciprocal regulation of cell adhesion and signaling. *J. Biol. Chem.* **275**: 22607–22610.

Chia, C. P., 1996. A 130-kDa plasma membrane glycoprotein involved in *Dictyostelium* phagocytosis. *Exp. Cell Res.* **227**: 182–189.

Chia, C. P. and Luna, E. J., 1989. Phagocytosis in *Dictyostelium discoideum* is inhibited by antibodies directed primarily against common carbohydrate epitopes of a major cell-surface plasma membrane glycoprotein. *Exp. Cell Res.* **181**: 11–26.

Choi, A. H. and Siu, C. H., 1987. Filopodia are enriched in a cell cohesion molecule of Mr 80,000 and participate in cell–cell contact formation in *Dictyostelium discoideum*. *J. Cell Biol.* **104**: 1375–1387.

Coates, J. C. and Harwood, A. J., 2001. Cell-cell adhesion and signal transduction during *Dictyostelium* development. *J. Cell Sci.* **114**: 4349–4358.

Cornillon, S., Pech, E., Benghezal, M., Ravanel, K., *et al.*, 2000. Phg1p is a nine-transmembrane protein superfamily member involved in *Dictyostelium* adhesion and phagocytosis. *J. Biol. Chem.* **275**: 34287–34292.

Cox, D., Berg, J. S., Cammer, M., Chinegwundoh, J. O., *et al.*, 2002. Myosin X is a downstream effector of PI(3)K during phagocytosis. *Nature Cell Biol.* **4**: 469–477.

Denker, S. P., Huang, D. C., Orlowski, J., Furthmayr, H. and Barber, D. L., 2000. Direct binding of the Na–H exchanger NHE1 to ERM proteins regulates the cortical cytoskeleton and cell shape independently of H^+ translocation. *Molecular Cell* **6**: 1425–1436.

Devreotes, P. N., 1994. G protein-linked signaling pathways control the developmental program of *Dictyostelium*. *Neuron* **12**: 235–241.

Di Palma, F., Holme, R. H., Bryda, E. C., Belyantseva, I. A., *et al.*, 2001. Mutations in *Cdh23*, encoding a new type of cadherin, cause stereocilia disorganization in waltzer, the mouse model for Usher syndrome type 1D. *Nature Genetics* **27**: 103–107.

Di Paolo, G., Pellegrini, L., Letinic, K., Cestra, G., *et al.*, 2002. Recruitment and regulation of phosphatidylinositol phosphate kinase type 1 gamma by the FERM domain of talin. *Nature* **420**: 85–89.

Ernest, S., Rauch, G. J., Haffter, P., Geisler, R., *et al.*, 2000. *Mariner* is defective in *myosin VIIA*: a zebrafish model for human hereditary deafness. *Human Molec. Genetics* **9**: 2189–2196.

Fey, P., Stephens, S., Titus, M. A. and Chisholm, R. L., 2002. SadA, a novel adhesion receptor in *Dictyostelium*. *J. Cell Biol.* **159**: 1109–1119.

Fontana, D. R. (1993) Two distinct adhesion systems are responsible for EDTA-sensitive adhesion in *Dictyostelium discoideum*. *Differentiation* **53**: 139–147.

Friedl, P., Borgmann, S. and Brocker, E. B., 2001. Amoeboid leukocyte crawling through extracellular matrix: lessons from the *Dictyostelium* paradigm of cell movement. *J. Leukocyte Biol.* **70**: 491–509.

García-Alvarez, B., de Pereda, J. M., Calderwood, D. A., Ulmer, T. S., *et al.*, 2003. Structural determinants of integrin recognition by talin. *Molecular Cell* **11**: 49–58.

Geiger, B., Bershadsky, A., Pankov, R. and Yamada, K. M., 2001. Transmembrane crosstalk between the extracellular matrix–cytoskeleton crosstalk. *Nature Rev. Molecular Cell. Biol.* **2**: 793–805.

Gibson, F., Walsh, J., Mburu, P., Varela, A., *et al.*, 1995. A type VII myosin encoded by the mouse deafness gene *shaker-1*. *Nature* **374**: 62–64.

Gingell, D., Todd, I. and Owens, N., 1982. Interaction between intracellular vacuoles and the cell suface analysed by finite aperture theory interference reflection microscopy. *J. Cell Sci.* **54**: 287–298.

Grimson, M. J., Coates, J. C., Reynolds, J. P., *et al.*, 2000. Adherens junctions and beta-catenin-mediated cell signalling in a non-metazoan organism. *Nature* **408**: 727–731.

Hamada, K., Shimizu, T., Matsui, T., Tsukita, S. and Hakoshima, T., 2000. Structural basis of the membrane-targeting and unmasking mechanisms of the radixin FERM domain. *EMBO J.* **19**: 4449–4462.

Han, Y. H., Chung, C. Y., Wessels, D., Stephens, S., *et al.*, 2002. Requirement of a vasodilator-stimulated phosphoprotein family member for cell adhesion, the formation of filopodia, and chemotaxis in *Dictyostelium*. *J. Biol. Chem.* **277**: 49877–49887.

Harloff, C., Gerisch, G. and Noegel, A. A., 1989. Selective elimination of the contact site A protein of *Dictyostelium discoideum* by gene disruption. *Genes Dev.* **3**: 2011–2019.

Harris, T. J. and Siu, C. H., 2002. Reciprocal raft–receptor interactions and the assembly of adhesion complexes. *Bioessays* **24**: 996–1003.

Harris, T. J., Ravandi, A., Awrey, D. E. and Siu, C. H., 2003. Cytoskeleton interactions involved in the assembly and function of glycoprotein-80 adhesion complexes in *Dictyostelium*. *J. Biol. Chem.* **278**: 2614–2623.

Hasson, T., 1999. Sensing a function for myosin-VIIa. *Current Biology* **9**: R838–R841.

Hasson, T., Heintzelman, M. B., Santos-Sacchi, J., Corey, D. P. and Mooseker, M. S., 1995. Expression in the cochlea and retina of myosin VIIa, the gene product defective in Usher syndrome type 1B. *Proc. Natl. Acad. Sci.* **92**: 9815–9819.

Hasson, T., Gillespie, P. G., Garcia, J. A., MacDonald, R. B., *et al.*, 1997. Unconventional myosins in inner-ear sensory epithelia. *J. Cell Biol.* **137**: 1287–1307.

Hemmings, L., Rees, D. J., Ohanian, V., Bolton, S. J., *et al.*, 1996. Talin contains three actin-binding sites each of which is adjacent to a vinculin-binding site. *J. Cell Sci.* **109**: 2715–2726.

Hibi, M., Nagasaki, A., Takahashi, M., Yamagishi, A. and Uyeda, T. Q. P., 2003. *Dictyostelium discoideum* talin A is crucial for myosin-II independent and adhesion-dependent cytokinesis. *J. Muscle Res. Cell Motility* **23**: in press.

Hirao, M., Sato, N., Kondo, T., Yonemura, S., *et al.*, 1996. Regulation mechanism of ERM (ezrin/radixin/moesin) protein/plasma membrane association: possible involvement of phosphatidylinositol turnover and Rho-dependent signaling pathway. *J. Cell Biol.* **135**: 37–51.

Hitt, A. L., Hartwig, J. H. and Luna, E. J., 1994. Ponticulin is the major high affinity link between the plasma membrane and the cortical actin network in *Dictyostelium*. *J. Cell Biol.* **126**: 1433–1444.

Hodge, T. and Cope, M. J. T. V., 2000. A myosin family tree. *J. Cell Sci.* **113**: 3353–3354.

Isakoff, S. J., Cardozo, T., Andreev, J., Li, Z., *et al.*, 1998. Identification and analysis of PH domain-containing targets of phosphatidylinositol 3-kinase using a novel *in vivo* assay in yeast. *EMBO J.* **17**: 5374–5387.

Knecht, D. A., Fuller, D. L. and Loomis, W. F., 1987. Surface glycoprotein, gp24, involved in early adhesion of *Dictyostelium discoideum*. *Dev. Biol.* **121**: 277–283.

Krause, M., Bear, J. E., Loureiro, J. J. and Gertler, F. B., 2002. The Ena/VASP enigma. *J. Cell Sci.* **115**: 4721–4726.

Kreitmeier, M., Gerisch, G., Heizer, C. and Müller-Taubenberger, A., 1995. A talin homologue of *Dictyostelium* rapidly assembles at the leading edge of cells in response to chemoattractant. *J. Cell Biol.* **129**: 179–188.

Küssel-Andermann, P., El-Amraoui, A., Safieddine, S., Nouaille, S., *et al.*, 2000. Vezatin, a novel transmembrane protein, bridges myosin VIIA to the cadherin-catenins complex. *EMBO J.* **19**: 6020–6029.

Ling, K., Doughman, R. L., Firestone, A. J., Bunce, M. W. and Anderson, R. A., 2002. Type I gamma phosphatidylinositol phosphate kinase targets and regulates focal adhesions. *Nature* **420**: 89–93.

Loomis, W. F., 1982. *The development of Dictyostelium discoideum*. Academic Press, New York.

Maniak, M., Rauchenberger, R., Albrecht, R., Murphy, J. and Gerisch, G., 1995. Coronin involved in phagocytosis: dynamics of particle-induced relocalization visualized by a green fluorescent protein tag. *Cell* **83**: 915–924.

Monkley, S. J., Zhou, X. H., Kinston, S. J., Giblett, S. M., *et al.*, 2000. Disruption of the talin gene arrests mouse development at the gastrulation stage. *Dev. Dynamics* **219**: 560–574.

Niewöhner, J., Weber, I., Maniak, M., Müller-Taubenberger, A. and Gerisch, G., 1997. Talin-null cells of *Dictyostelium* are strongly defective in adhesion to particle and substrate surfaces and slightly impaired in cytokinesis. *J. Cell Biol.* **138**: 349–361.

Niggli, V., Andreoli, C., Roy, C. and Mangeat, P., 1995. Identification of a phosphatidylinositol-4,5-bisphosphate-binding domain in the N-terminal region of ezrin. *FEBS Lett* **376**: 172–176.

Paglini, G., Kunda, P., Quiroga, S., Kosik, K. and Caceres, A., 1998. Suppression of radixin and moesin alters growth cone morphology, motility, and process formation in primary cultured neurons. *J. Cell Biol.* **143**: 443–455.

Parent, C. A., Blacklock, B. J., Froehlich, W. M., Murphy, D. B. and Devreotes, P. N., 1998. G protein signaling events are activated at the leading edge of chemotactic cells. *Cell* **95**: 81–91.

Pearson, M. A., Reczek, D., Bretscher, A. and Karplus, P. A., 2000. Structure of the ERM protein moesin reveals the FERM domain fold masked by an extended actin binding tail domain. *Cell* **101**: 259–270.

Petit, C., 2001. Usher syndrome: from genetics to pathogenesis. *Ann. Rev. Genom. Hum. Genetics* **2**: 271–297.

Ponte, E., Bracco, E., Faix, J. and Bozzaro, S., 1998. Detection of subtle phenotypes: the case of the cell adhesion molecule csA in *Dictyostelium. Proc. Natl. Acad. Sci.* **95**: 9360–9365.

Reczek, D., Berryman, M. and Bretscher, A., 1997. Identification of EBP50: A PDZ-containing phosphoprotein that associates with members of the ezrin-radixin-moesin family. *J. Cell Biol.* **139**: 169–179.

Schaller, M. D., Otey, C. A., Hildebrand, J. D. and Parsons, J. T., 1995. Focal adhesion kinase and paxillin bind to peptides mimicking beta integrin cytoplasmic domains. *J. Cell Biol.* **130**: 1181–1187.

Self, T., Mahony, M., Fleming, J., Walsh, J., *et al.*, 1998. Shaker-1 mutations reveal roles for myosin VIIA in both development and function of cochlear hair cells. *Development* **125**: 557–566.

Sesaki, H. and Siu, C. H., 1996. Novel redistribution of the Ca2+-dependent cell adhesion molecule DdCAD-1 during development of *Dictyostelium discoideum. Dev. Biol.* **177**: 504–516.

Sesaki, H., Wong, E. F. S. and Siu, C.H., 1997. The cell adhesion molecule DdCAD-1 in *Dictyostelium* is targeted to the cell surface by a nonclassical transport pathway involving contractile vacuoles. *J. Cell Biol.* **138**: 939–951.

Simson, R., Wallraff, E., Faix, J., Niewohner, J., *et al.*, 1998. Membrane bending modulus and adhesion energy of wild-type and mutant cells of *Dictyostelium* lacking talin or cortexillins. *Biophys. J.* **74**: 514–522.

Takeuchi, K., Sato, N., Kasahara, H., Funayama, N., *et al.*, 1994. Perturbation of cell adhesion and microvilli formation by antisense oligonucleotides to ERM family members. *J. Cell Biol.* **125**: 1371–1384.

Titus, M. A., 1999. A class VII unconventional myosin is required for phagocytosis. *Current Biol.* **9**: 1297–1303.

Tsujioka, M., Machesky, L. M., Cole, S. L., Yahata, K. and Inouye, K., 1999. A unique talin homologue with a villin headpiece-like domain is required for multicellular morphogenesis in *Dictyostelium. Current Biol.* **9**: 389–392.

Tuxworth, R. I., Weber, I., Wessels, D., Addicks, G. C., *et al.*, 2001. A role for myosin VII in dynamic cell adhesion. *Current Biol.* **11**: 318–329.

Varney, T. R., Casademunt, E., Ho, H. N., Petty, C., *et al.*, 2002. A novel *Dictyostelium* gene encoding multiple repeats of adhesion inhibitor-like domains has effects on cell–cell and cell–substrate adhesion. *Dev. Biol.* **243**: 226–248.

Vogel, G., Thilo, L., Schwarz, H. and Steinhart, R., 1980. Mechanism of phagocytosis in *Dictyostelium discoideum*: phagocytosis is mediated by different recognition sites as disclosed by mutants with altered phagocytotic properties. *J. Cell Biol.* **86**: 456–465.

Wang, J., Hou, L. S., Awrey, D., Loomis, W. F., *et al.*, 2000. The membrane glycoprotein gp150 is encoded by the *lagC* gene and mediates cell–cell adhesion by heterophilic binding during *Dictyostelium* development. *Dev. Biol.* **227**: 734–745.

Weber, I., Wallraff, E., Albrecht, R. and Gerisch, G., 1995. Motility and substratum adhesion of *Dictyostelium* wild-type and cytoskeletal mutant cells: a study by RICM/bright-field double-view image analysis. *J. Cell Sci.* **108**: 1519–1530.

Wehrle-Haller, B. and Imhof, B., 2002. The inner lives of focal adhesions. *Trends Cell Biol.* **12**: 382–389.

Weil, D., Blanchard, S., Kaplan, J., Guliford, P., *et al.*, 1995. Defective myosin VIIA gene responsible for Usher syndrome type 1B. *Nature* **374**: 60–61.

Wong, E., Yang, C., Wang, J., Fuller, D., *et al.*, 2002. Disruption of the gene encoding the cell adhesion molecule DdCAD-1 leads to aberrant cell sorting and cell-type proportioning during *Dictyostelium* development. *Development* **129**: 3839–3850.

Wong, E. F. S., Brar, S. K., Sesaki, H., Yang, C. Z. and Siu, C. H., 1996. Molecular cloning and characterization of DdCAD-1, a Ca^{2+}-dependent cell–cell adhesion molecule, in *Dictyostelium discoideum*. *J. Biol. Chem.* **271**: 16399–16408.

Yonemura, S. and Tsukita, S., 1999. Direct involvement of ezrin/radixin/moesin (ERM)-binding membrane proteins in the organization of microvilli in collaboration with activated ERM proteins. *J. Cell Biol.* **145**: 1497–1509.

3

Do Class I Myosins exert their Functions through Regulation of Actin Dynamics?

Thierry Soldati and **Claudia Kistler**

Since the discoveries of class I myosins, the first non-muscle myosins, about 30 years ago, their history has been linked both to the organization and working of actin filaments and to their connection to membranes, especially the plasmalemma. Indeed, the early biochemical characterization highlighted the capacity of these mechanochemical enzymes to crosslink actin filaments and to bind phospholipids. Recent findings shed new light on these relationships and reveal a more active role of class I myosins in regulating the spatial and temporal nucleation and elongation of actin filaments in close proximity of membrane sites determined by signalling and cell polarity mechanisms. The understanding of the role of the actin cytoskeleton in cell motility has recently made a giant leap forward with the discovery and dissection of the major actin nucleator, the Arp2/3 complex, and some of its activators. To investigate the molecular and cellular bases of the integration of myosin motor activity and actin dynamics in cortical management and cell motility, we make use of a genetically and biochemically tractable model organism, *Dictyostelium discoideum*. As the components of these complex machineries are evolutionarily conserved, their molecular and cellular dissection in *D. discoideum* is directly relevant to unravel their functional importance in higher organisms.

Cell Motility. Edited by Anne Ridley, Michelle Peckham and Peter Clark.
 ISBN 0 470 84872 3

Introduction

A brief summary of the process of actin dynamics

In the cytosol, ATP-G-actin is kept in a readily polymerizable pool bound to profilin. Profilin-actin is thought to be recruited to the site of polymerization via the poly-Pro motifs of activator proteins like VASP or WASp. Because *de novo* nucleation of filaments is rate-limiting, the Arp2/3 complex, the major cellular F-actin nucleator, lays at the core of cytoskeleton dynamics. Recent advances propose that extracellular signals are relayed through effector proteins like Rho family GTPases to the WASp or Scar family proteins, which in turn lead to the localized recruitment and activation of the Arp2/3 complex (Pollard *et al.*, 2000). Nucleation occurs mainly by branching off existing filaments, and polymerization of ATP-G-actin is thought directly to produce the force necessary to push membranes. Rapidly, capping of the newly generated filament is followed by stabilization through a variety of actin binding proteins. Binding of ADF/cofilin to ADP-F-actin initiates depolymerization. The pool of ATP-G-actin is reconstituted by the nucleotide exchange factor profilin (Blanchoin *et al.*, 2000). It has been recently demonstrated that this Arp2/3-dependent mechanism, possibly being the major one, nevertheless does not account for all engagements of the actin polymerization machinery. For example, PAK kinase can recruit and phosphorylate filamin proteins and together generate orthogonally cross-linked actin meshworks (Vadlamudi *et al.*, 2002). As another alternative, to generate unbranched actin filaments, proteins of the formin family work downstream of Rho GTPases and appear able to nucleate filaments and stay associated with their growing barbed ends (Pruyne *et al.*, 2002). The dendritic nucleation model offers a solid framework to dissect further the mechanisms of actin dynamics and their involvement in a vast array of cellular processes, and some questions are beginning to find an answer (Figure 3.1).

The myosin superfamily

The myosin superfamily of mechanoenzymes comprises 18 classes (Berg *et al.*, 2001). For example, the human genome encodes about 40 myosin genes of which about 25 are unconventional and come from at least 11 classes; *D. discoideum* appears to have 13 myosins from about six classes (Glockner *et al.*, 2002) and *Saccharomyces cerevisiae* has five myosins from three classes (Berg *et al.*, 2001). All members of the myosin family share a common structure; they are composed of three modules, the head, neck and tail domains. The N-terminal region harbours the motor unit, utilizing ATP to power movement along actin filaments. Almost all myosins follow the

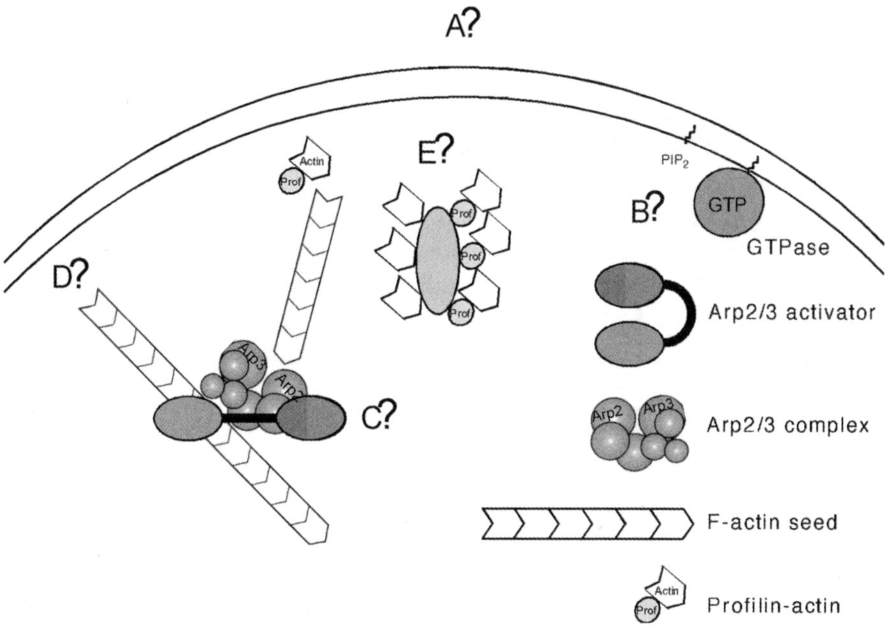

Figure 3.1 A model for actin nucleation and elongation, and some important questions. The popular dendritic nucleation model implies that upon reception of a stimulus (A) that leads to cell polarity through a signalling cascade that includes GTPases and lipid kinases (B), specific proteins can be relieved from an autoinhibitory conformation, recruit and activate the Arp2/3 complex (C) in close proximity to the membrane (D). During elongation, ATP-G actin is added at the growing barbed end of nascent actin filaments (E). These major steps are under intense investigation to solve the following questions. What is the range of stimuli that can lead to cell polarity and engagement of the actin dynamics machinery? (A?). Which GTPases, trimeric Gs and small GTPases of the Rho/Rac/Cdc42 family are involved and by which GEF are they activated? (B?). Beside the best characterized Arp2/3 activators of the WASp family, what other proteins can act in this pathway? (C?). How is the seed F-actin positioned close to the plasma membrane? (D?). Is the 'fuel' for elongation, ATP-G actin, preconcentrated at the growing end and is it bound by WH2 domain proteins and/or in the form of profilin-actin complex? (E?). Other factors, including capping proteins and cofilin, are crucial for motility but are not discussed here

so-called TEDS rule (Bement and Mooseker, 1995) by having a negatively charged amino acid residue (Asp (D), Glu (E) or phosphorylatable Ser (S) or Thr (T)) at a position of the head domain known as the cardiomyopathy loop, and crucial for activity. In class I myosins of lower eukaryotes, this site is phosphorylated by kinases of the PAK/Ste20 family, which are regulated by small GTPases of the Rho/Rac/Cdc42 family (reviewed in de la Roche and Cote (2001)). The neck or middle domain acts as a lever (Geeves, 2002) stiffened by the binding of light chains belonging to the superfamily of calmodulin-like EF hand proteins (Wolenski, 1995). Finally, via binding to specific proteins and cargoes, the tail is responsible for the specific function and location of myosins. In addition to the classical involvement of myosin II in producing contraction power, unconventional myosins are active as cortical managers, organelle and mRNA transporters, and as regulators of signal transduction. In particular, class I myosins (Figure 3.2A) form the second largest group after conventional myosin II, and appear to be essential players in the establishment and maintenance of cortical tension and related functions such as motility, endocytosis and exocytosis (Mermall *et al.*, 1998). Finally, the relevance of myosin for mammalian physiology and pathology was recently emphasized by the finding that many human and murine genetic diseases are associated with mutations in myosins (Kabaeva *et al.*, 2002; Berg, 2001; Mermall *et al.*, 1998).

Dictyostelium discoideum as a powerful model organism

D. discoideum is a eukaryotic social amoeba belonging to the crown group of organisms. Recent studies based on protein phylogeny clearly indicates that the Amoebozoa group is the closest relative of Metazoa and Fungi, and more distantly of Plantae (Baldauf *et al.*, 2000). In contrast to yeast, *D. discoideum* is famous as a simple model of multicellularity and performs many of the tasks typical of a higher eukaryote such as chemotaxis, motility, differentiation, cell–cell contact and efficient endocytosis. This organism also lives as a free cellular amoeba, feeding by phagocytosis and endocytosis. It possesses an endomembrane organization similar to higher organisms, and has already greatly contributed to our understanding of cytoskeleton and cell motility as well as of developmental and signal transduction pathways (Kessin, 2001). It is also a potent expression system; easy growth conditions of axenic laboratory strains give access to high quantities of proteins. As a haploid it is amenable to molecular genetics, and its relatively large size makes it a

Figure 3.2 (*opposite*) The family of class I myosins and the structure of members of subclass I. (A) Class I myosins form the second biggest class after conventional myosin II and are represented in most organisms except plants and some primitive protozoa. It is subdivided in four subclasses according to the structure of their neck and tail domains (see main text for details). (B) Biochemical analysis of class I myosins has demonstrated their potential function as regulated integrators of membrane and cytoskeleton interactions. The motor domain of non-metazoan class I myosins is regulated by phosphorylation of the TEDS site found in the hypertrophic cardiomyopathy loop that contacts actin. The neck domain binds a variable number of calmodulin-like light chains. The tail domain contains a polybasic TH1 domain that binds negatively charged phospholipids (+ + +); a Gly and Pro-rich TH2 domain that contains secondary actin-binding sites (GPA); an SH3 containing TH3 domain that link to a network of proteins involved in signalling, actin dynamics and endocytosis

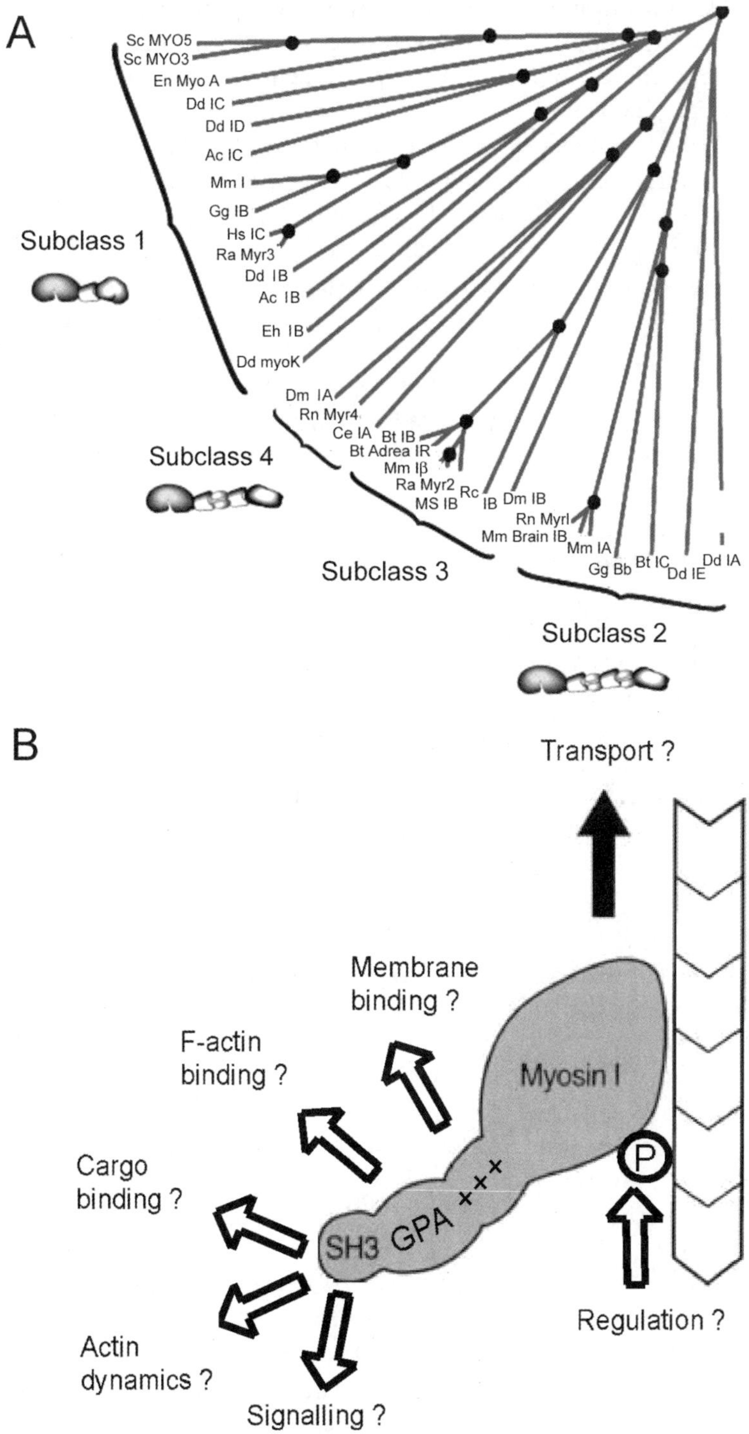
A
Sc MYO5
Sc MYO3
En Myo A
Dd IC
Dd ID
Ac IC
Mm I
Gg IB
Hs IC
Ra Myr3
Dd IB
Ac IB
Eh IB
Dd myoK
Subclass 1
Dm IA
Rn Myr4
Ce IA
Bt IB
Bt Adrea IR
Mm Iβ
Ra Myr2
MS IB
Rc IB
Dm IB
Rn Myrl
Mm Brain IB
Mm IA
Gg Bb
Bt IC
Dd IE
Dd IA
Subclass 4
Subclass 3
Subclass 2
B
Transport ?
Membrane binding ?
F-actin binding ?
Cargo binding ?
Myosin I
SH3 GPA
P
Actin dynamics ?
Signalling ?
Regulation ?

suitable object for morphological observations. Most importantly, the ease of obtaining mutant strains by random 'restriction enzyme mediated insertion' (REMI) mutagenesis, the recent establishment of regulated expression (Blaauw *et al.*, 2000) and high copy suppression systems (Robinson and Spudich, 2000), combined with the fact that the coordinated international effort to sequence and assemble its 35 Mb genome is reaching completion (http://www.sanger.ac.uk/Projects/D_discoideum/genomic_sequence.shtml), are enhancing its position as an excellent model organism.

In this chapter, I present a brief review of the recent literature concerned with the link between class I myosins and actin dynamics, and also proceed to discuss our recent unpublished results about *D. discoideum* MyoK in this context. Finally, as the components of these complex machineries are evolutionarily conserved, the proposed model is extended to the function of class I myosins in endocytic membrane trafficking in mammals.

Structure function analysis of Class I myosins

Class I myosins are divided into about four phylogenetic subclasses (Figure 3.2A; Mooseker and Cheney, 1995; Coluccio, 1997). The subclass I of ameboid-type myosins can be further subdivided into long-tailed and short-tailed. All class I myosins identified in most lower eukaryotes, including *Acanthamoeba*, *S. cerevisiae*, *S. pombe* and *Aspergillus* have long tails, but in *D. discoideum* class I myosins have either long tails (MyoB, MyoC and MyoD), short tails (MyoA, MyoE and MyoF) or virtually no tail at all (MyoK) (de la Roche and Cote, 2001). The long tails are 400–450 residues in size and comprise three tail homology (TH) domains. TH1 is rich in basic residues, TH2 exhibits a high content of glycine and proline while TH3 is more commonly referred to as an Src homology 3 (SH3) domain. The short tails are 300–350 residues in size and solely TH1 is recognizable. MyoK is a highly divergent 94 kDa type I myosin with a very short neck region and a tail only 38 residues in length, but has an insert of $\sim$150 amino acids within the motor domain that bears similarity to TH2 (Yazu *et al.*, 1999; Schwarz *et al.*, 2000). Phylogenetic analysis shows that *D. discoideum* MyoC is more closely related to *Acanthamoeba* myosin IA, the two *S. cerevisiae* type I myosins (Myo3p and Myo5p) and *Aspergillus* MYOA, while *D. discoideum* MyoB and MyoD occupy branches of the phylogenetic tree that contain *Acanthamoeba* myosin IB and myosin IC, respectively (Lee *et al.*, 1999). At the moment, only two homologues of long-tailed amoeboid class I myosins, myosin IE (Stöffler *et al.*, 1995) and IF (Crozet *et al.*, 1997) have been described in mammals, but the high degree of sequence relatedness is widely accepted as to reflect a high level of functional conservation. In contrast, whereas short-tailed myosins have not been identified in other lower eukaryotes than *D. discoideum*, many examples of such molecules have been extensively investigated in mammalian cells (Mooseker and Cheney, 1995; Coluccio, 1997).

It has been generally well established that TH1 binds anionic phospholipids, while TH1 and TH2, bind F-actin in an ATP-independent manner (de la Roche and Cote, 2001). As a consequence, all class I myosins probably have the potential to associate to a membrane and to translocate F-actin along its surface, as well as to bind two actin filaments and slide them relative to each other (Figure 3.2B). Intriguingly, *D. discoideum* MyoK has many of the same properties but uses a different molecular architecture. It has no TH1, TH2 and TH3 tail domains, but has a C-terminal farnesylation in place of phospholipid-binding TH2, and harbours a TH2-like insert within the motor domain (Schwarz *et al.*, 2000; and see below). The long tails of ameboid class I myosins may fold so as to position TH1 and TH2 side by side and the SH3 domain adjacent to the neck (Lee *et al.*, 1999). This location of the SH3 domain close to the motor might explain why the MgATPase activity of rat myosin IE is stimulated by proteolytic removal of the SH3 domain or by binding of an antibody to the C-terminus (Stöffler and Bähler, 1998).

Investigation of *D. discoideum* MyoK

MyoK was discovered as a result of an exhaustive screening for completion of the myosin repertoire in *D. discoideum,* which also revealed MyoM (Schwarz *et al.*, 1999). MyoM is the first myosin to carry a Rac GTPase activator domain in its tail and is involved in actin cytoskeleton reorganization (Geissler *et al.*, 2000). MyoK was characterized at the levels of cDNA and genomic sequences and organization, and protein expression and localization. MyoK shares structural features with class I myosins, but is distinguished by an unusual architecture (Figure 3.3; Schwarz *et al.*, 2000). MyoK is one of the smallest myosins found to date, and its most striking feature is the virtual absence of a tail or cargo-binding domain. MyoK ends with a -CLIQ sequence, corresponding to a so-called -CAAX motif, a known protein farnesylation signal (Figure 3.3A) that was shown to be sufficient to target a GFP fusion protein to the plasma membrane (E. Schwarz and T. Soldati, unpublished results). It is unclear whether, as all other myosins studied so far, MyoK carries a light chain bound to its short neck that does not carry any canonical IQ motif. As an apparent form of 'compensation', MyoK has a 150 residues insertion in its surface loop 1. This insertion is extremely rich in Gly, Pro and Arg (GPR), a composition similar to the TH2 tail domains of other class I myosin (Figure 3.3B), where it has been shown to contain F-actin binding sites. Moreover, the GPR-loop has about 40% identity (over 60% homology) with the Pro-rich domain of some WASp proteins (Figure 3.3C). Closer analysis revealed that it contains a variety of Pro-rich motifs

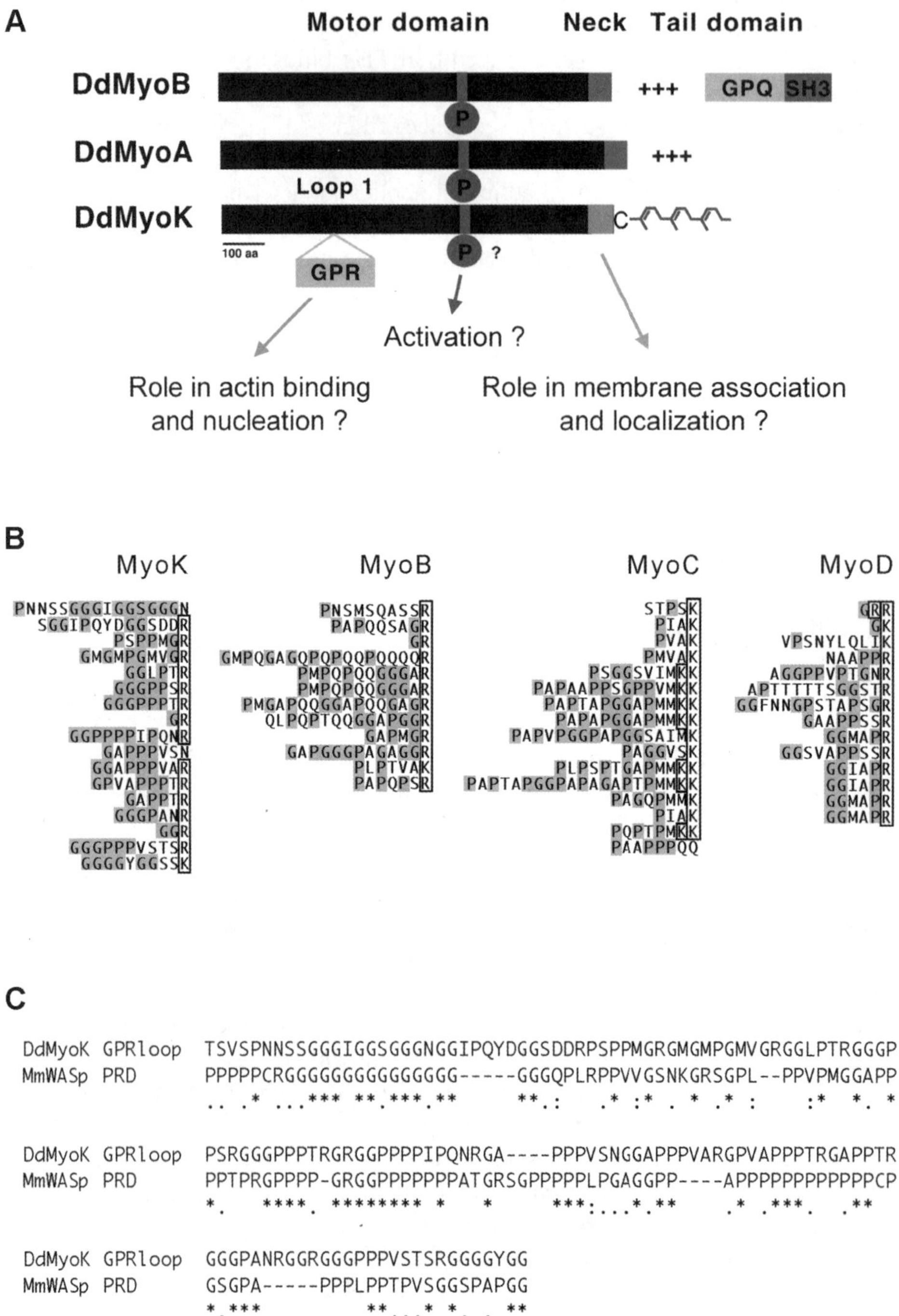
A
Motor domain
Neck
Tail domain
DdMyoB
+++
GPQ
SH3
P
DdMyoA
+++
Loop 1
P
DdMyoK
C
100 aa
GPR
P ?
Activation ?
Role in actin binding and nucleation ?
Role in membrane association and localization ?
B
MyoK
MyoB
MyoC
MyoD
C
DdMyoK GPRloop TSVSPNNSSGGGIGGSGGGNGGIPQYDGGSDDRPSPPMGRGMGMPGMVGRGGLPTRGGGP
MmWASp PRD PPPPPCRGGGGGGGGGGGGGG-----GGGQPLRPPVVGSNKGRSGPL--PPVPMGGAPP
DdMyoK GPRloop PSRGGGPPPTRGRGGPPPPIPQNRGA----PPPVSNGGAPPPVARGPVAPPPTRGAPPTR
MmWASp PRD PPTPRGPPPP-GRGGPPPPPPPATGRSGPPPPPLPGAGGPP----APPPPPPPPPPPPCP
DdMyoK GPRloop GGGPANRGGRGGGPPPVSTSRGGGGYGG
MmWASp PRD GSGPA-----PPPLPPTPVSGGSPAPGG

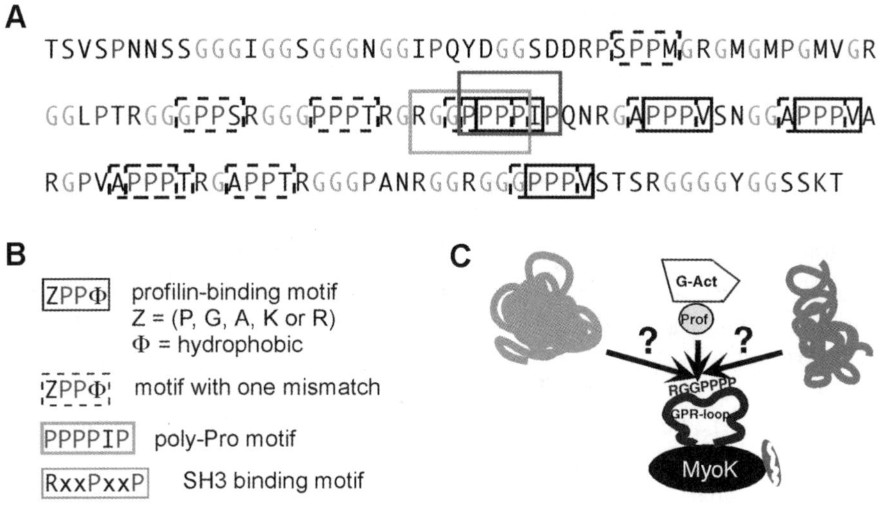

Figure 3.4 The MyoK GPR loop contains multiple protein–protein interaction motifs. (A) The 148 residues domain is extremely rich in Gly, Pro and Arg (64% total). (B) Profilin has been shown to interact with poly-Pro stretches and ZPPΦ motifs (Witke *et al.*, 2001) and SH3 domain containing proteins are known to interact with motifs of either RxxPxxP (class I) or PxxPxR (class II) motifs. These motifs are boxed in (A). Note the overlapping distribution of some motifs in the middle of the GPR domain, that could potentially lead to competition for and/or hierarchy of binding (C)

(Figure 3.4A) that have been shown to work as profilin-binding sites (for example ZPPΦ, Figure 3.4B (Witke *et al.*, 2001)). Finally, it contains a canonical class I SH3 binding motif (RxxPxxP). The GPR-loop might therefore serve as a multifunctional protein–protein interaction domain (Figure 3.4C) and it is not known yet whether the different, sometimes overlapping, binding sites lead to competition and/or hierarchy of binding events. First biochemical analysis indicated that MyoK has two independent actin-binding sites (one ATP-dependent and one salt-sensitive), suggesting that it may act as a novel

---

Figure 3.3 (*opposite*) Structure of class I myosins in *D. discoideum*. (A) The ameboid myosins are either long-tailed (like MyoB) or short-tailed (like MyoA), but MyoK has a more exotic architecture. It has a very short neck/tail domain, its C-terminus is carboxymethyl-farnesylated, and it bears a 150-residue insertion in a surface loop of the motor domain otherwise well conserved in length. This extended loop 1 is extremely rich in Gly, Pro organized in pseudo repeats punctuated by the positive charges of Arg, and is thus called GPR loop. This composition is very similar to the TH2 domains of other class I myosins (B), even though the relative densities of Gly and Pro change, and Arg is sometimes replaced by double Lys (as in MyoC). In addition, the GPR loop of MyoK bears high sequence similarity with the Pro-rich domains (PRD) of WASp family members, for example, it has about 40% identity and over 60% homology with the PRD of mouse WASp (C) regulated actin-crosslinker. Nevertheless, as MyoK is not abundant, we hypothesized that it works less mechanically than catalytically and might have a distinctive role in cortical management (Schwarz *et al.*, 2000).

Phenotypes resulting from manipulation of class I myosins

Immunolocalization studies of class I myosins in amoeba (de la Roche and Cote, 2001; Ma *et al.*, 2001) and in higher eukaryotes (Kalhammer and Bahler, 2000; Berg *et al.*, 2001) visualized their association with dynamic regions of the cell cortex, such as the leading pseudopod of migrating cells and endocytic structures, microvillosities, cell–cell junctions (Stöffler *et al.*, 1998) and sites of particle ingestion (Diakonova *et al.*, 2002). In less plastic fungal cells, the major site of myosin I localization is within the actin cortical patches (Win *et al.*, 2002).

Gene disruption studies have implicated *D. discoideum* class I myosins in a palette of actin-based processes, many of which were interpreted as a result of the lipid-binding and actin-binding properties of TH1 and TH2, leading to a contraction of the cortical actin network or transport of vesicles along actin filaments (Ma *et al.*, 2001; de la Roche and Cote, 2001). Cells lacking either MyoA or MyoB moved with reduced velocity, formed more pseudopods, and turned more frequently than wild-type (Wessels *et al.*, 1991, 1996; Titus *et al.*, 1993). Double mutants (A-/B-, B-/C-) showed that these myosins contribute to the generation of cortical tension (Dai *et al.*, 1999), suggesting that both long-tailed and short-tailed myosins play a role in the organization of the actin cytoskeleton. Cell movement, macropinocytosis and phagocytosis all use the actin cytoskeleton to extend membrane protrusions and it has been shown that myosin motors are also involved in these processes. *D. discoideum* MyoB has been located to the phagocytic cup (Fukui *et al.*, 1989), and membrane ruffles (Novak *et al.*, 1995). Cells deficient in MyoB have a reduced rate of phagocytosis (Jung and Hammer, 1990; Jung *et al.*, 1996), whereas overexpression of MyoB resulted in decreased macropinocytosis (Novak and Titus, 1997). Single MyoC mutants have a decreased initial rate of fluid-phase uptake (Jung *et al.*, 1996), and the slow growth of various double and triple mutants (A-/B-, B-/C-, B-/D-, B-/C-/D-) was interpreted as additive impairments of myosin I function in fluid-phase uptake (Jung *et al.*, 1996; Novak *et al.*, 1995). These results led to the proposition that class I myosins share partially overlapping but mainly non-redundant functions in endocytosis. Findings in *Acanthamoeba* (Baines *et al.*, 1992) and *Entamoeba histolytica* (Voigt *et al.*, 1999) are in excellent agreement with these observations.

To study the function of MyoK *in vivo*, we analysed the phenotypic alterations in *myoK* null cells and MyoK overexpressing cells. We observed morphological changes of the cortical actin cytoskeleton accompanied by impairments in the rate of phagocytosis and in chemotactic motility. Cells deficient in one or more myosin Is extend more lateral pseudopods during motility, exhibit decreased rates of pinocytosis and phagocytosis and have a lower cortical tension (Jung *et al.*, 1996; Novak *et al.*, 1995; Schwarz *et al.*, 2000; Dai *et al.*, 1999) while overexpression of some class I myosins augments cortical tension and restricts extension of actin-filled protrusions (Novak and Titus, 1997; Schwarz *et al.*, 2000). The phenotype of *myoK* null cells was the more dramatic of all single myosin I knock outs reported so far (Dai *et al.*, 1999), especially with regard to the significant effects on cortical tension. Indeed, *myoK* null cells had a 24% lower cortical tension than wild-type cells, and myoK overexpresses a 70% higher cortical tension (Schwarz *et al.*, 2000). This is remarkable considering the high potential degree of functional redundancy between the seven class I myosins of *D. discoideum*.

Because of their sheer number, it is practically impossible to disrupt all seven class I myosins genes in *D. discoideum*, but in *S. cerevisiae* it was shown that mutants lacking both Myo3p and Myo5p exhibit a strong growth defect, accumulate intracellular vesicles and are severely impaired in endocytosis (Geli and Riezman, 1996; Goodson *et al.*, 1996). MYOA is required for the viability of *Aspergillus* and is involved in the generation of cell polarity and focal secretion at the tip of growing hyphae (McGoldrick *et al.*, 1995; Osherov *et al.*, 1998). Several more recent studies highlight the importance of the SH3 domain and thereby emphasize another aspect of myosin I function. The defects in growth, endocytosis and actin organization exhibited by *D. discoideum* cells lacking MyoA and MyoB can be fully rescued by wild-type MyoB but not by MyoB missing its SH3 domain (MyoBΔSH3) (Novak and Titus, 1998). Furthermore, overexpression of MyoBΔSH3 does not generate the defects associated with overexpression of wild-type MyoB (Novak and Titus, 1997). In *S. cerevisiae*, the severe defects resulting from Myo3p/Myo5p double knockout can be complemented by either Myo3p or Myo5p lacking TH2 but not lacking their SH3 domain (Anderson *et al.*, 1998; Evangelista *et al.*, 2000). In contrast, *Aspergillus* MYOA appears to function even when its SH3 domain is deleted (Osherov *et al.*, 1998).

Class I myosins and the actin dynamics connection

SH3 domains in general are found in an impressive number of factors involved in signal transduction, actin dynamics and endocytic membrane trafficking (Schafer, 2002; Warren *et al.*, 2002; Dewar *et al.*, 2002; Mochida *et al.*, 2002; Soulard *et al.*, 2002; Bon *et al.*, 2000), and have been shown to interact specifically with either one or a few proline-rich motifs of either class I

(RxxPxxP) or class II (PxxPxR) (Mayer, 2001). Recently, ligands of *S. cerevisiae* and *D. discoideum* class I myosin SH3 domains have been identified (see below), and shown to link these molecular motors with the endocytic machinery, the Arp2/3 complex and the machinery responsible for actin dynamics. In addition, the discovery of complex networks of molecules associating through a cascade of interactions between SH3 and proline-rich domains (PRD) (Holt and Koffer, 2001; Tong *et al.*, 2002) sheds new light on the functional significance of TH2 and SH3 domains of class I myosins. For example, TH2 domains bear strong resemblance with the PRD of WASp, SCAR and formins, all involved in regulation of actin nucleation (Mullins, 2000; Pruyne *et al.*, 2002; Evangelista *et al.*, 2002). It is also important to note that, beside being able to recruit and stimulate Arp2/3 nucleating activity via a C-terminal acidic domain, the proteins of the WASp/Scar family have a WASp homology 2 (WH2) domain that can recruit G-actin (Mullins, 2000). In addition, in yeast and mammalian cells Las17p/WASp interacts with verprolin (Vrp1p)/WIP (WASp-interacting protein) (Vaduva *et al.*, 1997; Vaduva *et al.*, 1999). Many of the proteins shown to interact with WASp have both PRD and SH3 domains, emphasizing the potentially high complexity of the resulting networks (Tong *et al.*, 2002).

The two *S. cerevisiae* type I myosins, Myo3p and Myo5p, are linked to the Arp2/3 complex by multiple interactions. First, these myosins bind directly to proline-rich motifs in both Vrp1p and Las17p by means of their SH3 domain (Evangelista *et al.*, 2000; Lechler *et al.*, 2000; Geli *et al.*, 2000). The interaction with Vrp1p is required for their proper localization to cortical patches that mark sites of polarized cell growth (Anderson *et al.*, 1998). In addition, Vrp1p binds Las17p which recruits the Arp2/3 complex in a pathway of actin nucleation parallel to the one dependent on class I myosins (Lechler *et al.*, 2001). Moreover, both myosins directly interact with the Arp2/3 complex via a C-terminal acidic motif (Evangelista *et al.*, 2000; Lechler *et al.*, 2000; Lee *et al.*, 2000). Related acidic sequences are similarly located at the C-terminus of other fungal class I myosins (*Candida albicans*, *S. pombe* and *Aspergillus*) of Las17p and of human WASp. In *S. cerevisiae*, Myo3p, Myo5p and Las17p function in a redundant manner to activate the Arp2/3 complex, as removal of the acidic sequence from class I myosins or Las17p has little effect, but deletion of all acidic domains virtually eliminates actin filament assembly in cortical patches (Lee *et al.*, 2000). In an assay based on permeabilized urea-treated cells, Myo3p or Myo5p were shown to promote actin polymerization only when their motor activity was intact. Nucleation activity was impaired by dephosphorylation or mutation of the TEDS site (Lechler *et al.*, 2000). Contrary to WASp, yeast Las17p lacks a Cdc42-binding domain. Therefore, these latter results suggest that the primary pathway by which this GTPase induces actin polymerization in *S. cerevisiae* is through the activation of myosin I by Cdc42-dependent kinases of the PAK family.

The tail of Myo5p was shown to induce cytosol-dependent actin polymerization around Sepharose beads. This assay appears to mimick the formation of dynamic cortical actin patches (Geli *et al*., 2000; Idrissi *et al*., 2002), structures that have been involved in endocytosis. Importantly, the agreement between the biochemical requirements *in vitro* and the genetic requirements *in vivo* appears striking and corroborate the working model of an intricate link between actin dynamics and endocytosis. Both processes depend on the Arp2/3 complex, Vrp1p and cofilin (Geli *et al*., 2000; Idrissi *et al*., 2002). Interestingly, profilin seems to be dispensable, perhaps highlighting that, as the WH2 domain (present in Las17p and also Vrp1p) can bind G-actin directly, it may be bypassing the need for the recruitment of monomeric ATP-actin by profilin. These findings are similar but not identical to the requirements for the *in vitro* reconstitution of the motility of *Listeria* from purified components, which is commonly considered to be a model for lamellipodium extension (Loisel *et al*., 1999).

In amoeba, class I myosins are also able to recruit the Arp2/3 complex, but do so with a twist, involving an adapter protein. CARMIL and Acan125 were isolated from extracts using recombinant GST fusions with the SH3 domains of *D. discoideum* MyoB and MyoC (Jung *et al*., 2001), and *Acanthamoeba* MyoA (Lee *et al*., 1999) and MyoC (Xu *et al*., 1995) as affinity baits, respectively. These homologous adapters of 116 and 125 kDa bear two C-terminal PXXP motifs that are ligands of the SH3 domain and consist of multiple leucine-rich repeat sequences (Xu *et al*., 1997). The native CARMIL complex contains all seven components of the Arp2/3 complex and the α and β subunits of capping protein (Jung *et al*., 2001), which blocks the barbed end of actin filaments. Determining the significance of the presence of both an actin nucleator and a polymerization terminator in the same complex will require more investigation, but it invites further speculations about their functions in shaping the very architecture of the cortical actin meshwork. MyoB, MyoC, CARMIL and the Arp2/3 complex are concentrated in actin-driven cellular protrusions, especially crown-shaped macropinocytic cups and the leading edge of migrating cells. Most recently, preliminary data indicated that, when fused to GST, the GPR loop of MyoK is able to bind not only F-actin, but also the profilin–actin complex. In addition, a pull-down experiment with the GST-GPR loop construct in cytosolic extracts detected a protein of about 55 kD. MALDI-TOF analysis identified it as the *D. discoideum* homologue of the yeast and mammal Abp1p/SH3P7r2 protein (C. Kistler and T. Soldati, unpublished results). This latter protein is conserved from yeast to human and has been shown to recruit and activate the Arp2/3 complex (Goode *et al*., 2001; Kessels *et al*., 2001). These data are very suggestive that the function of MyoK in cortical management and regulation of motility and phagocytosis are exerted through the recruitment and modulation of the actin nucleation machinery (see model Figure 3.5).

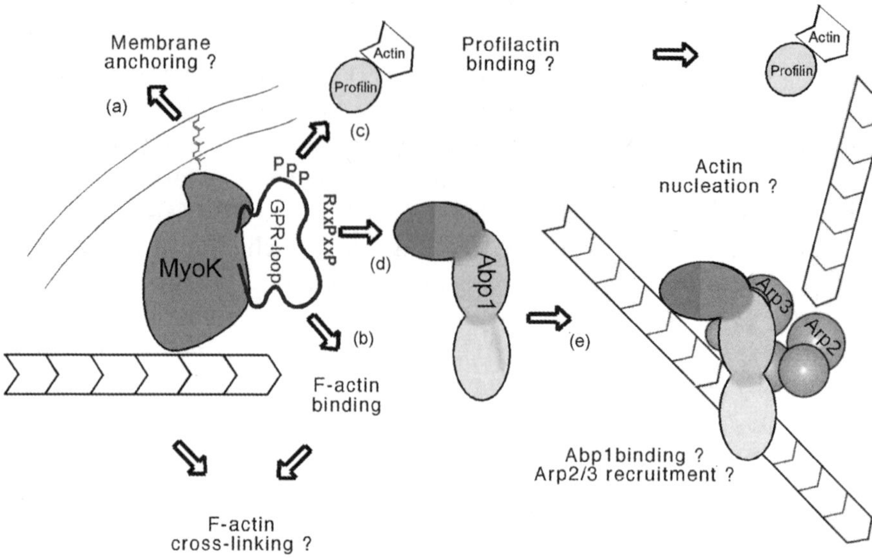

Figure 3.5 A working model of MyoK molecular functions. Results from the investigation of MyoK function *in vivo* and *in vitro* (Schwarz *et al.*, 2000) combined with our most recent unpublished experiments allow us to propose the following working hypothesis. First, MyoK is the first myosin carrying a C-terminal CAAX prenylation motif and preliminary experiments indicate that the last 55 residues of MyoK mediate prenylation and membrane attachment of a GFP fusion protein (a). Secondly, the GPR-loop appears able to bind not only F-actin (b), but also the profilin-actin complex (c). Thirdly, we identified the *D. discoideum* homologue of the yeast and mammal Abp1/SH3P7r2 protein as a binding partner for the SH3 binding motif (d). Therefore, we propose that MyoK, via its membrane anchor and affinity for F-actin, could position (and activate) the Arp2/3 complex (e) in close proximity of the plasma membrane. If MyoK motor domain gets activated by phosphorylation of its TEDS site, it could maintain contact with the barbed end of F-actin and participate in elevating the local concentration of the profilin-actin complex, the 'fuel' for F-actin elongation

Conclusions and outlook

The C-terminal acidic domain that mediates the direct association of yeast myosin I with the Arp2/3 complex is missing in the *D. discoideum* and *Acanthamoeba* myosins I, but the connection to the Arp2/3 complex has been maintained through the adaptor protein CARMIL. Homologues of CARMIL exist in *Caenorhabditis elegans* (Jung *et al.*, 2001), *Drosophila* and other higher eukaryotes including mammals, providing an important clue that the long-tailed class I myosins present in animals may be similarly linked to the Arp2/3 complex via their SH3 domain. Importantly, the example set by *D. discoideum* MyoK of an alternative mode for the indirect recruitment of the Arp2/3 complex by an adaptor bound via its GPR loop offers the attractive possibility that a comparable mechanism is active in higher eukaryotes where the PRD of class I myosins and other linker proteins might do the job. Finally, beside these three ways of linking to the Arp2/3 complex, class I myosins make use of two different ways to concentrate G-actin, the fuel for F-actin elongation. Class I myosins either bind profilin-actin through a Pro-rich domain (MyoK GPR loop) or bind G-actin indirectly via the WH2 domains of the adaptor proteins WASp, CARMIL and Vrp1p (Paunola *et al.*, 2002). Concentrating monomeric actin is not essential for actin-based motility (Loisel *et al.*, 1999) but greatly enhances the efficiency and speed of the process (Zalevsky *et al.*, 2001; Machner *et al.*, 2001; Cossart, 2000; Geese *et al.*, 2000; Loisel *et al.*, 1999).

It is attractive to speculate that actin filament assembly might involve clusters of membrane-anchored class I myosins and bound adaptors that link to the Arp2/3 complex. In such a position, class I myosins could stay associated both with the plasma membrane and with the barbed end of actin filaments, which usually point out to the cell periphery, and could facilitate addition of actin monomers. They might also be targeted to spots of actin dynamics/endocytosis through the complex interaction networks mentioned above and thereby increase tremendously the local concentration of both nucleation, elongation and capping activities, favouring a highly dynamic and branched meshwork.

Alternatively, myosins could transport Arp2/3 complexes towards the barbed ends of actin filaments, accounting for the apparent requirement for myosin I motor activity in actin assembly in the reconstituted system of Lechler *et al.* (2000). In the light of such a model, it is intriguing to note that in amoeba the complex might be kept at the pointed end of a very short and capped actin filament (Jung *et al.*, 2001). They might also capture Arp2/3 complexes at the rear of the ageing actin meshwork, where it is depolymerized, and recycle the nucleation activity to the front of the advancing polymerization zone. Such models potentially suffer from two conceptual problems: the lack of processivity of myosin I motors and the concurrent necessity for an uncapping activity (Falet *et al.*, 2002). It is well known that class I myosins are not processive because, unlike kinesin motors, they spend most of their time dissociated from the actin filament, and Ostap and Pollard (1996) predicted that clusters of 20 myosin I molecules would be needed for processive motility. On the other hand, a complex consisting for example of Las17p, Vrp1p and the Arp2/3 complex may be able to cluster sufficient myosin I molecules to support processive movement along an actin filament (Machesky, 2000). If class I myosins are used to recruit the Arp2/3 complex at its site of action, then the role of the WASp family proteins in this model is unclear. One possibility is that WASp family proteins are also transported there by class I myosins, or otherwise that they 'mark a spot' at the plasma membrane and dock the myosin I via its SH3 domain.

The evidence that class I myosins and actin dynamics play a crucial role in endocytosis in yeasts and other lower eukaryotes is overwhelming, however, their involvement in the endocytic machinery of mammalian cells is still acutely lagging. In this context, recent findings are extremely exciting and highlight the likely conservation of these mechanisms. For example, Myo1B is associated with endosomes and lysosomes (Raposo *et al.*, 1999) and cooperates with microtubules for the movement of lysosomes (Cordonnier *et al.*, 2001). Both calmodulin and Myo1C regulate membrane trafficking along the recycling pathway of MDCK cells (Huber *et al.*, 2000). The recycling of the glucose transporter in response to insulin is facilitated by Myo1c (Bose *et al.*, 2002). The major challenges for the future will be to determine the roles played by actin filaments at different steps in the internalization of proteins and fluid, and to determine how the interface of the endocytic machinery and the actin cytoskeleton is structured and regulated.

Acknowledgements

I thank all the laboratory members who along the years have contributed to a better understanding of myosin function, and especially to Eva Schwarz and Claudia Kistler, who have generated all the data on *D. discoideum* MyoK, including many exciting and still to be published results mentioned here. The work has been supported by the Max-Planck Society, the Deutsche Forschungsgemeinschaft and the BBSRC.

References

Anderson, B. L., Boldogh, I., Evangelista, M., Boone, C., *et al.*, 1998. The src homology domain-3 (SH3) of a yeast type-I myosin, myo5p, binds to verprolin and is required for targeting to sites of actin polarization. *J. Cell Biol.* **141**: 1357–1370.

Baines, I. C., Brzeska, H. and Korn, E. D., 1992. Differential localization of *Acanthamoeba* myosin I isoforms. *J. Cell Biol.* **119**: 1193–1203.

Baldauf, S. L., Roger, A. J., Wenk-Siefert, I. and Doolittle, W. F., 2000. A kingdom-level phylogeny of eukaryotes based on combined protein data. *Science* 972–977.

Bement, W. M. and Mooseker, M. S., 1995. TEDS rule: a molecular rationale for differential regulation of myosins by phosphorylation of the heavy chain head. *Cell Motil. Cytoskeleton* **31**: 87–92.

Berg, J. S., Powell, B. C. and Cheney, R. E., 2001. A millennial myosin census. *Mol. Biol. Cell.* **12**: 780–794.

Blaauw, M., Linskens, M. H. K. and van Haastert, P. J. M., 2000. Efficient control of gene expression by a tetracycline-dependent transactivator in single *Dictyostelium discoideum* cells. *Gene* **252**: 71–82.

Blanchoin, L., Pollard, T. D. and Mullins, R. D., 2000. Interactions of ADF/cofilin, Arp2/3 complex, capping protein and profilin in remodeling of branched actin filament networks. *Curr. Biol.* **10**: 1273–1282.

Bon, E., Recordon-Navarro, P., Durrens, P., Iwase, M., *et al.*, 2000. A network of proteins around Rvs167p and Rvs161p, two proteins related to the yeast actin cytoskeleton. *Yeast* **16**: 1229–1241.

Bose, A., Guilherme, A., Robida, S. I., Nicoloro, S. M., *et al.*, 2002. Glucose transporter recycling in response to insulin is facilitated by myosin Myo1c. *Nature* **420**: 821–824.

Coluccio, L. M., 1997. Myosin I. *Am. J. Physiol.* **273**: C347–C359.

Cordonnier, M. N., Dauzonne, D., Louvard, D. and Coudrier, E., 2001. Actin filaments and myosin I alpha cooperate with microtubules for the movement of lysosomes. *Mol. Biol. Cell* **12**: 4013–4029.

Cossart, P., 2000. Actin-based motility of pathogens: the Arp2/3 complex is a central player. *Cell Microbiol.* **2**: 195–205.

Crozet, F., el Amraoui, A., Blanchard, S., Lenoir, M., *et al.* (1997). Cloning of the genes encoding two murine and human cochlear unconventional type I myosins. *Genomics* **40**: 332–341.

Dai, J. W., TingBeall, H. P., Hochmuth, R. M., Sheetz, M. P. and Titus, M. A., 1999. Myosin I contributes to the generation of resting cortical tension. *Biophys. J.* **77**: 1168–1176.

de la Roche, M. A. and Cote, G. P., 2001. Regulation of *Dictyostelium* Myosin I and II. *Biochim. Biophys. Acta* **15**: 245–261.

Dewar, H., Warren, D. T., Gardiner, F. C., Gourlay, C. G., *et al.*, 2002. Novel proteins linking the actin cytoskeleton to the endocytic machinery in *Saccharomyces cerevisiae*. *Mol. Biol. Cell* **13**: 3646–3661.

Diakonova, M., Bokoch, G. and Swanson, J. A., 2002. Dynamics of cytoskeletal proteins during Fcgamma receptor-mediated phagocytosis in macrophages. *Mol. Biol. Cell* **13**: 402–411.

Evangelista, M., Klebl, B. M., Tong, A. H. Y., Webb, B. A., *et al.*, 2000. A role for myosin-I in actin assembly through interactions with Vrp1p, Bee1p, and the Arp2/3 complex. *J. Cell Biol.* **148**: 353–362.

Evangelista, M., Pruyne, D., Amberg, D. C., Boone, C. and Bretscher, A., 2002. Formins direct Arp2/3-independent actin filament assembly to polarize cell growth in yeast. *Nat. Cell Biol.* **4**: 32–41.

Falet, H., Hoffmeister, K. M., Neujahr, R., Italiano, J. E., Jr., *et al.*, 2002. Importance of free actin filament barbed ends for Arp2/3 complex function in platelets and fibroblasts. *Proc. Natl. Acad. Sci. USA* **99**: 16782–16787.

Fukui, Y., Lynch, T. J., Brzeska, H. and Korn, E. D., 1989. Myosin I is located at the leading edges of locomoting *Dictyostelium* amoebae. *Nature* **341**: 328–331.

Geese, M., Schluter, K., Rothkegel, M., Jockusch, B. M., *et al.*, 2000. Accumulation of profilin II at the surface of *Listeria* is concomitant with the onset of motility and correlates with bacterial speed. *J. Cell Sci.* **113**: 1415–1426.

Geeves, M. A., 2002. Stretching the lever-arm theory. *Nature* **415**: 129–131.

Geissler, H., Ullmann, R. and Soldati, T., 2000. The tail domain of myosin M catalyses nucleotide exchange on Rac1 GTPases and can induce actin-driven surface protrusions. *Traffic* **1**: 399–410.

Geli, M. I. and Riezman, H., 1996. Role of type I myosins in receptor-mediated endocytosis in yeast. *Science* **272**: 533–535.

Geli, M. I., Lombardi, R., Schmelzl, B. and Riezman, H., 2000. An intact SH3 domain is required for myosin I-induced actin polymerization. *EMBO J.* **19**: 4281–4291.

Glockner, G., Eichinger, L., Szafranski, K., Pachebat, J. A., *et al.*, 2002. Sequence and analysis of chromosome 2 of *Dictyostelium discoideum*. *Nature* **418**: 79–85.

Goode, B. L., Rodal, A. A., Barnes, G. and Drubin, D. G., 2001. Activation of the Arp2/3 complex by the actin filament binding protein Abp1p. *J. Cell Biol.* **153**: 627–634.

Goodson, H. V., Anderson, B. L., Warrick, H. M., Pon, L. A. and Spudich, J. A., 1996. Synthetic lethality screen identifies a novel yeast myosin I gene (MYO5): myosin I proteins are required for polarization of the actin cytoskeleton. *J. Cell Biol.* **133**: 1277–1291.

Holt, M. R. and Koffer, A., 2001. Cell motility: proline-rich proteins promote protrusions. *Trends Cell Biol.* **11**: 38–46.

Huber, L. A., Fialka, I., Paiha, K., Hunziker, W., *et al.*, 2000. Both calmodulin and the unconventional myosin myr4 regulate membrane trafficking along the recycling pathway of MDCK cells. *Traffic* **1**: 494–503.

Idrissi, F. Z., Wolf, B. L. and Geli, M. I., 2002. Cofilin, but not profilin, is required for Myosin-I-induced actin polymerization and the endocytic uptake in yeast. *Mol. Biol. Cell* **13**: 4074–4087.

Jung, G. and Hammer, J. A. D., 1990. Generation and characterization of *Dictyostelium* cells deficient in a myosin I heavy chain isoform. *J. Cell Biol.* **110**: 1955–1964.

Jung, G., Wu, X. and Hammer, J. A., 3rd, 1996. *Dictyostelium* mutants lacking multiple classic myosin I isoforms reveal combinations of shared and distinct functions. *J. Cell Biol.* **133**: 305–323.

Jung, G., Remmert, K., Wu, X. F., Volosky, J. M. and Hammer, J. A., 2001. The *Dictyostelium* CARMIL protein links capping protein and the Arp2/3 complex to type I myosins through their SH3 domains. *J. Cell Biol.* **153**: 1479–1497.

Kabaeva, Z. T., Perrot, A., Wolter, B., Dietz, R., *et al.*, 2002. Systematic analysis of the regulatory and essential myosin light chain genes: genetic variants and mutations in hypertrophic cardiomyopathy. *Eur. J. Hum. Genet.* **10**: 741–748.

Kalhammer, G. and Bahler, M., 2000. Unconventional myosins. *Essays Biochem.* **35**: 33–42.

Kessels, M. M., Engqvist-Goldstein, A. E., Drubin, D. G. and Qualmann, B., 2001. Mammalian Abp1, a signal-responsive F-actin-binding protein, links the actin cytoskeleton to endocytosis via the GTPase dynamin. *J. Cell Biol.* **153**: 351–366.

Kessin, R. H. (2001). *Dictyostelium – Evolution, cell biology, and the development of multicellularity*. Cambridge, UK, Cambridge University Press.

Lechler, T., Shevchenko, A. and Li, R., 2000. Direct involvement of yeast type I myosins in Cdc42-dependent actin polymerization. *J. Cell Biol.* **148**: 363–373.

Lechler, T., Jonsdottir, G. A., Klee, S. K., Pellman, D. and Li, R., 2001. A two-tiered mechanism by which Cdc42 controls the localization and activation of an Arp2/3-activating motor complex in yeast. *J. Cell Biol.* **155**: 261–270.

Lee, W. L., Ostap, E. M., Zot, H. G. and Pollard, T. D. 1999. Organization and ligand binding properties of the tail of *Acanthamoeba* myosin-IA – Identification of an actin-binding site in the basic (tail homology-1) domain. *J. Biol. Chem.* **274**: 35159–35171.

Lee, W. L., Bezanilla, M. and Pollard, T. D., 2000. Fission yeast myosin-I, Myo1p, stimulates actin assembly by Arp2/3 complex and shares functions with WASp. *J. Cell Biol.* **151**: 789–799.

Loisel, T. P., Boujemaa, R., Pantaloni, D. and Carlier, M. F., 1999. Reconstitution of actin-based motility of *Listeria* and *Shigella* using pure proteins. *Nature* **401**: 613–616.

Ma, S., Fey, P. and Chisholm, R. L., 2001. Molecular motors and membrane traffic in *Dictyostelium*. *Biochim. Biophys. Acta* **15**: 234–244.

Machesky, L. M., 2000. The tails of two myosins. *J. Cell Biol.* **148**, 219–221.

Machner, M. P., Urbanke, C., Barzik, M., Otten, S., *et al.*, 2001. ActA from *Listeria* monocytogenes can interact with up to four Ena/VASP homology 1 domains simultaneously. *J. Biol. Chem.* **276**: 40096–40103.

Mayer, B. J. (2001). SH3 domains: complexity in moderation. *J. Cell Sci.* **114**: 1253–1263.

McGoldrick, C. A., Gruver, C. and May, G. S., 1995. myoA of *Aspergillus nidulans* encodes an essential myosin I required for secretion and polarized growth. *J. Cell Biol.* **128**: 577–587.

Mermall, V., Post, P. L. and Mooseker, M. S., 1998. Unconventional myosins in cell movement, membrane traffic, and signal transduction. *Science* **279**: 527–533.

Mochida, J., Yamamoto, T., Fujimura-Kamada, K. and Tanaka, K., 2002. The novel adaptor protein, Mti1p, and Vrp1p, a homolog of Wiskott–Aldrich syndrome protein-interacting protein (WIP), may antagonistically regulate type I myosins in *Saccharomyces cerevisiae*. *Genetics* **160**: 923–934.

Mooseker, M. S. and Cheney, R. E., 1995. Unconventional myosins. *Ann. Rev. Cell Dev. Biol.* **11**: 633–675.

Mullins, R. D., 2000. How WASP-family proteins and the Arp2/3 complex convert intracellular signals into cytoskeletal structures. *Curr. Opin. Cell Biol.* **12**: 91–96.

Novak, K. D. and Titus, M. A., 1997. Myosin I overexpression impairs cell migration. *J. Cell Biol.* **136**: 633–647.

Novak, K. D. and Titus, M. A., 1998. The myosin I SH3 domain and TEDS rule phosphorylation site are required for in vivo function. *Mol. Biol. Cell* **9**: 75–88.

Novak, K. D., Peterson, M. D., Reedy, M. C. and Titus, M. A., 1995. *Dictyostelium* myosin I double mutants exhibit conditional defects in pinocytosis. *J. Cell Biol.* **131**: 1205–1221.

Osherov, N., Yamashita, R. A., Chung, Y. S. and May, G. S., 1998. Structural requirements for in vivo myosin I function in *Aspergillus nidulans*. *J. Biol. Chem.* **273**: 27017–27025.

Ostap, E. M. and Pollard, T. D., 1996. Biochemical kinetic characterization of the *Acanthamoeba* myosin-I ATPase. *J. Cell Biol.* **132**: 1053–1060.

Paunola, E., Mattila, P. K. and Lappalainen, P., 2002. WH2 domain: a small, versatile adapter for actin monomers. *FEBS Lett.* **513**: 92–97.

Pollard, T. D., Blanchoin, L. and Mullins, R. D., 2000. Molecular mechanisms controlling actin filament dynamics in nonmuscle cells. *Ann. Rev. Biophys. Biomol. Struct.* **29**: 545–576.

Pruyne, D., Evangelista, M., Yang, C., Bi, E., *et al.*, 2002. Role of formins in actin assembly: nucleation and barbed-end association. *Science* **297**: 612–615.

Raposo, G., Cordonnier, M. N., Tenza, D., Menichi, B., *et al.*, 1999. Association of myosin I alpha with endosomes and lysosomes in mammalian cells. *Mol. Biol. Cell* **10**: 1477–1494.

Robinson, D. N. and Spudich, J. A., 2000. Dynacortin, a genetic link between equatorial contractility and global shape control discovered by library complementation of a *Dictyostelium discoideum* cytokinesis mutant. *J. Cell Biol.* **150**: 823–838.

Schafer, D. A., 2002. Coupling actin dynamics and membrane dynamics during endocytosis. *Curr. Opin. Cell Biol.* **14**: 76–81.

Schwarz, E. C., Geissler, H. and Soldati, T. 1999. A potentially exhaustive screening strategy reveals two novel divergent myosins in *Dictyostelium*. *Cell Biochem. Biophys.* **30**: 413–435.

Schwarz, E., Neuhaus, E. M., Kistler, C., Henkel, A. W. and Soldati, T. 2000. *Dictyostelium* myosin IK is involved in the maintenance of cortical tension and affects motility and phagocytosis. *J. Cell Sci.* **113**: 413–435.

Soulard, A., Lechler, T., Spiridonov, V., Shevchenko, A., *et al.*, 2002. *Saccharomyces cerevisiae* Bzz1p is implicated with type I myosins in actin patch polarization and is able to recruit actin-polymerizing machinery in vitro. *Mol. Cell Biol.* **22**: 7889–7906.

Stöffler, H. E. and Bähler, M., 1998. The ATPase activity of Myr3, a rat myosin I, is allosterically inhibited by its own tail domain and by Ca2+ binding to its light chain calmodulin. *J. Biol. Chem.* **273**: 14605–14611.

Stöffler, H. E., Ruppert, C., Reinhard, J. and Bähler, M., 1995. A novel mammalian myosin I from rat with an SH3 domain localizes to Con A-inducible, F-actin-rich structures at cell–cell contacts. *J. Cell Biol.* **129**: 819–830.

Stöffler, H.-E., Hönnert, U., Bauer, C. A., Hofer, D., *et al.*, 1998. Targeting of the myosin-I myr3 to intercellular adherens type junctions induced by dominant active Cdc42 in HeLa cells. *J. Cell Sci.* **111**: 2779–2788.

Titus, M. A., Wessels, D., Spudich, J. A. and Soll, D., 1993. The unconventional myosin encoded by the myoA gene plays a role in *Dictyostelium* motility. *Mol. Biol. Cell* **4**: 233–246.

Tong, A. H., Drees, B., Nardelli, G., Bader, G. D., *et al.*, 2002. A combined experimental and computational strategy to define protein interaction networks for peptide recognition modules. *Science* **295**: 321–324.

Vadlamudi, R. K., Li, F., Adam, L., Nguyen, D., *et al.*, 2002. Filamin is essential in actin cytoskeletal assembly mediated by p21-activated kinase 1. *Nat. Cell Biol.* **4**: 681–690.

Vaduva, G., Martin, N. C. and Hopper, A. K., 1997. Actin-binding verprolin is a polarity development protein required for the morphogenesis and function of the yeast actin cytoskeleton. *J. Cell Biol.* **139**: 1821–1833.

Vaduva, G., Martinez-Quiles, N., Anton, I. M., Martin, N. C., *et al.*, 1999. The human WASP-interacting protein, WIP, activates the cell polarity pathway in yeast. *J. Biol. Chem.* **274**: 17103–17108.

Voigt, H., Olivo, J. C., Sansonetti, P. and Guillen, N., 1999. Myosin IB from *Entamoeba histolytica* is involved in phagocytosis of human erythrocytes. *J. Cell Sci.* **112**: 1191–1201.

Warren, D. T., Andrews, P. D., Gourlay, C. W. and Ayscough, K. R., 2002. Sla1p couples the yeast endocytic machinery to proteins regulating actin dynamics. *J. Cell Sci.* **115**: 1703–1715.

Wessels, D., Murray, J., Jung, G., Hammer, J. A. I. and Soll, D. R., 1991. Myosin IB null mutants of *Dictyostelium* exhibit abnormalities in motility. *Cell Motil. Cytoskeleton* **20**: 301–315.

Wessels, D., Titus, M. and Soll, D. R., 1996. A *Dictyostelium* myosin I plays a crucial role in regulating the frequency of pseudopods formed on the substratum. *Cell Motil. Cytoskeleton* **33**: 64–79.

Win, T. Z., Mulvihill, D. P. and Hyams, J. S., 2002. Take five: A myosin class act in fission yeast. *Cell Motil. Cytoskeleton* **51**: 53–56.

Witke, W., Sutherland, J. D., Sharpe, A., Arai, M. and Kwiatkowski, D. J., 2001. Profilin I is essential for cell survival and cell division in early mouse development. *Proc. Natl. Acad. Sci. USA* **98**: 3832–3836.

Wolenski, J. S., 1995. Regulation of Calmodulion binding myosins. *Trends Cell Biol.* **5**: 310–316.

Xu, P., Zot, A. S. and Zot, H. G. (1995). Identification of Acan125 as a myosin-I-binding protein present with myosin-I on cellular organelles of *Acanthamoeba*. *J. Biol. Chem.* **270**: 25316–25319.

Xu, P., Mitchelhill, K. I., Kobe, B., Kemp, B. E. and Zot, H. G., 1997. The myosin-I-binding protein Acan125 binds the SH3 domain and belongs to the superfamily of leucine-rich repeat proteins. *Proc. Natl. Acad. Sci. USA* **94**: 3685–3690.

Yazu, M., Adachi, H. and Sutoh, K., 1999. Novel *Dictyostelium* unconventional myosin MyoK is a class I myosin with the longest loop-1 insert and the shortest tail. *Biochem. Biophys. Res. Commun.* **255**: 711–716.

Zalevsky, J., Grigorova, I. and Mullins, R. D., 2001. Activation of the Arp2/3 complex by the *Listeria* ActA protein – ActA binds two actin monomers and three subunits of the Arp2/3 complex. *J. Biol. Chem.* **276**: 3468–3475.

4

Ephrin-regulated Contact Repulsion of Growth Cones

Lene K. Harbott, Daniel J. Marston and **Catherine D. Nobes**

Eph receptor activation by membrane-tethered ephrin ligands is critical for establishing correct neural projections, for creating boundaries between tissues, and for patterning the developing vasculature. The intracellular signals that regulate these diverse, cell contact-dependent biological responses are largely unknown. We show that activation of two Eph receptors, EphB2 and EphA7, by soluble ephrins stimulate the assembly of protrusive actin structures (filopodia and lamellipodia) and contractile actin stress fibres in Swiss 3T3 fibroblasts through the regulation of Rho GTPases. We have examined actin structures and the morphology of retinal ganglion cell (RGC) growth cones treated with soluble ephrin-A molecules. Ephrin-A5-Fc treatment causes loss of the lamella of the RGC growth cone, and retraction of the axon. The specific Rho kinase inhibitor, Y27632, prevents ephrin-induced axon retraction, but not loss of growth cone lamella. Co-culturing RGCs with fibroblasts expressing ephrin molecules on their surface allows us to examine the dynamic responses triggered by Eph receptor–ephrin interaction at sites of cell–cell contact. Besides growth cone collapse of the neurons, we show that fibroblasts also respond to Eph-ephrin signalling by lamella collapse at sites of contact and subsequent directional changes in migration.

Cell Motility. Edited by Anne Ridley, Michelle Peckham and Peter Clark.
 ISBN 0 470 84872 3

Introduction

Eph receptor tyrosine kinases and their surface-bound ligands, the ephrins, play a key role in the patterning of many tissues during embryonic development (Kullander and Klein, 2002; Wilkinson, 2000). Their importance in vascular development, tissue-border formation, cell migration and axon guidance is well described. In contrast with most receptor tyrosine kinases, Eph receptor activation occurs through direct cell–cell contact since the ephrins are cell surface molecules. The cellular response to Eph receptor/ephrin interaction can be repulsion, for example between neighbouring epithelial or endothelial cells or as the neuronal growth cone navigates its way towards its target. In other circumstances the Eph receptor/ephrin interaction can lead to an attractive or adhesive response (Holmberg and Frisen, 2002). It has recently been shown that upon Eph receptor binding, a response is induced in the ephrin-expressing cell (Kullander and Klein, 2002; Wilkinson, 2000). This capacity for bi-directional signalling allows both receptor- and ligand-expressing cells to be influenced by the mutual contact. Although the signalling pathways downstream of Eph receptors and ephrins that regulate cell repulsion/attraction responses are unclear, it appears that these pathways converge to regulate the actin cytoskeleton and cell adhesion.

Eph receptor and ephrin families

The Eph receptors and ephrins are numerous; currently 14 Eph receptors and eight ephrins have been identified. Eph receptors comprise the largest receptor tyrosine kinase family, accounting for one-third of all tyrosine kinase receptors. The ligands and the receptors each exist in one of two classes, A or B, based on structural similarity within the class, and the affinity of their binding to a cognate partner (Gale *et al.*, 1996). The ephrin-A ligands are attached to the outer leaflet of the plasma membrane via a glycosylphosphatidylinositol (GPI) anchor, whereas proteins of the ephrin-B class have a transmembrane region and a short cytoplasmic tail. Generally A-class receptors bind A-class ligands and B-class receptors bind B-class ligands. The exception is EphA4, which can bind both A-class and some B-class ligands.

Eph receptor/ephrin regulation of axon guidance

The most studied situation in which Eph receptor/ephrin function plays a critical role is axon path-finding. During embryonic development many different neuronal cell types respond to ephrins. Collectively ephrins play important roles in setting up the topology of nerve connections between the retina and the brain, between the various layers of the cortex, and also appear to be critical in preventing some midline crossing events during corticospinal tract formation (Kullander and Klein, 2002; Wilkinson, 2001). The predominant model system has been the projection of retinal ganglion cells (RGCs) from the retina through the optic chiasm to the tectum in the developing embryo (Drescher *et al.*, 1997; Flanagan and Vanderhaeghen, 1998).

Ephrins were initially isolated as repulsive axon guidance molecules expressed in the tectum, which cause collapse of RGC growth cones *in vitro* (Drescher *et al.*, 1995; Nakamoto *et al.*, 1996; Monschau *et al.*, 1997; see Figure 4.2A). Ephrin molecules are also repulsive to RGC axons *in vivo*, since disruption of tectal ephrin-A molecules causes RGC axons to overshoot their target positions (Frisen *et al.*, 1998; Feldheim *et al.*, 2000). Taken together with expression studies, this strongly suggests a function for ephrins as contact-cues that might repel axons.

In the embryonic tectum ephrins-A5 and -A2 are expressed in an increasing anterior-to-posterior gradient. Temporal RGC growth cones express high levels of receptor (particularly EphA3) and stop at positions in the anterior tectum, where they experience sufficient repulsive signals from these cells to trigger contact-repulsion. Nasal axons, expressing lower levels of EphA receptors, are able to navigate further into the posterior tectum before they reach the position expressing appropriate levels of ephrin ligand to trigger contact-repulsion (Drescher *et al.*, 1995; Monschau *et al.*, 1997; Cheng *et al.*, 1995).

Eph receptor/ephrin mediated control of cell segregation

Segment-restricted expression of Eph receptors and ephrins in the developing hindbrain prevents intermingling of sub-populations of neural epithelial cells, promoting the formation of sharp rhombomeric compartment-boundaries (Xu *et al.*, 1999). The mutual contact repulsion of Eph receptor/ephrin expressing cells causes them to sort into separate domains. It has been demonstrated that bi-directional signalling is necessary for the restriction of cell intermingling (Mellitzer *et al.*, 1999). Likewise, fibroblast-like cells of the pre-somitic mesoderm in a number of vertebrate species express alternating stripes of Eph receptors and ephrins that may block cell intermingling (Durbin *et al.*, 1998).

Developmental patterning of the vasculature is also critically dependent on interactions between Eph receptors and ephrins. Transgenic mice which are null for ephrin-B2 or EphB4 (the cognate receptor for ephrin-B2 in endothelial cells) support a model whereby bidirectional, repulsive signalling between EphB4- and ephrin-B2-expressing cells leads to the segregation of these two endothelial cell populations. This segregation appears to underlie the separation of arteries and veins, which is crucial for the correct development of the embryonic vasculature (Pandey *et al.*, 1995; Wang *et al.*, 1998; Adams *et al.*, 1999).

Interestingly, ephrins can also stimulate endothelial sprouting inducing the assembly of capillary networks both *in vitro* and *in vivo* (Adams *et al.*, 1999; Stein *et al.*, 1998; Wang *et al.*, 1998). These findings suggest that a combination of repulsive and attractive responses to Eph/ephrin signalling in endothelial cells might regulate the assembly, branching and remodelling of the primitive vascular network during embryogenesis.

Eph receptor/ephrin signalling

The exact nature of the cellular response underlying Eph receptor/ephrin mediated regulation of axon pathfinding, tissue boundary formation, vasculogenesis and angiogenesis is unknown. Cell–cell repulsion responses suggest de-adhesion mechanisms whereby an Eph receptor expressing cell will decrease its affinity for, and thus detach from, its ephrin expressing neighbour. Indeed, there is some evidence for ephrin-stimulated modulation of cell adhesions in tissue culture cells expressing Eph receptors, although these are cell–substrate adhesion effects rather than cell–cell adhesion effects (Huynh-Do *et al.*, 1999; Zou *et al.*, 1999; Miao *et al.*, 2000; Carter *et al.*, 2002). The collapse response of nerve growth cones to ephrins is proposed to be via disassembly of actin structures and thus a loss of the dynamic protrusive actin structures, filopodia and lamellipodia, required for growth cone migration (Meima *et al.*, 1997a,b).

Binding ephrins induces Eph receptors to form higher order clusters, leading to receptor autophosphorylation on several intracellular tyrosine residues. Like other receptor tyrosine kinases, activated Eph receptors can recruit adaptor molecules including Nck, p85-PI3 kinase, Src and Grb2 (Kullander and Klein, 2002) although, to date, none of these adaptors has been shown to be critically important.

Interestingly, in addition to phosphorylation-dependent recruitment of signalling effectors in the receptor-expressing cells it has been shown that B-class ephrins become phosphorylated on their intracellular tail following Eph receptor/ephrin interaction (Holland *et al.*, 1996; Bruckner *et al.*, 1997). One adaptor, recruited to ephrin-B1, is Grb4, which appears to trigger loss of polymerized actin structures and the disassembly of focal adhesions (Cowan and Henkemeyer, 2001).

Eph receptor activation by soluble ephrins rapidly stimulates the assembly of filamentous actin structures in fibroblast cells

To analyse early changes in actin organization following activation of Eph receptor kinases, we microinjected eukaryotic expression vectors encoding EphB2 and EphA7 into the nuclei of quiescent, serum-starved Swiss 3T3 cells. These fibroblast cells are frequently used to study cellular cytoskeletal changes induced by growth factors (Nobes and Hall, 1995; Kozma *et al.*, 1995; Ridley and Hall, 1992), since serum starvation induces loss of filamentous actin structures such as lamellipodia and stress fibres.

We find a diffuse surface localization of EphB2 or EphA7 can be detected 2 h after microinjection of constructs encoding these receptors (data not shown). Expression of EphB2 or EphA7 alone does not induce actin filament assembly within this time (data not shown).

Activation of EphB2 by the addition of soluble, dimeric ephrin-B1-Fc results in a rapid increase in tyrosine phosphorylation (data not shown) and a concomitant accumulation of polymerized actin in broad sheet-like lamellipodia (Figure 4.1A) beginning within 5–7 minutes of ligand presentation. Lamellipodial activity is downregulated after 30 min and EphB2 receptors internalized (data not shown).

Activation of EphA7 leads to a pattern of actin assembly distinct from that stimulated by activating EphB2, with cells forming localized filopodial protrusions accompanied by associated lamellipodial spreading, within 5–7 min of stimulation by dimeric ephrin-A4-Fc (Figure 4.1B). The phosphotyrosine immunofluorescence pattern is also different for the two receptors, with activated EphA7-expressing cells displaying small clusters of phosphotyrosine localized to zones of filopodial and lamellipodial activity (data not shown). Following the initial increase in actin accumulation at the cell periphery, actin stress fibres are frequently observed in both EphB2 and EphA7 expressing cells, beginning 10–15 min after stimulation with ephrin (Figure 4.1A and 4.1B).

EphB2 and EphA7 induced lamellipodial protrusion is mediated by the small GTPase Rac

Small GTPases of the Rho family – the three best characterized being Rho, Rac and Cdc42 – control the assembly of filamentous actin structures and are implicated in the regulation of axon guidance. Cdc42 induces actin filament assembly leading to filopodia protrusion; Rac triggers protrusion of lamellipodia, and Rho regulates the assembly of contractile actin:myosin

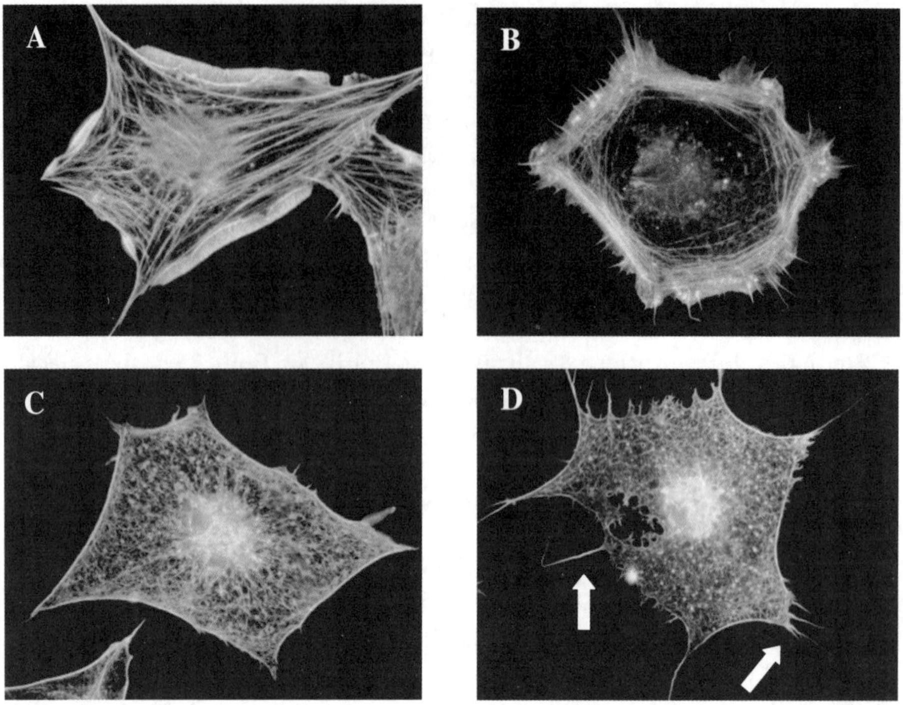

Figure 4.1 Activation of EphB2 and EphA7 triggers actin reorganization in fibroblasts. Serum-starved subconfluent Swiss 3T3 cells were microinjected with plasmids encoding EphB2 (A), EphB2 + N17Rac (C), EphA7 (B), EphA7 + N17Rac (D). After 3 h cells were treated with 1 μg ephrin-B1-Fc (A, C) or ephrin-A4-Fc (B, D) for 15 min before fixation. Actin filaments (A–D) were visualized with TRITC-phalloidin. Arrows in D indicate filopodia

filaments (Nobes and Hall, 1995; Kozma *et al.*, 1995; Ridley and Hall, 1992; Ridley *et al.*, 1992).

To determine whether the lamellipodia induced by stimulated EphB2 and EphA7 are a result of activation of the small GTPase Rac, we have co-injected cells with a eukaryotic expression construct encoding myc-tagged dominant negative Rac(N17Rac). Expression of N17Rac protein, confirmed by immunofluorescence after fixing cells, does not alter the increase in phosphotyrosine immunofluorescence observed after addition of ephrins to Eph receptor expressing cells (data not shown). Expression of N17Rac completely blocks lamellipodial protrusions stimulated by ephrin-B1 activation of EphB2, and by ephrin-A4 activation of EphA7 (Figure 4.1C and 4.1D). However, Rac inhibition does not interfere with the assembly of filopodia in activated EphA7-expressing cells (see arrows in Figure 4.1D). N17Rac also inhibits the assembly of actin stress fibres after EphB2 and EphA7 activation, suggesting that stress fibres are produced as a result of Rac-mediated Rho activation (Ridley *et al.*, 1992). This we have confirmed by co-injecting Eph receptor expressing cells with C3 transferase (a specific inhibitor of Rho) and observing a failure of actin stress fibre formation without inhibition of lamellipodial protrusions (data not shown).

Role of Rho GTPases in ephrin induced growth cone collapse

Several different model systems have demonstrated a role for members of the Rho family proteins in axon guidance. From the classic studies of Rho GTPase manipulation in fibroblasts, it might be predicted that activation of the small GTPases Rac and Cdc42 would mediate growth cone responses to attractive cues, whereas Rho would likely underlie growth cone collapse and axon retraction events via Rho-associated-kinase signalling and consequent actin-myosin based contraction events (Dickson, 2001). A novel guanine nucleotide exchange factor for Rho GTPases named ephexin has been identified as an EphA receptor interacting protein through a two-hybrid screen (Shamah *et al.*, 2001). Transfection of a dominant negative mutant version of ephexin into RGCs inhibited ephrin-A induced growth cone collapse, implying a link between Eph receptors, Rho GTPases and the actin cytokeleton. While it has been shown that growth cone collapse induced by another repulsive cue, Semaphorin3A, requires Rac activity (Jin and Strittmatter, 1997; Kuhn *et al.*, 1999; Vastrik *et al.*, 1999), studies of ephrin-induced RGC growth cone collapse have demonstrated a reduction in the levels of active Rac in the neurons after ephrin treatment, concurrent with an increase in Rho activity (Wahl *et al.*, 2000). Indeed, Mueller and colleagues (Wahl *et al.*, 2000) have demonstrated that inhibiting Rho or Rho associated kinase (a Rho effector molecule) prevents ephrin-induced growth cone collapse in RGCs.

We have repeated this experiment on cultured RGCs, by stimulating with soluble ephrin molecules. RGC axons growing on laminin have a characteristic growth cone, consisting filopodia (Figure 4.2A first panel, arrowheads) with a veil-like lamella between them (Figure 4.2A, arrow). Addition of 1 μg/ml ephrin-A5 to the bathing medium of cultured RGCs causes rapid collapse of the growth cone lamella and filopodia, followed by axon retraction within 5 min (Figure 4.2A). Using a phosphorylation-specific Eph receptor antibody we observe that Eph receptors on the RGC growth cone and along the length of the axon rapidly become phosphorylated in response to ephrin-A5 (Figure 4.2B, right-hand panels). This antibody reveals a single band after immunoblotting whole cell lysates from RGCs stimulated with ephrin-A5, confirming that there is a significant increase in Eph receptor phosphorylation in these cells in response to ephrin-A5 (Figure 4.2C).

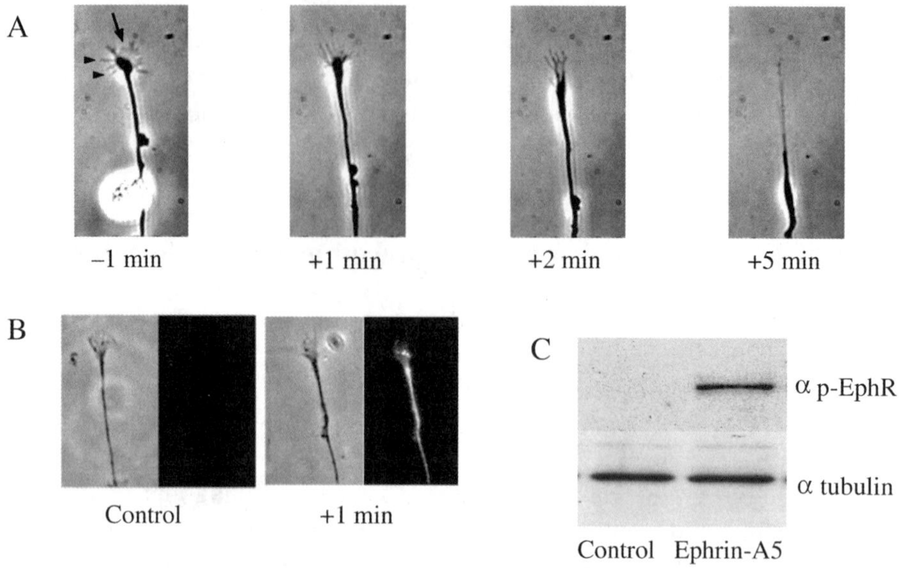

Figure 4.2 Ephrin-A5 causes collapse of Chick RGC growth cones via activation of Eph receptors. (A) Ephrin-A5-fc dimers were clustered with an anti-human antibody and added to the bathing medium of cultured RGCs. Times indicated are times after addition of ephrin. (B) Addition of soluble ephrin-A5 leads to phosphorylation of Eph receptors. The anti-pEph receptor antibody (right-hand panels) recognizes the phosphorylation of two highly conserved juxtamembrane tyrosines in the Eph receptor intracellular tail. Times shown are times after ephrin addition. (C) Isolated and purified RGCs show phosphorylation of Eph receptors after 15 min stimulation with soluble clustered ephrin A5. Tubulin is shown as a loading control

In the presence of a specific inhibitor of Rho associated kinase, Y27632, soluble ephrin-A5-Fc induces the loss of lamellae seen in the control situation (compare Figure 4.2A and 4.3A). However, the collapse of filopodia and the dramatic axon retraction are not observed. RGC growth cones fixed and stained with phalloidin can be placed into one of three categories according to their morphology and filamentous actin distribution (Figure 4.3B). We consider a full growth cone to have lamella and filopodia and a partial growth cone to have no lamellae but at least three or more filopodia. Full collapse is defined as loss of both lamellae and filopodia. Although the proportion of growth cones which have fully collapsed in response to ephrin-A5 stimulation is returned to near control levels by the presence of Y27632, the majority of growth cones show a partially collapsed morphology with no lamellae (Figure 4.3B). Since inhibition of Rho associated kinase prevents the ephrin-induced collapse of filopodia and dramatic axon retraction, but not loss of lamella, it seems the Rho-Rho kinase pathway must not be the only one activated during ephrin-A5 induced growth cone collapse.

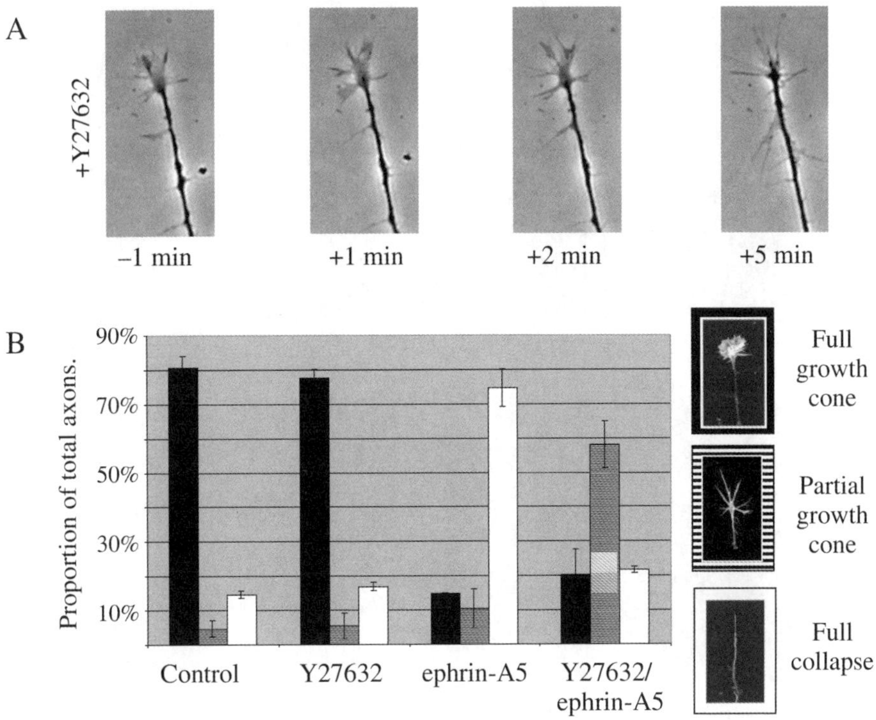

Figure 4.3 Inhibiting Rho associated kinase partially rescues ephrin-A5 induced growth cone collapse. (A) Time-lapse stills of a RGC growth cone in the presence of 10 μM Y27632. Times shown are times after addition of ephrin. (B) RGC growth cones fixed and stained with phalloidin can be placed into one of three categories according to their filamentous actin distribution, as shown in the right-hand panels. A full growth cone is defined as consisting full lamella and filopodia. A partial growth cone has no lamella but has three or more filopodia. Full collapse is defined as loss of both lamella and filopodia. Cells were fixed after 10 min stimulation with soluble ephrin-A5

As mentioned previously, ephrin molecules are membrane bound limiting Eph receptor–ephrin interaction to sites of cell–cell contact. We have developed a co-culture assay in which we can study the interaction of RGC growth cones with cells expressing ephrin molecules on their surface. When a RGC growth cone encounters an ephrin-A-expressing fibroblast cell in culture the growth cone withdraws its lamella and the axon retracts within 10 min (Figure 4.4A). The fibroblast appears also to respond to the cell contact by rapid lamella collapse (see Figure 4.4B). In other examples, a migrating fibroblast will respond to the encounter by changing the direction of its migration (data not shown). This demonstrates the bi-directional signalling upon Eph receptor–ephrin activation at sites of cell–cell contact observed in other systems.

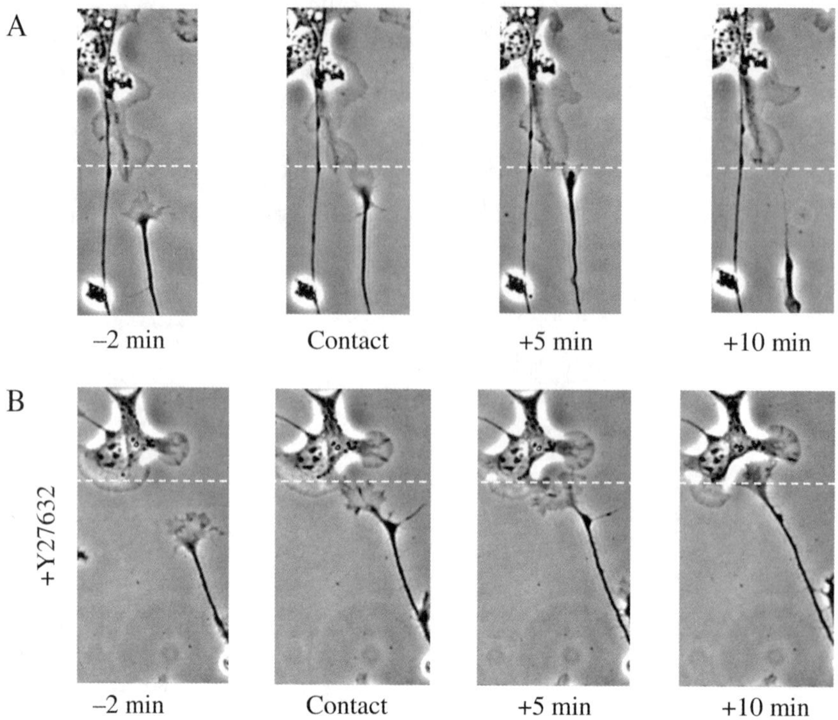

Figure 4.4 Ephrin-A ligands on the surface of Swiss 3T3 fibroblasts induce Rho kinase mediated growth cone collapse and axon retraction. (A) A RGC growth cone in culture encounters an ephrin-A expressing fibroblast. Times shown are times after contact of the cells. (B) Time-lapse stills of a co-culture experiment in which 10 μM Y27632 has been added to the bathing medium. Times indicated are times after contact is seen to occur between the two cells

By addition of soluble extracellular portions of A-class Eph receptors we block the interaction between ephrin-A ligands expressed on the surface of the fibroblast cell, and EphA receptors on the RGC growth cone. Addition of soluble EphA5 prevents the growth cone collapse and axon retraction seen in the RGC, and the repulsive response in the fibroblast cell (data not shown), confirming that these responses are directly mediated by Eph receptor–ephrin interaction.

The presence of the Rho associated kinase inhibitor, Y27632, in the co-culture assay dramatically prevents the rapid retraction of the axon we observe in response to contact with a cell expressing surface ephrin, and delays the loss of lamella (Figure 4.4B). In the presence of Y27632 the lamella is lost on average 11.6 ± 4.8 minutes after contact with the ephrin-expressing fibroblast ($n = 13$), as compared with 5.9 ± 2.6 minutes in the control situation (n = 24). The response of the Y27632 treated RGC to the ephrin-expressing fibroblast contrasts with the response to soluble ephrin. The lamella of Y27632 treated RGC is lost within 5 minutes after stimulation with soluble ephrin-A5 (see Figure 4.3A). Following contact with the fibroblast cell the lamella of Y27632 treated RGC is retained for longer and the growth cone continues to advance across the substrate (Figure 4.4B). These contrasting results might reflect a difference in concentration or complement of ephrin-A ligands that stimulate RGC Eph receptors. Soluble ephrin-A5 causes activation of Eph receptors all over the growth cone and along the axon (Figure 4.2B) whereas only the receptors at the leading edge of the growth cone will be activated by direct contact with the fibroblast cell, which may express ephrin-A ligands other than ephrin-A5.

Inhibiting Rho associated kinase does not prevent the repulsive response of the ephrin-expressing fibroblast cell after interaction with the RGC (Figure 4.4B). This suggests that Rho kinase may not be involved in the ephrin-signalling pathway.

Conclusions

It is clear that, *in vitro*, ephrin stimulation of Eph receptor expressing cells has a profound effect on the actin cytoskeleton, and that this in turn can influence the dynamic behaviour of the cells involved. In fibroblast cells we see two temporally distinct actin based responses: first a protrusive response in which lamellae and/or filopodia are induced, and later the formation of contractile actin stress fibres which are associated with a cell's withdrawal or retraction.

In neuronal axons in culture, soluble and membrane bound ephrin causes axonal retraction, which is mediated by Rho associated kinase. It is therefore possible that contractile actin filaments analogous to fibroblast stress fibres are assembled in the neuronal axon following ephrin stimulation. In the co-culture assay it seems that the growth cone may continue to advance a small distance across the substrate after contact with the ephrin-expressing cell. It will be interesting to analyse this further to determine whether the full compliment of actin responses seen in the Eph receptor expressing fibroblast cells are reflected in the neuronal response to ephrins; both the early protrusive and the later retraction behaviours.

In the case of stimulation with membrane-bound ligand, not only does inhibiting Rho associated kinase prevent axonal retraction, but also the loss of lamella is significantly delayed and the axon continues to advance across the substrate. The differences in response to soluble versus membrane bound ligand may reflect differences in the complement of ephrin molecules presented, or the level of neuronal Eph receptor activation. Ongoing work involves introducing ephrin-A molecules into cells which do not endogenously express ephrins. In this way we can precisely control the levels of ephrins involved in the interaction, and investigate possible differences in response to different A-ephrins.

It is interesting to view the interaction of Eph receptor-expressing neurons with ephrin-expressing fibroblasts in the light of the work of Michael Abercrombie. During the phenomenon of contact inhibition of locomotion in cultured cells, he described how dynamic actin structures such as lamellipodia are paralysed at new cell contact sites. Overall this can lead to the migration of contacting cells away from one another. The signalling pathways involved must therefore converge on the actin cytoskeleton and cell adhesion systems in both cells involved in the interaction. In the co-culture assay described here both cells respond by remodelling their actin based motility machineries, and ultimately pull away from the site of contact. It may turn out to be that similar mechanisms are involved in Eph receptor/ephrin regulated cell repulsion and Abercrombie's contact inhibition of locomotion.

References

Adams, R. H., Wilkinson, G. A., Weiss, C., Diella, F., *et al.*, 1999. Roles of ephrinB ligands and EphB receptors in cardiovascular development: demarcation of arterial/venous domains, vascular morphogenesis, and sprouting angiogenesis. *Genes Dev* **13**: 295–306.

Bruckner, K., Pasquale, E. B. and Klein, R., 1997. Tyrosine phosphorylation of transmembrane ligands for Eph receptors. *Science* **275**: 1640–1643.

Carter, N., Nakamoto, T., Hirai, H. and Hunter, T., 2002. EphrinA1-induced cytoskeletal re-organization requires FAK and p130(cas). *Nat. Cell Biol.* **4**: 565–573.

Cheng, H. J., Nakamoto, M., Bergemann, A. D. and Flanagan, J. G., 1995. Complementary gradients in expression and binding of ELF-1 and Mek4 in development of the topographic retinotectal projection map. *Cell* **82**: 371–381.

Cowan, C. A. and Henkemeyer, M., 2001. The SH2/SH3 adaptor Grb4 transduces B-ephrin reverse signals. *Nature* **413**: 174–179.

Dickson, B. J., 2001. Rho GTPases in growth cone guidance. *Curr. Opin. Neurobiol.* **11**: 103–110.

Drescher, U., Kremoser, C., Handwerker, C., Loschinger, J., *et al.*, 1995. In vitro guidance of retinal ganglion cell axons by RAGS, a 25 kDa tectal protein related to ligands for Eph receptor tyrosine kinases. *Cell* **82**: 359–370.

Drescher, U., Bonhoeffer, F. and Muller, B. K., 1997. The Eph family in retinal axon guidance. *Curr. Opin. Neurobiol.* **7**: 75–80.

Durbin, L., Brennan, C., Shiomi, K., Cooke, J., *et al.*, 1998. Eph signaling is required for segmentation and differentiation of the somites. *Genes Dev.* **12**: 3096–3109.

Feldheim, D. A., Kim, Y. I., Bergemann, A. D., Frisen, J., *et al.*, 2000. Genetic analysis of ephrin-A2 and ephrin-A5 shows their requirement in multiple aspects of retinocollicular mapping. *Neuron* **25**: 563–574.

Flanagan, J. G. and Vanderhaeghen, P., 1998. The ephrins and Eph receptors in neural development. *Annu. Rev. Neurosci.* **21**: 309–345.

Frisen, J., Yates, P. A., McLaughlin, T., Friedman, G. C., *et al.*, 1998. Ephrin-A5 (AL-1/RAGS) is essential for proper retinal axon guidance and topographic mapping in the mammalian visual system. *Neuron* **20**: 235–243.

Gale, N. W., Holland, S. J., Valenzuela, D. M., Flenniken, A., *et al.*, 1996. Eph receptors and ligands comprise two major specificity subclasses and are reciprocally compartmentalized during embryogenesis. *Neuron* **17**: 9–19.

Holland, S. J., Gale, N. W., Mbamalu, G., Yancopoulos, G. D., *et al.*, 1996. Bidirectional signalling through the EPH-family receptor Nuk and its transmembrane ligands. *Nature* **383**: 722–725.

Holmberg, J. and Frisen, J., 2002. Ephrins are not only unattractive. *Trends Neurosci* **25**: 239–243.

Huynh-Do, U., Stein, E., Lane, A. A., Liu, H., *et al.*, 1999. Surface densities of ephrin-B1 determine EphB1-coupled activation of cell attachment through alphavbeta3 and alpha5beta1 integrins. *Embo. J.* **18**: 2165–2173.

Jin, Z. and Strittmatter, S. M., 1997. Rac1 mediates collapsin-1-induced growth cone collapse. *J. Neurosci.* **17**: 6256–6263.

Kozma, R., Ahmed, S., Best, A. and Lim, L., 1995. The Ras-related protein Cdc42Hs and bradykinin promote formation of peripheral actin microspikes and filopodia in Swiss 3T3 fibroblasts. *Mol. Cell Biol.* **15**: 1942–1952.

Kuhn, T. B., Brown, M. D., Wilcox, C. L., Raper, J. A. and Bamburg, J. R., 1999. Myelin and collapsin-1 induce motor neuron growth cone collapse through different pathways: inhibition of collapse by opposing mutants of rac1. *J. Neurosci.* **19**: 1965–1975.

Kullander, K. and Klein, R., 2002. Mechanisms and functions of Eph and ephrin signalling. *Nat. Rev. Mol. Cell Biol.* **3**: 475–486.

Meima, L., Kljavin, I. J., Moran, P., Shih, A., *et al.*, 1997a. AL-1-induced growth cone collapse of rat cortical neurons is correlated with REK7 expression and rearrangement of the actin cytoskeleton. *Eur. J. Neurosci.* **9**: 177–188.

Meima, L., Moran, P., Matthews, W. and Caras, I. W., 1997b. Lerk2 (ephrin-B1) is a collapsing factor for a subset of cortical growth cones and acts by a mechanism different from AL-1 (ephrin-A5). *Mol. Cell Neurosci.* **9**, 314–328.

Mellitzer, G., Xu, Q. and Wilkinson, D. G., 1999. Eph receptors and ephrins restrict cell intermingling and communication. *Nature* **400**: 77–81.

Miao, H., Burnett, E., Kinch, M., Simon, E. and Wang, B., 2000. Activation of EphA2 kinase suppresses integrin function and causes focal-adhesion-kinase dephosphorylation. *Nat. Cell Biol.* **2**: 62–69.

Monschau, B., Kremoser, C., Ohta, K., Tanaka, H., *et al.*, 1997. Shared and distinct functions of RAGS and ELF-1 in guiding retinal axons. *Embo. J.* **16**: 1258–1267.

Nakamoto, M., Cheng, H. J., Friedman, G. C., McLaughlin, T., *et al.*, 1996. Topographically specific effects of ELF-1 on retinal axon guidance in vitro and retinal axon mapping *in vivo*. *Cell* **86**: 755–766.

Nobes, C. D. and Hall, A., 1995. Rho, rac, and cdc42 GTPases regulate the assembly of multimolecular focal complexes associated with actin stress fibers, lamellipodia, and filopodia. *Cell* **81**: 53–62.

Pandey, A., Shao, H., Marks, R. M., Polverini, P. J. and Dixit, V. M., 1995. Role of B61, the ligand for the Eck receptor tyrosine kinase, in TNF-alpha-induced angiogenesis. *Science* **268**: 567–569.

Ridley, A. J. and Hall, A., 1992. The small GTP-binding protein rho regulates the assembly of focal adhesions and actin stress fibers in response to growth factors. *Cell* **70**: 389–399.

Ridley, A. J., Paterson, H. F., Johnston, C. L., Diekmann, D. and Hall, A., 1992. The small GTP-binding protein rac regulates growth factor-induced membrane ruffling. *Cell* **70**: 401–410.

Shamah, S. M., Lin, M. Z., Goldberg, J. L., Estrach, S., *et al.*, 2001. EphA receptors regulate growth cone dynamics through the novel guanine nucleotide exchange factor ephexin. *Cell* **105**: 233–244.

Stein, E., Lane, A. A., Cerretti, D. P., Schoecklmann, H. O., *et al.*, 1998. Eph receptors discriminate specific ligand oligomers to determine alternative signaling complexes, attachment, and assembly responses. *Genes Dev.* **12**: 667–678.

Vastrik, I., Eickholt, B. J., Walsh, F. S., Ridley, A. and Doherty, P., 1999. Sema3A-induced growth-cone collapse is mediated by Rac1 amino acids 17-32. *Curr. Biol.* **9**: 991–998.

Wahl, S., Barth, H., Ciossek, T., Aktories, K. and Mueller, B. K., 2000. Ephrin-A5 induces collapse of growth cones by activating Rho and Rho kinase. *J. Cell Biol.* **149**: 263–270.

Wang, H. U., Chen, Z. F. and Anderson, D. J., 1998. Molecular distinction and angiogenic interaction between embryonic arteries and veins revealed by ephrin-B2 and its receptor Eph-B4. *Cell* **93**: 741–753.

Wilkinson, D. G., 2000. Eph receptors and ephrins: regulators of guidance and assembly. *Int. Rev. Cytol.* **196**: 177–244.

Wilkinson, D. G., 2001. Multiple roles of EPH receptors and ephrins in neural development. *Nat. Rev. Neurosci.* **2**: 155–164.

Xu, Q., Mellitzer, G., Robinson, V. and Wilkinson, D. G., 1999. *In vivo* cell sorting in complementary segmental domains mediated by Eph receptors and ephrins. *Nature* **399**: 267–271.

Zou, J. X., Wang, B., Kalo, M. S., Zisch, A. H., *et al.*, 1999. An Eph receptor regulates integrin activity through R-Ras. *Proc. Natl. Acad. Sci. USA* **96**: 13 813–13 818.

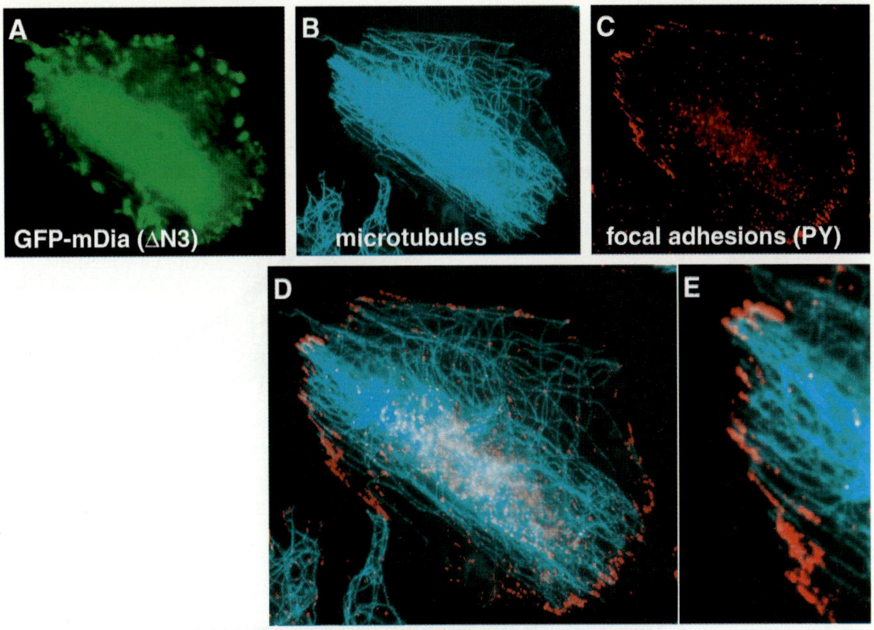

Figure 5.5

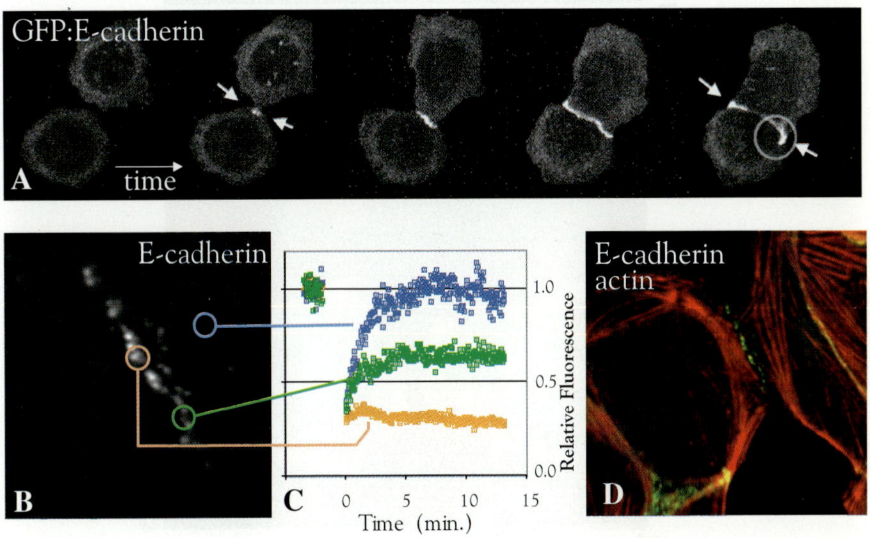

Figure 6.1

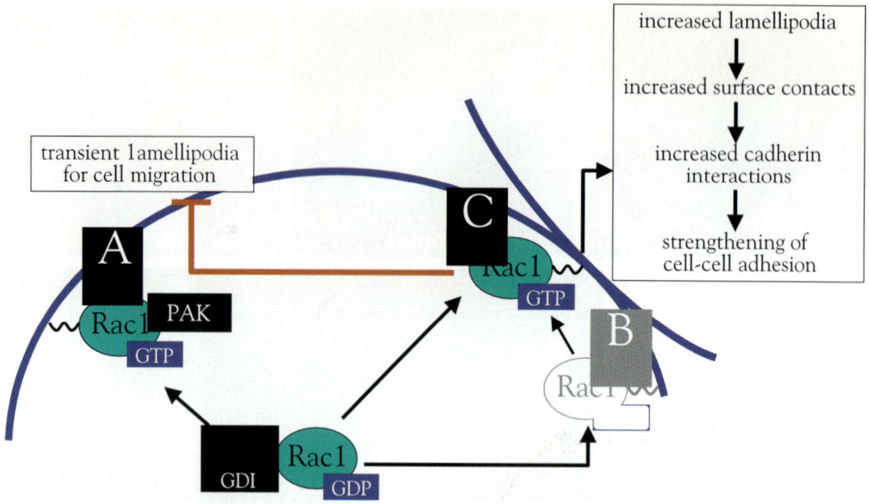

Figure 6.2

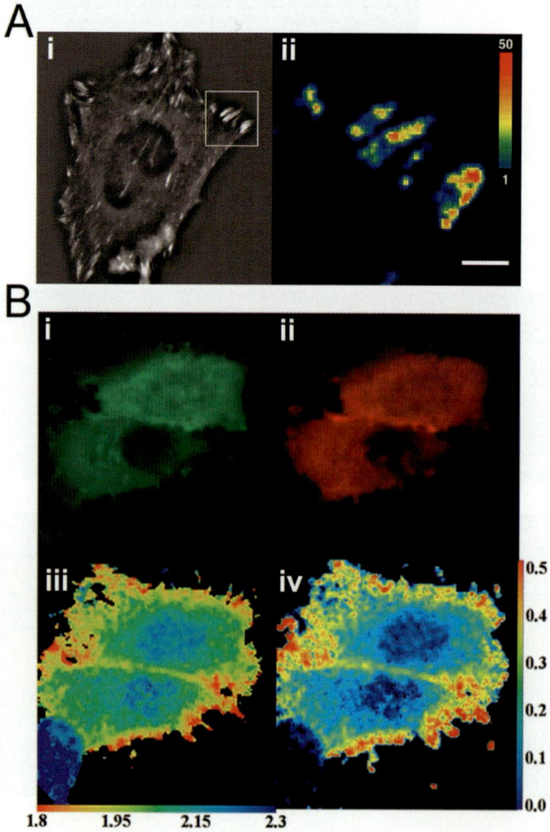

Figure 7.1

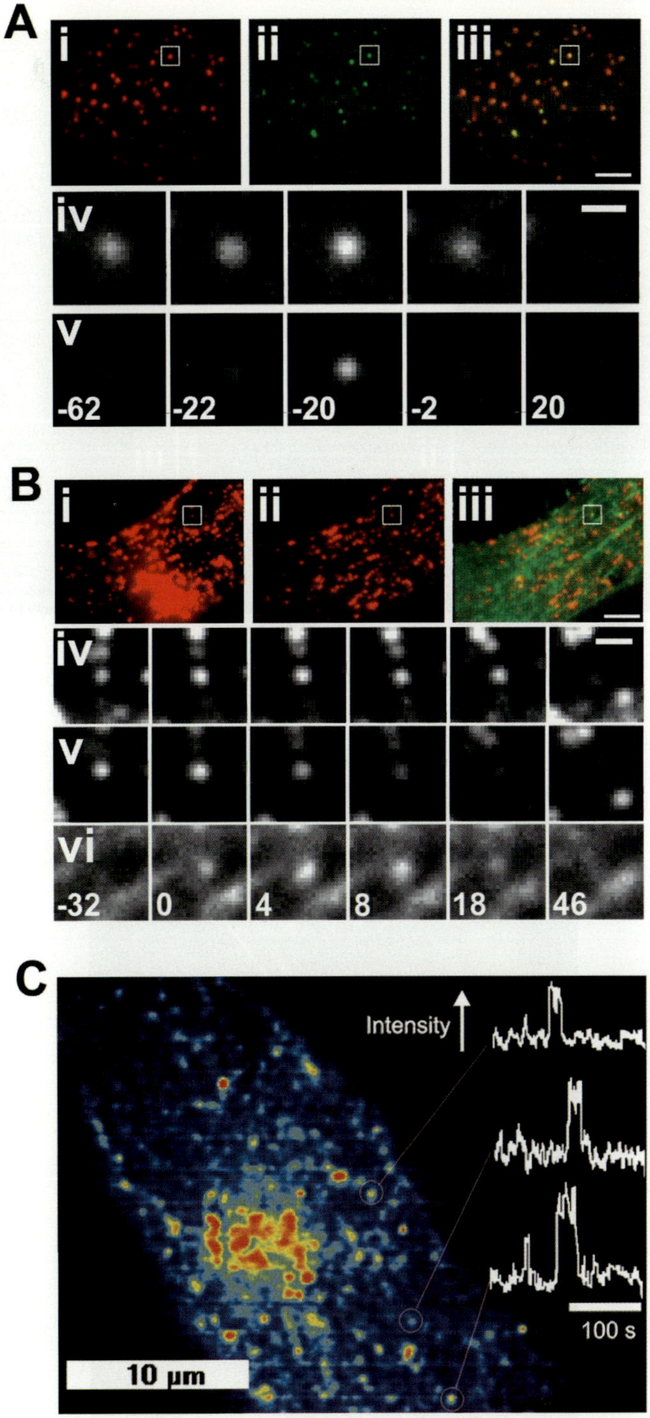
A
i
ii
iii
iv
v
-62
-22
-20
-2
20
B
i
ii
iii
iv
v
vi
-32
0
4
8
18
46
C
Intensity
100 s
10 µm

Figure 7.2

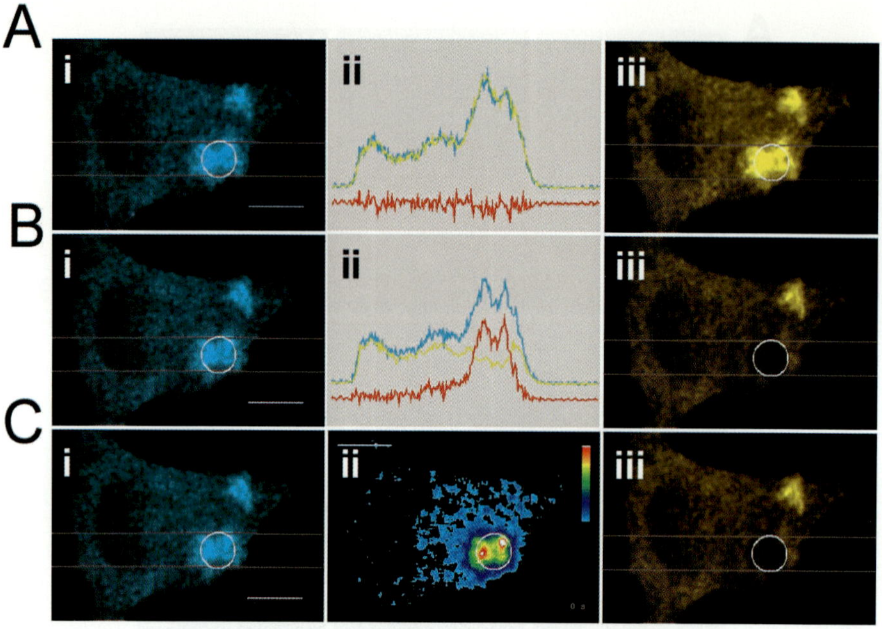

Figure 7.5

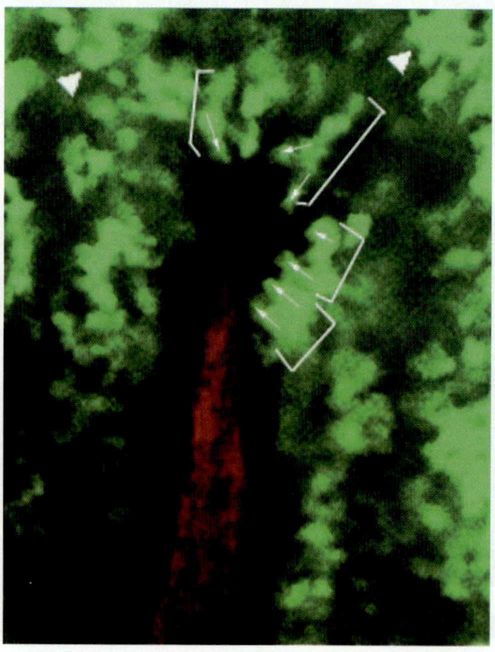

Figure 11.1

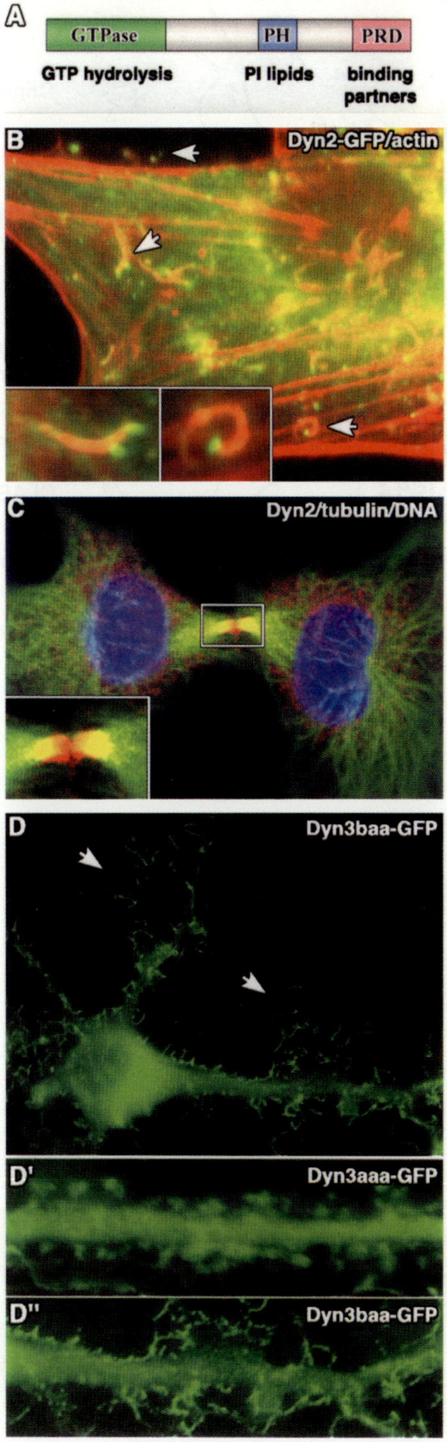
A
GTPase
PH
PRD
GTP hydrolysis
PI lipids
binding partners
B
Dyn2-GFP/actin
C
Dyn2/tubulin/DNA
D
Dyn3baa-GFP
D'
Dyn3aaa-GFP
D''
Dyn3baa-GFP

Figure 12.1

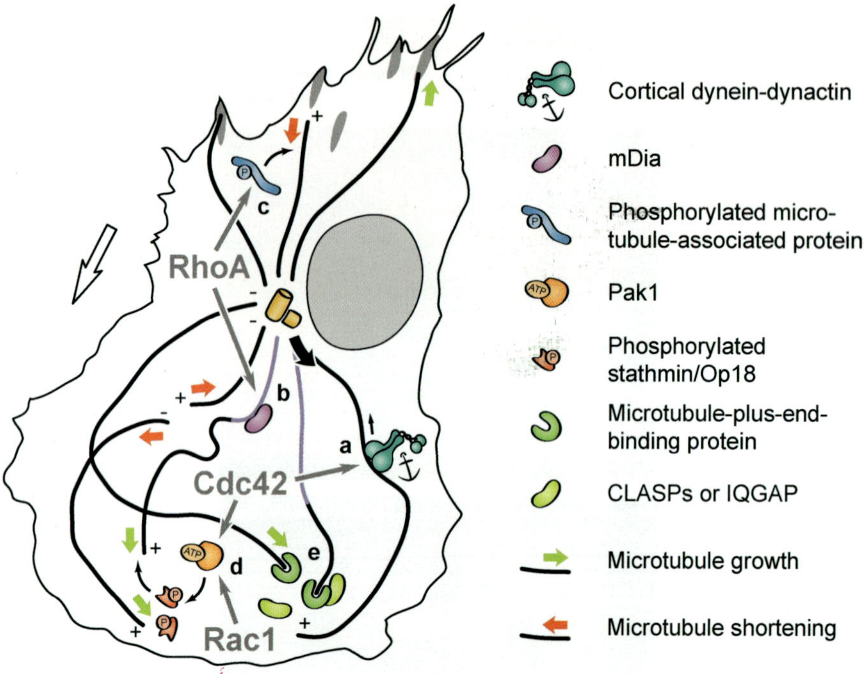

Figure 13.3

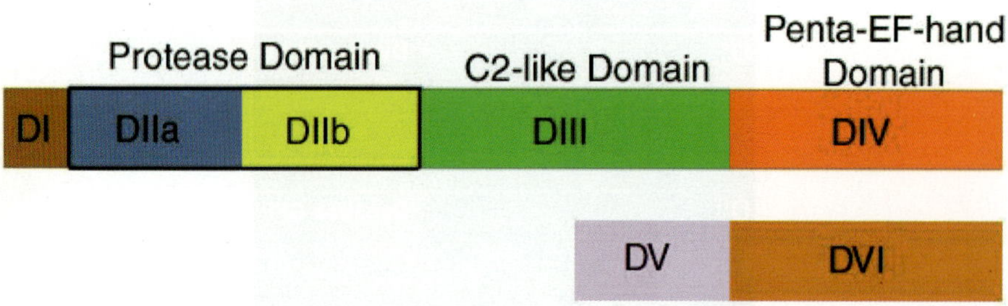

Figure 14.1

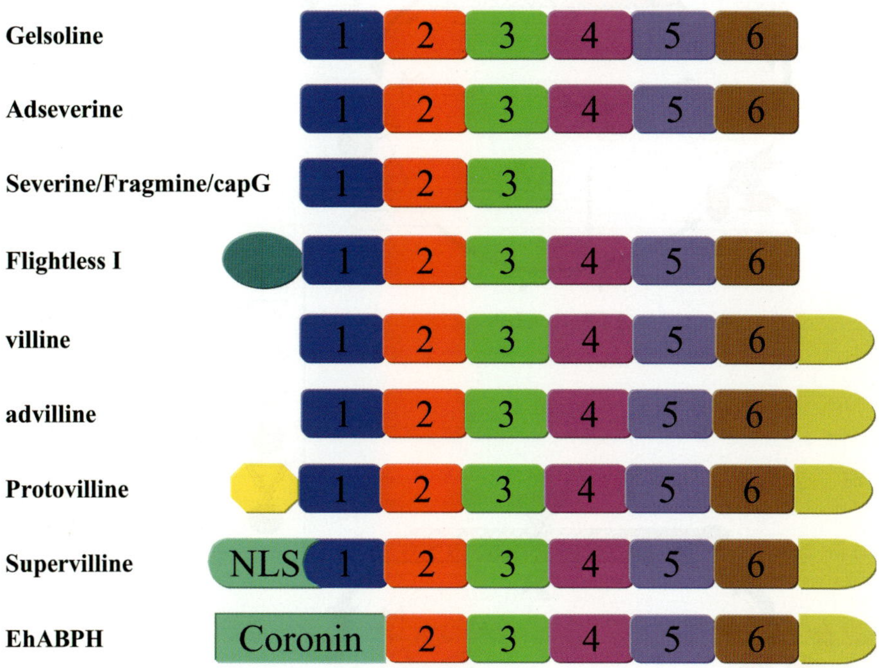
Gelsoline
1
2
3
4
5
6
Adseverine
Severine/Fragmine/capG
Flightless I
villine
advilline
Protovilline
Supervilline
NLS
EhABPH
Coronin

Figure 15.1

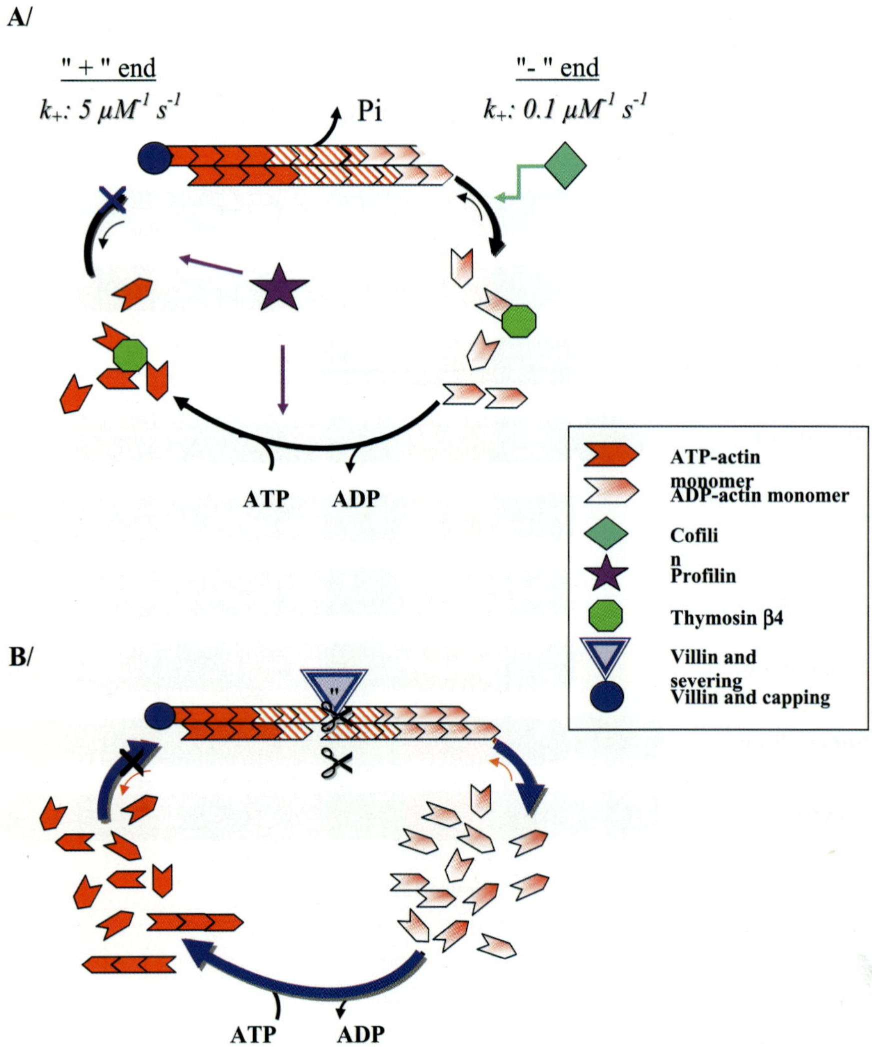
A/
" + " end
k+: 5 μM-1 s-1
"- " end
k+: 0.1 μM-1 s-1
Pi
ATP
ADP
ATP-actin monomer
ADP-actin monomer
Cofilin
Profilin
Thymosin β4
Villin and severing
Villin and capping
B/
ATP
ADP

Figure 15.3

5

Interplay between the Actin Cytoskeleton, Focal Adhesions and Microtubules

Christoph Ballestrem, Natalia Magid, Julia Zonis, Michael Shtutman and **Alexander Bershadsky**

Focal adhesions (FAs) are dynamic molecular complexes associated with integrin-family transmembrane receptors, which connect the actin cytoskeleton with the extracellular matrix. FA assembly is induced by tension either applied to these structures externally, or resulting from myosin II-driven cell contractility. Thus, FAs function as mechanosensors, 'reporting' to the cell information about the physical properties of the surrounding environment. Rho, a principal molecular switch triggering FA formation, operates by activating Rho associated kinase (ROCK) and the formin homology protein, mDia1. ROCK is required for the activation of the myosin II-driven contractility, and its function can be bypassed by externally applied force. Under these conditions, mDia1 remains necessary for FA assembly. mDia1nucleates actin filaments, and affects microtubule dynamics at both the plus and minus ends, which might facilitate microtubule concentration near maturing FAs. Since microtubules interfere with myosin II-driven contractility, this may create a negative feedback loop controlling FA growth.

Cell Motility. Edited by Anne Ridley, Michelle Peckham and Peter Clark.
 ISBN 0 470 84872 3

Introduction

Cell movement results from the coordinated remodelling of the cytoskeleton and adhesion structures. The cellular machinery involved in this process is very complex and includes not only the structural components (such as cytoskeletal fibres, molecular motors and transmembrane adhesion receptors), but also a plethora of adapter, regulatory and signalling proteins that control dynamics and interactions of the core structural elements. The complexity of the cell locomotory system is high and includes many regulatory loops, providing proper coordination between different structural components, and enabling this system to respond correctly to a variety of external stimuli. The major types of external signals affecting cell motility, besides soluble chemotactic factors and 'motogenic' ligands, are the adhesion contacts with other cells and the extracellular matrix (ECM). In particular, integrin-mediated cell-matrix adhesions not only physically support cell locomotion, but also generate signals that essentially determine the character of cell motility, its direction, velocity and persistence (Cary *et al.*, 1999; Holly *et al.*, 2000).

Despite the complexity of the external stimuli and the types of locomotory responses, the general model for the cell motility regulation, as it emerges from the studies of the last 10 years, is surprisingly uniform (Figure 5.1). The external signals, including the signals from the adhesion receptors, are not transduced directly to the effector mechanisms, but rather to an integrating system based mainly on the small G-proteins of the Rho family (Kjoller and Hall, 1999; Ridley, 2001; Schmidt and Hall, 2002) and perhaps some other small G-proteins such as ARFs (Casanova, 2003; Turner and Brown, 2001). Each specific stimulus is 'translated' into a specific combination of activities/localizations of these G-proteins. The G-proteins, in turn, modulate the activity and localization of downstream effectors, initiating cascades of events that lead to cytoskeletal reorganization, and ultimately to alterations in adhesion and locomotory behaviour (Bishop and Hall, 2000; Etienne-Manneville and Hall, 2002; Ridley, 2001).

The two main cytoskeletal effectors of the small Rho GTPases are the actin microfilaments and the microtubule system. Organization of these major systems of cellular fibres have a surprising number of common features (Mitchison, 1992). Both actin filaments and microtubules are polar and demonstrate an assembly-prone, fast growing 'plus end' and a slow growing 'minus end' that favours dissasembly. Both types of fibres are associated with molecular motors (kinesins and dyneins for microtubules, myosins for actin filaments) that directionally move along these fibres (Cross and Carter, 2000). Apparent structural similarity was found between the myosin and kinesin motor domains (Vale and Milligan, 2000). The dynamics of the processes of assembly–disassembly are in both cases rich and complex, and permit different modes of subunit turnover at the steady state, including 'treadmilling' and

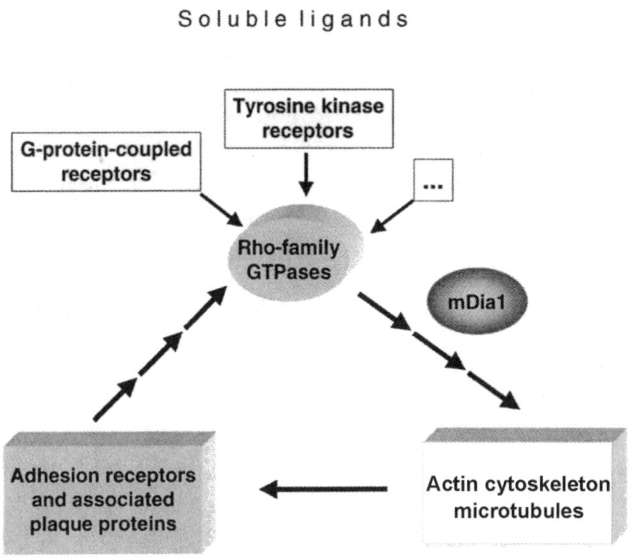

Figure 5.1 A general scheme of cell motility regulation and the role of mDia1. External 'motogenic' factors stimulate tyrosine kinase receptors, G-protein coupled receptors, or other signalling pathways leading to an alteration of activities of a subset of Rho family GTPases. This, in turn, triggers a subset of targets, which induce cytoskeletal re-organization. The cytoskeletal changes are required for formation/release of cell-matrix adhesions, and cytoskeletal and adhesion reorganization together produce cell movements. Adhesion-dependent signals, in turn, affect the Rho GTPase activities, closing the feedback loop. mDia1, one of immediate Rho targets, is discussed in detail in the present review. It affects both actin and microtubules, resulting in reorganization of both the cytoskeleton and focal adhesions

'dynamic instability' (see for actin, Littlefield and Fowler (2002), and for microtubules, Waterman-Storer and Salmon (1997)). Fibre assembly and disassembly are regulated by a variety of specific proteins, and the principles of this regulation have many common features (see Pollard and Borisy, 2003 for actin, Heald and Nogales, 2002 for microtubules). The existence of a particular mechanism of actin regulation usually indicates that a parallel mechanism is used by microtubules, and vice versa. For example, the molecular complexes that nucleate microtubules and actin filaments include, as essential components, proteins that are highly homologous to the subunits of the corresponding fibres. Thus, the principal component of microtubule-nucleating complexes is gamma-tubulin (Moritz and Agard, 2001), while the core of the complex that nucleates actin filaments (Arp2/3 complex) comprises the actin-related proteins, Arp2 and Arp3 (Pollard and Beltzner, 2002).

Despite the apparent similarity in the general organization of the two cytoskeletal systems, cells use actin and microtubule units to perform different, albeit complementary functions. An illuminating example of cooperation between the actin and microtubule system is provided by the process of assembly, maturation and disassembly of matrix adhesions, molecular complexes connecting ECM with the actin cytoskeleton via transmembrane integrin receptors (Geiger *et al.*, 2001; Small *et al.*, 2002). The interplay between the actin cytoskeleton, microtubule system and integrin-mediated matrix adhesions is the main subject of the present review.

Coordination of the functions of the actin and microtubule systems is achieved via a variety of cross-talk mechanisms operating at several levels. There are, for example, several types of cross-linker proteins, which appear directly to connect microtubules and actin filaments (Fuchs and Karakesisoglou, 2001). However, coordination of the fibre dynamics in time and space requires more sophisticated regulation than simply direct physical linkage. The Rho family GTPases were recently shown to control dynamics and organization not only of the actin cytoskeleton but also of microtubules (Cook *et al.*, 1998; Daub *et al.*, 2001; Fukata *et al.*, 2002; Ishizaki *et al.*, 2001; Nakano *et al.*, 2002; Palazzo et *al.*, 2001). One of the mechanisms underlying such parallel regulation is based on the unique features of a direct Rho target, Diaphanous related formin homology protein, mDia1 (Alberts, 2002). In the coming pages, we will discuss the activities of this formin and its possible role in the coordination of microtubules and the actin cytoskeleton in the course of focal adhesion remodelling.

Actin, microtubules and cell–matrix adhesions in crawling cell locomotion

Crawling locomotion of cells can be viewed as a periodically repeating sequence of events that includes: formation of pseudopodial protrusions, their attachment and translocation of the cell body in the direction of the new attachment sites (Lauffenburger and Horwitz, 1996). All these events can, in principle, be served by the actin cytoskeleton, in the absence of microtubules. In particular, formation of filopodial and lamellipodial protrusions is based exclusively on the regulated polymerization of actin filament superstructures (Borisy and Svitkina, 2000). Moreover, fragments of fish epidermal keratocytes, which lack nuclei, centrosomes and microtubules can nevertheless move with remarkable speed and persistence (Euteneuer and Schliwa, 1984; Verkhovsky *et al.*, 1999). However, beginning with the pioneering work of Vasiliev (Vasiliev *et al.*, 1970) it has been shown that, in practice, a majority of cell types do require the microtubule system for directional locomotion.

Many studies have been performed, using specific microtubule-disrupting drugs (Peterson and Mitchison, 2002), to elucidate the role of microtubules in cell migration. Comparisons of cell types, whose migration is sensitive to microtubule disruption, with those retaining their ability to migrate in the absence of functional microtubules, revealed one clear distinction between them: these cell types differ in their mode of the attachment to the ECM. More precisely, we must consider different types of integrin-mediated matrix adhesions. Even under simplified conditions of monolayer culture, at least four types of structures linking the ECM with the actin cytoskeleton via transmembrane integrin molecules have been described: ubiquitous 'classical' focal adhesions (or focal contacts) and focal complexes, and more specialized fibrillar adhesions, and podosomes (Geiger *et al.*, 2001). In three-dimensional matrices derived from tissues or cell culture, additional variants of integrin adhesions appear (Cukierman *et al.*, 2001). Besides morphological distinctions, different classes of integrin-mediated matrix adhesions vary in their dynamics, protein composition and adhesion strength (Adams, 2001; Geiger *et al.*, 2001; Wehrle-Haller and Imhof, 2002). Focal complexes begin initially as dot-like adhesions (less than 1 μm in size) originating at the tips of lamellipodia or filopodia. They are short-lived structures and either disappear, or develop into classical focal adhesions (Figure 5.2), elongated oval plaques several micrometers in size associated with the bundles of actin filaments ('stress fibres'). Fibrillar adhesions evolve from focal adhesions and participate in the assembly of extracellular fibronectin fibres, while podosomes are small (0.5 μm) ring-shaped adhesion structures found in specialized cell types, such as macrophages and osteoclasts (Geiger *et al.*, 2001).

The dependence on microtubules for the locomotion of a particular cell type correlates with the presence of classical focal adhesions. Cells that form focal complex type adhesions but do not convert them into focal adhesions can move successfully without microtubules. These microtubule-independent cell types include the fish keratocytes mentioned above (Euteneuer and Schliwa, 1984), 'professional' migrating cells such as polymorphonuclear leukocytes (which actually move even faster after microtubule disruption (Keller and Niggli, 1993)), and perhaps some poorly adherent cancer cells (Ivanova *et al.*, 1980; Sroka *et al.*, 2002). Detailed studies of fish keratocytes, for example, revealed mainly punctate adhesions with an area of less than 1 μm^2 (Lee and Jacobson, 1997); very few of these adhesions transform into typical focal adhesions (Anderson and Cross, 2000). In contrast, cell types that require microtubules for migration are well-attached cells that in addition to focal complexes demonstrate typical focal adhesions. Microtubule-dependence of migration was observed in fibroblasts and fibroblast-like cells (Gail and Boone, 1971; Goldman, 1971; Vasiliev *et al.*, 1970), which move relatively slowly, but also in fast migrating cells, such as Melb-a melanoblasts or melanoma B16 cells (Ballestrem *et al.*, 2000). Disruption of microtubules in such cell types leads to a further increase in the number and size of the focal adhesions (Bershadsky *et al.*, 1996; Kirchner *et al.*, 2003), which impedes cell migration by increasing their adhesion to the substrate to the level

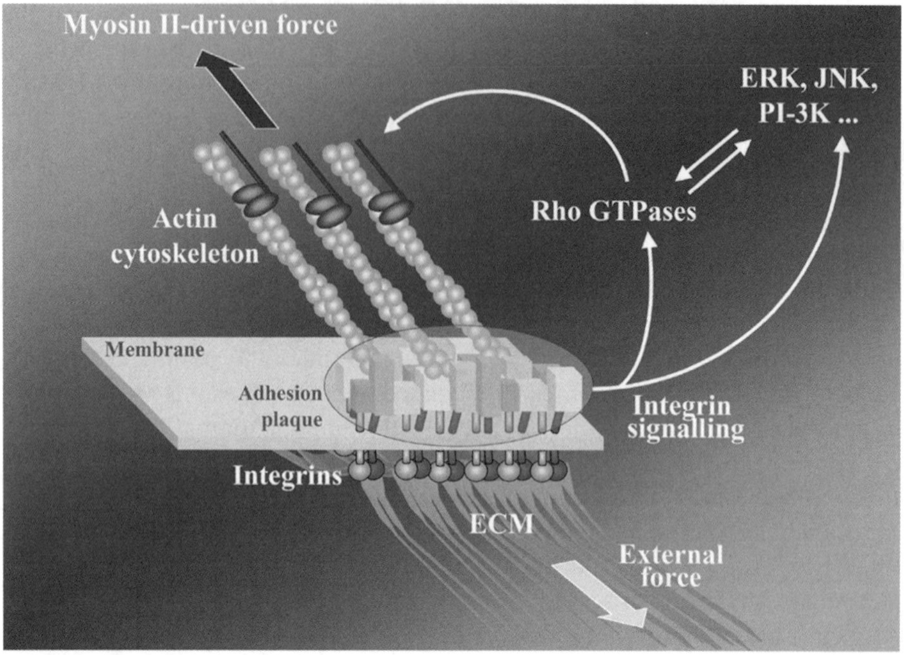

Figure 5.2 Focal adhesion as a mechanosensor. The focal adhesion is a multi-molecular complex connecting the extracellular matrix with the actin cytoskeleton. Heterodimeric transmembrane integrin receptors bind matrix proteins via their extracellular domains, while their cytoplasmic domains are associated with a dense submembrane plaque containing more than 50 different proteins ('boxes' enclosed in the oval area), including structural elements, as well as proteins involved in signal transduction. The plaque, in turn, is connected to the termini of actin filament bundles. The assembly and maintenance of focal adhesions depend on local mechanical forces. This force may be generated by myosin II-driven isometric contraction of the actin cytoskeleton, or by extracellular perturbations such as matrix stretching or fluid shear stress. Force-induced assembly of the adhesion plaque leads to the activation of a variety of signalling pathways that control cell proliferation, differentiation and survival (e.g. MAP kinase and PI 3-kinase pathways) as well as the organization of the cytoskeleton (e.g. Rho family GTPase pathways). Rho, in particular, is an indispensable regulator of focal adhesion assembly affecting actin polymerization and myosin II driven contractility. Reproduced with permission from Geiger and Bershadsky (2002)

incompatible with locomotion (Ballestrem *et al.*, 2000). Thus, the requirement for microtubules in cell migration depends on the mode of formation of integrin-dependent cell-matrix contacts, and applies mainly to cells that form classical focal adhesions.

A second important fact, which sheds light on the role of microtubules in cell locomotion, is their involvement in the regulation of myosin II-driven contractility. Since the studies of Danowski (1989), it has become increasingly clear that microtubule systems share the ability to suppress cell contractility.

Disruption of microtubules was shown to activate contractility in many different cell types (Elbaum *et al.*, 1999; Small *et al.*, 2002). The mechanism of microtubule-mediated suppression of contractility is not yet clear, and could be based both on delivery and removal of contractility regulating signalling molecules by the microtubule system, and on mechanical resistance of microtubule arrays (with associated motors and crosslinking proteins) to actomyosin contractility (see, for more detailed discussion Elbaum *et al.*, 1999; Ingber, 2002; Small *et al.*, 2002; Wittmann and Waterman-Storer, 2001). Thus, to understand possible role of microtubules in the regulation of cell migration, it is necessary first to explain how formation of focal adhesions is related to the regulation of cell contractility.

Mechanosensory function of focal adhesions and its modulation by microtubules

Adherent cells apply physical forces to the extracellular matrix, generated through the contractile activity of the actin cytoskeleton (Balaban *et al.*, 2001; Dembo and Wang, 1999; Galbraith and Sheetz, 1997; Harris *et al.*, 1980; Tien *et al.*, 2002) and reviewed in Beningo and Wang (2002). The forces are produced by myosin II and are transmitted via the focal adhesions. The average magnitude of the force for fibroblast-like cells is about 5 nN per square micrometre of focal adhesion plaque. In the course of these studies, a surprising feature of focal adhesions was discovered. These structures appear to assemble and maintain their integrity only under conditions in which forces of this magnitude are applied to them. Any treatment that inhibits myosin II-driven contractility – including incubation with chemical inhibitors of myosin light chain kinase (ML-7 and KT5926), myosin ATPase (BDM), or transfection with caldesmon, a protein that also inhibits actin-dependent myosin II ATPase activity, prevents formation of new focal adhesions, and, moreover, leads to rapid disassembly of the existing ones (Chrzanowska-Wodnicka and Burridge, 1996; Helfman *et al.*, 1999). In cells attached to a flexible substrate, which can be deformed by the cell, the tension force acting at the adhesion plaques are smaller than in an identical cell attached to the solid substrate. Thus, the average size of the focal adhesions that cells form upon attachment to flexible substrates is smaller than those formed in cells attached to solid substrate (Pelham and Wang, 1997). Finally, the assembly of the focal adhesions can be induced by externally applied tension forces, using either a micropipette attached to the cell surface, or the local stretching of a flexible substrate (Kaverina, 2002; Riveline *et al.*, 2001). External force applied to the focal adhesion can completely substitute the cell-generated force, so that micropipette manipulation leads to focal adhesion growth even in cells treated by BDM or transfected with caldesmon (Riveline *et al.*, 2001). Thus, individual focal adhesions function as miniature mechanosensors: they respond to an increase of applied tension force by activation of assembly, and to a reduction of such force by disassembly (Figure 5.2).

The mechanism of such unusual behaviour of focal adhesions is not yet clear, and its understanding requires more detailed knowledge of the organization and dynamics of these structures than we presently have. Perhaps mechanical force applied to the focal adhesion induces alteration(s) in the conformation of some of its components and/or in their mutual positions (Geiger and Bershadsky, 2001, 2002). This stretching could increase the probability of incorporation of new components in such a way that the focal adhesion assembles, preserving its self-similarity. The observation that relaxation of tension by a variety of myosin II inhibitors leads not just to cessation of growth, but to a rapid disassembly of focal adhesions suggests that these structures are intrinsically dynamic: they continuously loose 'old' subunits and incorporate the 'new' ones in a tension-dependent fashion. Recent direct observations of β3-integrin dynamics in focal adhesions using the technique of fluorescence recovery after photobleaching (FRAP), indeed, revealed high turnover rate (Ballestrem *et al.*, 2001) and dependence of fluorescence recovery on the myosin II activity (Tsuruta *et al.*, 2002). Among other factors, the protease calpain specifically localized at the focal adhesions might be involved in their rapid disassembly (Bhatt *et al.*, 2002). Unlike focal adhesions, other types of integrin-mediated adhesion structures such as focal complexes and fibrillar adhesions do not disassemble following inhibition of myosin II contractility (Riveline *et al.*, 2001; Zamir *et al.*, 2000).

The dependence of focal adhesion assembly on applied force creates a positive feedback loop amplifying the growth of these structures in cells attached to a solid substrate. In fact, since several components of focal adhesions are actin-associated proteins that can bind and perhaps even nucleate actin filaments, an increase in the focal adhesion plaque size will lead to an increase in the number of actin filaments associated with it. Increasing the actin filament number will lead, via their interactions with myosin II motors, to augmentation of the centripetal tension force applied to this focal adhesion. Force augmentation will in turn promote further focal adhesion growth (via additional activation of the focal adhesion mechanosensor), further increasing the bundle of associated actin filaments, and so on. Thus, if this mechanism were not regulated, focal adhesions would grow indefinitely, being limited only by a possible rupture (if the final force applied to it by the cell exceeds a certain threshold).

Obviously, the cell needs a physiological mechanism(s) that can attenuate this positive feedback loop and can weaken or disrupt focal adhesions, as necessary. In particular, such mechanisms are needed to enable focal adhesion-forming cells to migrate along the substrate (otherwise they simply will not be able to release once-formed adhesions). Cross-talk between focal adhesions and microtubules could provide an effective local mechanism to interfere with the self-accelerating growth of focal adhesions.

As described above, microtubules oppose cell contractility; their disruption induces cell contraction accompanied by myosin II light chain phosphorylation (Kolodney and Elson, 1995). It is likely that the inhibitory effect of microtubules on myosin II activity is local. Several experiments have dissected the interplay between myosin II, focal adhesions and microtubules. Microtubule disruption leads to significant augmentation of focal adhesions, which can be prevented by suppression of myosin II-contractility (Bershadsky *et al.*, 1996; Helfman *et al.*, 1999). Moreover, local application of myosin II inhibitors by micropipette allows the restoration of normal migration ability of the cells lacking microtubules, most probably by normalizing the size and distribution of the focal adhesions (Kaverina *et al.*, 2000). Observations of microtubule dynamics in living cells suggest that tension developing in the regions of cell attachment to the substrate promotes rapid microtubule ingrowth, and subsequent cortical targeting (Kaverina, 2002; Suter *et al.*, 1998). Direct targeting of microtubule ends to the focal adhesion plaques was observed in fibroblasts (Kaverina *et al.*, 1998); this targeting was accompanied by cessation of focal adhesion growth, and often by their disassembly (Kaverina *et al.*, 1999). These studies suggest that microtubules are attracted to focal adhesions, locally suppress the tension forces applied to these structures, and thereby interrupt the positive feedback loop described above (for a more detailed discussion, see Small *et al.* (2002)).

Which factors coordinate these processes of tension-dependent growth of focal adhesions with the subsequent advent of microtubules relaxing the tension? To put this question in the context of signalling, it is important to remember that formation and maintenance of classical focal adhesions depends on the activity of Rho (Ridley and Hall, 1992; Rottner *et al.*, 1999). Two targets of Rho were shown to be necessary and sufficient to mediate Rho's function in this process: Rho associated kinase (ROCK), and the formin homology protein, mDia1 (Tominaga *et al.*, 2000; Watanabe *et al.*, 1999). ROCK is known to activate myosin II (Fukata *et al.*, 2001; Kimura *et al.*, 1996); our experiments revealed that the function of ROCK in focal adhesion formation is through the activation of myosin II-driven forces applied to these structures (Riveline *et al.*, 2001). Function of the second Rho target, mDia1, appears to be more complex. Under conditions of external force application, mDia1 is necessary and sufficient to mediate the Rho-signal, dictating focal adhesion assembly (Riveline *et al.*, 2001). Moreover, a model explaining the potential mechanism of coordination between microtubules and focal adhesions is suggested by our findings that mDia1 can participate in the regulation of microtubule dynamics and targeting.

mDia1 as a possible coordinator of actin, focal adhesions and microtubule assembly

Diaphanous related formin homology proteins (DRF) are ubiquitous, and are found in mice, humans, *Drosophila*, *C. elegans* (Severson *et al.*, 2002), *Aspergillus* (Sharpless and Harris, 2002), and fission and budding yeast. A mutation of this protein in *Drosophila*, called Diaphanous, affects cytokinesis (Afshar *et al.*, 2000; Castrillon and Wasserman, 1994). Structural organization of DRF proteins is similar in different species and includes a small Rho-family G-protein-binding domain at the N-terminus, several formin-homology domains (FH1, FH2 and sometimes FH3) in the central part of the molecule, and an auto-inhibitory domain (DAD) at the C-terminus (Alberts, 2001, 2002). In its 'closed' conformation, when DAD is bound to the N-terminus, the molecule is inactive. Occupation of the DRF protein by Rho-GTP (or by Cdc42-GTP in some DRF), in its binding site at the N-terminus, releases DAD and converts the 'closed' conformation into an 'open' one, in which formin homology domains are accessible to their ligands, rendering the entire molecule biologically active (Figure 5.3). Truncated mutants of DRF preserving the formin homology domains in the open conformation are constitutively active (Tominaga *et al.*, 2000; Watanabe *et al.*, 1999).

One apparent function of DRF proteins is in the regulation of actin assembly. In budding yeast, formins Bni1 and Bnr1 are necessary and sufficient for the assembly of elongated, tropomyosin-containing cables of

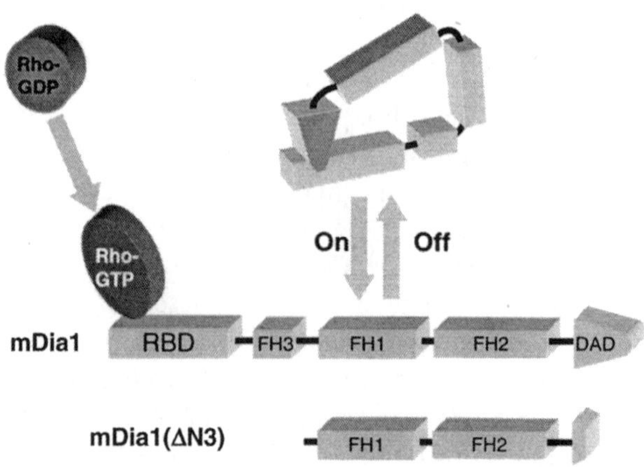

Figure 5.3 Activation of mDia by active RhoA. In its inactive state, mDia is thought to exist in a closed conformation where its C-terminal auto-inhibitory domain (DAD) is bound to the N-terminus. Upon binding of Rho-GTP to the Rho-binding domain (RBD) at the N-terminus, the whole molecule unfolds and the binding sites for downstream signalling molecules localized at the FH1 and FH2 domains become exposed. A constitutively active form of mDia1 (mDia1 ΔN3), lacking the N-terminal domain, contains intact FH1 and FH2 domains (see Watanabe *et al.*, 1999)

actin filaments (Evangelista *et al.*, 2002; Sagot *et al.*, 2002a). The classic actin nucleator, Arp2/3 complex (see Introduction), is not necessary for the cable formation, even though it is indispensable for the assembly of yeast actin structures of another type, so called actin 'patches' (Winter *et al.*, 1997). *In vitro* experiments demonstrated that truncated constitutively active formin nucleates the assembly of actin filaments (Pruyne *et al.*, 2002; Sagot *et al.*, 2002b). While the nucleating Arp2/3 complexes are attached to the 'minus' filament ends and connect them with other actin filaments creating an integrated branching filament array (Pollard and Beltzner, 2002), the nucleating formins remain at the 'plus' filament ends (Pruyne *et al.*, 2002) and promote the assembly of long non-branching filaments (Pruyne *et al.*, 2002; Sagot *et al.*, 2002b). Both FH1 and FH2 formin domains are required for actin filament nucleation. Formin-induced actin polymerization can be strongly enhanced by cooperation with profilin (Sagot *et al.*, 2002b), an actin monomer binding protein that can associate with formins via a proline-rich FH1 domain (Watanabe *et al.*, 1997).

In mammalian cells, expression of a truncated, constitutively active mutant of mDia1 formin also strongly augments the level of actin polymerization (Watanabe *et al.*, 1999). Moreover, in the cells expressing this mutant, polymerized actin forms numerous bundles, all orientated in a single direction, and the cells acquire a characteristic elongated bipolar morphology (Figure 5.4). Since FH1 and FH2 formin domains are highly conserved, it is most likely that mDia1, in a manner similar to yeast formins, strongly promotes actin filament nucleation. However, the complex phenotypic effect induced by active mDia1 in mammalian cells is difficult to explain only by enhanced actin nucleation. Enhancement of Arp2/3 complex-dependent actin nucleation (by overexpression of the Arp2/3 complex activator, scar, or its C-terminal VCA domain) leads not to formation, but rather to disruption of all normal actin structures and complete loss of cell polarization (Machesky and Insall, 1998). At the same time, expression of active mDia1 leads to development of a highly organized, albeit exaggerated, polarized phenotype.

Thus, it appears that the active mDia1, in fact, triggers not only actin filament nucleation, but a series of events that together produce the polarized phenotype. Among relevant events, the most interesting one is a possible effect on microtubule dynamics and organization. In fact, formins were shown to affect microtubules in addition to actin. For example, in fission yeast, either knockout or overexpression of For3 formin leads to significant alterations in the cytoplasmic microtubule array (Nakano *et al.*, 2002). There are several indications that, in higher eukaryotes, formins may also be involved in microtubule regulation. First, it appears that in cells expressing active mDia1, microtubules are aligned along the same direction as the numerous actin bundles (Figure 5.4), and this effect depends on the FH2 domain of mDia1 (Ishizaki *et al.*, 2001). It was further shown that activation of mDia leads to an

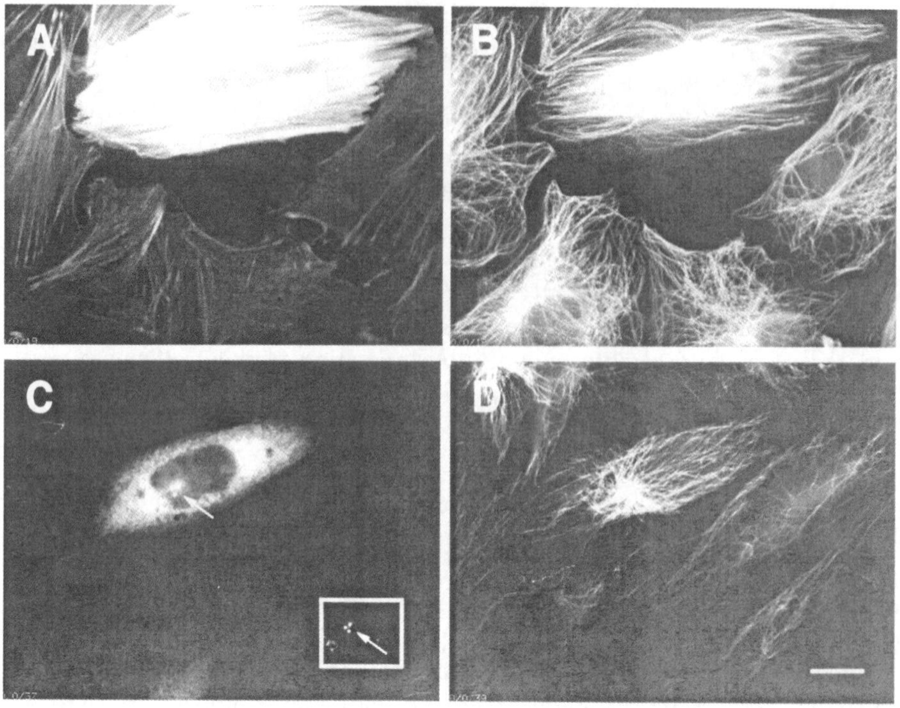

Figure 5.4 Cytoskeletal alterations produced by active mDia1. (A) Cells expressing mDia1 ΔN3 contain an elevated level of actin filaments, as revealed by Cy3-phalloidin staining. These actin filaments (A) together with microtubules (B,D) align along the axis of the bipolar cell. As visualized using a GFP-fusion construct of mDia1 ΔN3 (C) together with γ-tubulin antibody staining (insert in C), the active form of mDia1 localizes to the centrosome. Bar 10 μm

increase in the fraction of microtubules in which α-tubulin undergoes a specific post-translational modification, the removal of C-terminal tyrosine by tubulin-carboxypeptidase (Palazzo *et al.*, 2001). The increased fraction of the detyrosinated α-tubulin usually correlates with a longer microtubule lifetime (Khawaja *et al.*, 1988; Webster *et al.*, 1987), so that active mDia might induce microtubule stabilization. We investigated the effect of mDia1 on microtubules in more detail, using direct observations of microtubule dynamics in living cells and measurements of microtubule polymer mass. Our studies revealed that active mDia1 profoundly affects microtubule assembly–disassembly processes, both at the 'plus' and 'minus' microtubule end (Ballestrem *et al.*, 2003; unpublished).

In control cells, expressing normal mDia1, microtubules nucleated at the centrosome grow toward the cell periphery at a constant speed. When the microtubule plus end approaches the cell edge, it demonstrates 'dynamic instability', an oscillatory behaviour with alternating periods of growth and shortening, and finally shrinks back to the centrosome (Komarova *et al.*, 2002b). Some microtubules detach from the centrosome (Abal *et al.*, 2002; Keating *et al.*, 1997), which usually leads to their rapid shortening from the released minus end (Komarova *et al.*, 2002b). Enucleated cell fragments (cytoplasts) lacking centrosomes provide a convenient model for the studies of such non-centrosomal microtubules (Chausovsky *et al.*, 2000; Rodionov *et al.*, 1999). These microtubules demonstrate 'treadmilling' dynamics, in which rapid depolymerization at the free minus end is balanced by the growth at the plus end (Rodionov *et al.*, 1999; Rodionov and Borisy, 1997). The steady state is possible if the concentration of the non-polymerized tubulin is sufficiently high. Therefore the mass of the microtubule polymer in the centrosome-free cytoplasts is substantially smaller than in centrosome-containing cytoplasts or in intact cells (Rodionov *et al.*, 1999) (see also Figure 5.6).

Examining effects of mDia1 on microtubule dynamics, we have shown that in cells transfected with the active form of mDia1 microtubules grow more slowly and their growth is less tenacious than in control cells (the average value of the growth speed was twofold lower, and the variance about twofold higher). However, the velocity of microtubule shortening during the oscillation phase at the cell periphery was also strongly reduced, while the frequencies of the transitions from growth to shrinking and vice versa were not significantly changed. Thus, the amplitude of the oscillations of the microtubule plus ends significantly decreased in cells expressing the active mDia1. These dynamic changes correlated with an apparent increase of overlap between microtubule ends and focal adhesions at the cell leading edge (Figure 5.5). The effect of mDia1 on the minus end microtubule dynamics was also dramatic. Microtubule polymer level in the centrosome-free cytoplasts produced from the cells expressing constitutively active mDia1 did not decrease, as in control cytoplasts, but remained almost as high, as in intact cells (Figure 5.6). This suggests that active mDia1 protects microtubule minus-ends from rapid disassembly in the absence of the centrosome. In this connection, it is interesting that both transfected and endogenous mDia1 was enriched at the centrosomes of intact cells (Figure 5.4). Effect of mDia1 on the microtubule minus-end required cooperation with another Rho target, ROCK, which is also localized to the centrosome (Chevrier *et al.*, 2002).

What could the mechanism of mDia1-mediated microtubule regulation be, and what role might it play in the coordination of microtubule and actin functions in the processes of cell polarization and directional migration? The answer to the first question depends on revealing possible partners of mDia1 among proteins that directly control microtubule dynamics. As mentioned above, there are several groups of such proteins, including proteins that sequester non-polymerized tubulin dimers, 'classical' MAPs that bind to the microtubules along their entire length, and microtubule tip-binding proteins (Heald and Nogales, 2002). The last group of proteins is perhaps most

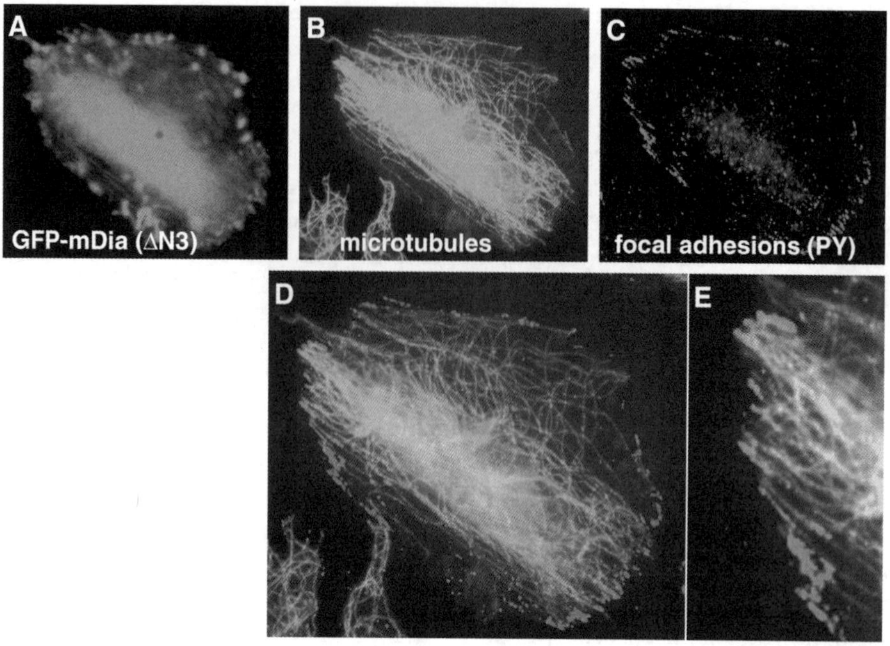

Figure 5.5 Microtubule ends at the periphery of active mDia1-expressing cells co-localize with focal adhesions. CHO-K1 cells expressing mDia1 ΔN3-GFP (A) have a bipolar shape with microtubules aligned along the axis (B). Many of the microtubule ends at the periphery of the cell overlap with focal adhesions (visualized by anti-phosphotyrosine antibody staining) (C). Frames (D) and (E) represent the merged microtubule and focal adhesion images; E shows a part of D at higher magnification. (A colour reproduction of this figure can be found in the colour plate section)

relevant, since their primary role in the regulation of microtubule dynamics and targeting to the cell cortex both in yeast and in higher eukaryotic cells is becoming increasingly clear (Gundersen, 2002). This group includes the evolutionarily conserved proteins EB1, CLIP-170 and LIS1 and their respective partners APC, CLASPs and the dynein–dynactin complex (McNally, 2001; Schuyler and Pellman, 2001). There are certainly additional links between the members of this group: LIS1 can bind CLIP-170 (Coquelle *et al.*, 2002), while EB1 binds p150Glued, an essential component of dynactin (Askham *et al.*, 2002), etc. Effects of several proteins from this group on microtubule dynamics *in vitro* and *in vivo* are well documented (Komarova *et al.*, 2002a; Tirnauer *et al.*, 2002). *Xenopus* XMAP215 and the kinesin family microtubule-depolymerizing protein XKCM1 also regulate dynamics at the microtubule ends (Kinoshita *et al.*, 2001). Relationships between these proteins and the proteins of the EB1/CLIP-170/LIS1 group are not yet clear. While the microtubule tip-binding proteins are (by definition) localized to the microtubule plus ends, it is interesting that many (if not all) members of

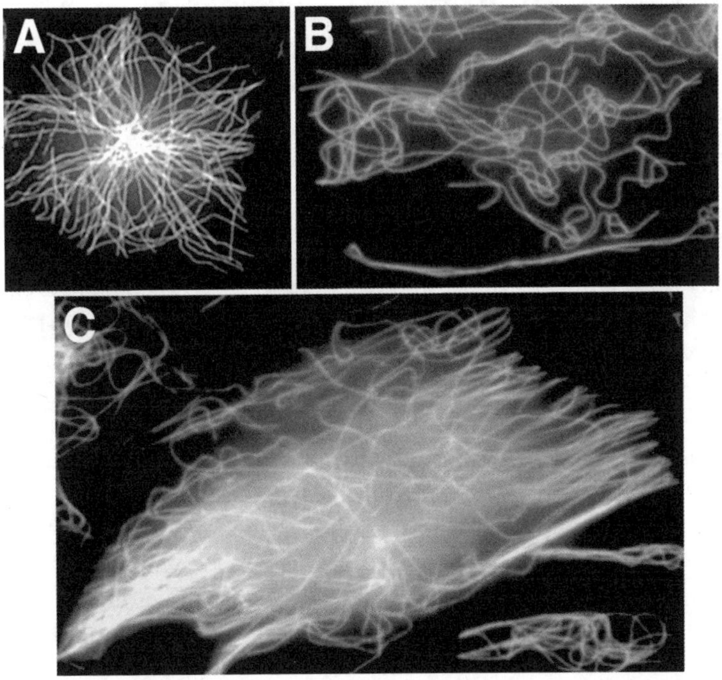

Figure 5.6 mDia1 stabilizes non-centrosomal microtubules. (A) In centrosome-containing cytoplasts, microtubules form a radial array with minus-ends attached to the centrosome. (B) Cytoplasts lacking a centrosome contain only a few microtubules; the amount of tubulin polymer is much less than in centrosome-containing cytoplasts. (C) Centrosome-free cytoplasts expressing active mDia1 preserve their bipolar shape and contain many microtubules aligned along their axes. The amount of tubulin polymer is of the same order of magnitude as in centrosome-containing cytoplasts. Staining with anti-tubulin antibody. Photograph (A) is reproduced with permission from Chausovsky *et al.* (2000)

this group can also be found in association with the centrosome (see, for example Rehberg and Graf (2002)). This indicates possible additional functions of these proteins in the control of minus end dynamics and/or their attachment to the centrosome. This is consistent with the centrosomal localization of mDia1, as mentioned above.

There is no direct information yet concerning possible interactions of mDia1 with any of these microtubule-associated proteins. One interesting possibility can be inferred from data on the composition of the dynein–dynactin complex. An essential component of this complex, is a filament-forming actin-related protein 1 (Arp1) known also as centractin, which is more homologous to actin than other Arps (Holleran *et al.*, 1998). Moreover, Arp1 not only binds to other components of dynactin complex, like p150Glued (Waterman-Storer *et al.*, 1995), but also to several types of genuine actin-binding proteins such as barbed end capping protein (Schafer *et al.*, 1994) and spectrin (Holleran *et al.*, 2001). Since mDia1 is an actin-binding protein, it might, in principle, bind Arp1. This could provide a mechanism for a possible link between mDia1 and dynactin, and, via dynactin, to other microtubule-end-tracking proteins (CLIP-170, LIS1, EB1, etc.), that may affect microtubule dynamics.

Of course, less direct mechanisms of mDia1 effects on microtubules could also be envisioned. For example, it was suggested in a recent study that activation of mDia1 leads to activation of the Rac1 protein (Tsuji *et al.*, 2002). Rac1 activation could stabilize microtubules via PAK1-mediated phosphorylation (and inactivation) of a tubulin-sequestering protein, Op18/stathmin (Daub *et al.*, 2001) or via direct effect on IQGAP protein that binds CLIP-170 (Fukata *et al.*, 2002). All these possibilities deserve to be tested experimentally.

Finally, we must consider how the effects of mDia on microtubule dynamics described above could be integrated into the general scheme of the interplay between contractile actin cytoskeleton, mechanosensory focal adhesions and contraction-suppressing microtubules outlined in the previous section. The first question that should be addressed in this connection is, where and when does the activation of mDia1 occur? Since the only known mDia1 activator is Rho, the spatial and temporal distribution of active mDia1 should be similar to that of active Rho. While no studies have been reported relating to the distribution of active Rho, there are some data describing gradients of Rac and Cdc42 activity, obtained by various modifications of the fluorescence resonance energy transfer (FRET) technique (Gardiner *et al.*, 2002; Itoh *et al.*, 2002; Kraynov *et al.*, 2000). Rac activation can occur as a result of integrin signalling (Price *et al.*, 1998), and there is good evidence that formation of the new cell-ECM adhesions at the cell leading edge is responsible for establishing the spatial gradient of Rac activity (Del Pozo *et al.*, 2002; Tzima *et al.*, 2002). Integrin signalling can also activate Rho, although the mechanism is less clear; this activation depends in a complex way on the integrin type (Danen *et al.*, 2002; Miao *et al.*, 2002), and is biphasic over time (activation is preceded by a decay) (Ren *et al.*, 1999). Nevertheless, it is reasonable to suggest that the integrin-mediated signal at the focal adhesion can at some point activate Rho, and the gradient of Rho activity (and therefore mDia1 activity) will depend on focal adhesion distribution.

Based on our model, gradients of mDia1 activity should locally remodel microtubule behaviour (Figure 5.7). Minus end stabilization will lead to selective protection of microtubules released from the centrosome. Alterations in plus end dynamics may increase the average time these ends will spend in the cell region with high mDia1 activity. Thus, mDia1 should promote an elevated concentration of microtubules (especially their plus ends) in proximity to the source of integrin signalling. This, in turn, may trigger some of the following effects: (1) suppression of actomyosin contractility in that region of the cell; (2) increase of microtubule-dependent transport to and

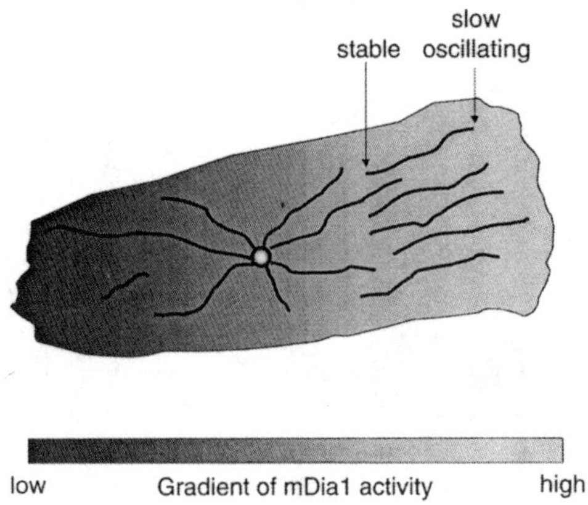

Figure 5.7 A scheme depicting a possible role of mDia1 in the organization of the microtubule system. In areas with high mDia1 activity, microtubules slow down their plus-end dynamics, while their minus-ends, even those not associated with the centrosome, are protected from depolymerization. This leads to accumulation of microtubules in the regions with higher mDia1 activity, presumably those regions adjacent to the source of integrin signalling, e.g., focal adhesions (see text for detailed discussion)

from this area; (3) facilitation of direct microtubule targeting to the focal adhesion plaques. These effects should make it possible for the cell to orientate and migrate in the direction determined by the integrin-mediated ECM signals, which could be either a gradient of ECM molecule concentration (Brandley and Schnaar, 1989), or gradient of substrate rigidity (Lo *et al.*, 2000). In particular, accumulation of microtubules behind the focal adhesions inducing relaxation of contractility should brake the positive feedback loop of tension-induced focal adhesion growth (see previous section) allowing the cell to avoid both abrupt focal adhesion breakage, and their indefinite growth (Figure 5.8). Instead, the older focal adhesions, which have attracted microtubules and therefore reduced the actomyosin tension in their proximity, will weaken or disassemble, giving the cell the ability to relocate onto the new substrate area. Thus, microtubules facilitate directional migration of cells that are attached to a substrate with well-developed focal adhesions. Rapidly migrating cells forming mainly small focal complexes are less dependent on the microtubule-mediated regulation.

Conclusion and perspectives

In this review, we highlighted several pathways coupling formation and growth of focal adhesions with contractile activity and cytoskeletal dynamics. Focal adhesions function as mechanosensors, responding to physical forces

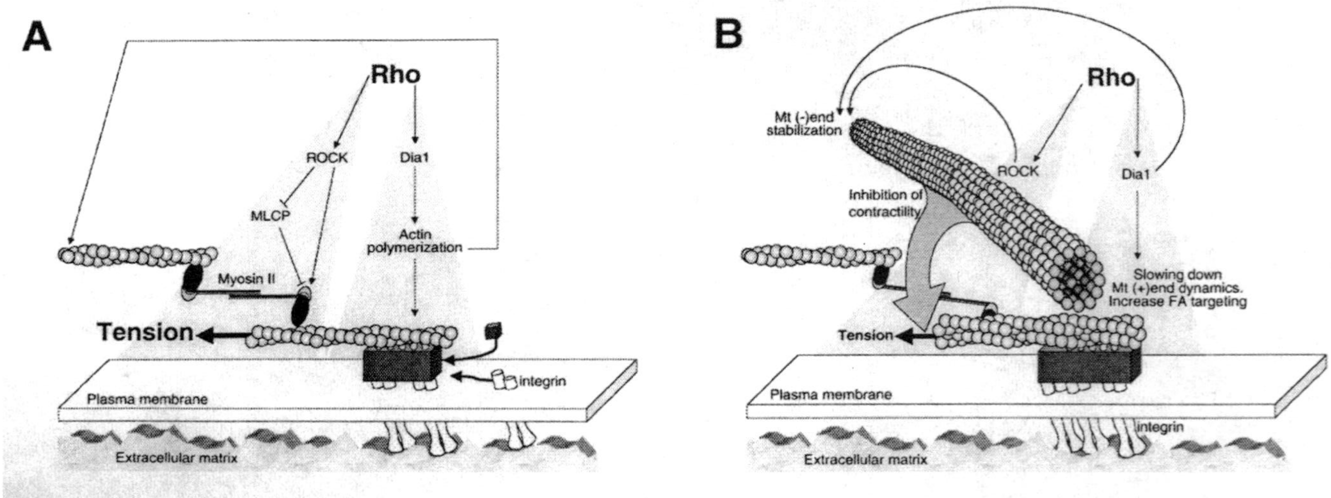

Figure 5.8 A unifying model presenting roles of actin and microtubule systems in the development of focal adhesions. (A) Tension-dependent assembly. Rho induces formation of focal adhesions by triggering ROCK and mDia1. The function of ROCK is to activate myosin II-driven cell contraction via inhibition of myosin-light chain phosphatase (MLCP) and perhaps direct phosphorylation of myosin II light chain (crescent symbols). Tension applied to mechanosensory focal adhesion plaque (black box) promotes incorporation of new components. Active mDia1 promotes nucleation of linear actin filaments. When tension is applied externally, mDia1 is the only Rho-target that is necessary for focal adhesion growth. (B) Microtubule-dependent turnover. High levels of active mDia1 in the proximity of the mature focal adhesion lead to alteration of microtubule dynamics at both plus- and minus-ends. ROCK activity is required for mDia1-induced microtubule minus-end stabilization. Alteration of the microtubule dynamics may facilitate targeting of their plus-ends to mature focal adhesions. Concentration of microtubules in the proximity of focal adhesion leads to local inhibition of the myosin-driven contractility in this region, and consequently to cessation of growth or even disassembly of focal adhesions. Frame (A) is based on the study by Riveline *et al.* (2001), and is reproduced with permission from Geiger and Bershadsky (2001)

(applied externally or generated by the cell) by directional assembly. The Rho target, formin homology protein mDia1, in addition to the nucleation of actin filament polymerization, was shown to affect microtubule dynamics both at the plus and minus ends, and, possibly, to facilitate microtubule growth in the direction of focal adhesions. Since microtubules interfere with myosin II-driven contractility, the mDia1-induced changes in microtubule dynamics and targeting may create a negative feedback loop controlling focal adhesion growth.

Obviously, this model requires further development and refinement. The molecular mechanism explaining the effect of force application on focal adhesion assembly has not been elucidated. We do not yet know how microtubules oppose cell contractility. Although bulk microtubule disruption induces cell contraction, it has not been demonstrated that the ends of individual microtubules can prevent contractility in their proximity; moreover, the possible mechanism of such local relaxation is not known. Finally, the molecular mechanism of mDia1's effects on microtubule dynamics also should be clarified. These questions provide attractive avenues for future studies.

References

Abal, M., Piel, M., Bouckson-Castaing, V., Mogensen, M., *et al.*, 2002. Microtubule release from the centrosome in migrating cells. *J. Cell Biol.* **159**: 731–737.

Adams, J. C., 2001. Cell-matrix contact structures. *Cell Mol. Life Sci.* **58**: 371–392.

Afshar, K., Stuart, B. and Wasserman, S. A., 2000. Functional analysis of the *Drosophila* diaphanous FH protein in early embryonic development. *Development* **127**: 1887–1897.

Alberts, A. S., 2001. Identification of a carboxyl-terminal diaphanous-related formin homology protein autoregulatory domain. *J. Biol. Chem.* **276**: 2824–2830.

Alberts, A. S., 2002. Diaphanous-related Formin homology proteins. *Curr. Biol.* **12**: R796.

Anderson, K. I. and Cross, R., 2000. Contact dynamics during keratocyte motility. *Curr. Biol.* **10**: 253–260.

Askham, J. M., Vaughan, K. T., Goodson, H. V. and Morrison, E. E., 2002. Evidence that an interaction between EB1 and p150(Glued) is required for the formation and maintenance of a radial microtubule array anchored at the centrosome. *Mol. Biol. Cell* **13**: 3627–3645.

Balaban, N. Q., Schwarz, U. S., Riveline, D., Goichberg, P., *et al.*, 2001. Force and focal adhesion assembly: a close relationship studied using elastic micropatterned substrates. *Nat. Cell Biol.* **3**: 466–472.

Ballestrem, C., Wehrle-Haller, B., Hinz, B. and Imhof, B. A., 2000. Actin-dependent lamellipodia formation and microtubule-dependent tail retraction control-directed cell migration. *Mol. Biol. Cell* **11**: 2999–3012.

Ballestrem, C., Hinz, B., Imhof, B. A. and Wehrle-Haller, B., 2001. Marching at the front and dragging behind: differential alphaVbeta3-integrin turnover regulates focal adhesion behavior. *J. Cell Biol.* **155**: 1319–1332.

Beningo, K. A. and Wang, Y. L., 2002. Flexible substrata for the detection of cellular traction forces. *Trends Cell Biol.* **12**: 79–84.

Bershadsky, A., Chausovsky, A., Becker, E., Lyubimova, A. and Geiger, B., 1996. Involvement of microtubules in the control of adhesion-dependent signal transduction. *Curr. Biol.* **6**: 1279–1289.

Bhatt, A., Kaverina, I., Otey, C. and Huttenlocher, A., 2002. Regulation of focal complex composition and disassembly by the calcium-dependent protease calpain. *J. Cell Sci.* **115**: 3415–3425.

Bishop, A. L. and Hall, A., 2000. Rho GTPases and their effector proteins. *Biochem. J.* **348** (2): 241–255.

Borisy, G. G. and Svitkina, T. M., 2000. Actin machinery: pushing the envelope. *Curr. Opin. Cell Biol.* **12**: 104–112.

Brandley, B. K. and Schnaar, R. L., 1989. Tumor cell haptotaxis on covalently immobilized linear and exponential gradients of a cell adhesion peptide. *Dev. Biol.* **135**: 74–86.

Cary, L. A., Han, D. C. and Guan, J. L., 1999. Integrin-mediated signal transduction pathways. *Histol. Histopathol.* **14**: 1001–1009.

Casanova, J. E., 2003. ARFs. *Curr. Biol.* **13**: R123.

Castrillon, D. H. and Wasserman, S. A., 1994. Diaphanous is required for cytokinesis in *Drosophila* and shares domains of similarity with the products of the limb deformity gene. *Development* **120**: 3367–3377.

Chausovsky, A., Bershadsky, A. D. and Borisy, G. G., 2000. Cadherin-mediated regulation of microtubule dynamics. *Nat. Cell Biol.* **2**: 797–804.

Chevrier, V., Piel, M., Collomb, N., Saoudi, Y., *et al.*, 2002. The Rho-associated protein kinase p160ROCK is required for centrosome positioning. *J. Cell Biol.* **157**: 807–817.

Chrzanowska-Wodnicka, M. and Burridge, K., 1996. Rho-stimulated contractility drives the formation of stress fibers and focal adhesions. *J. Cell Biol.* **133**: 1403–1415.

Cook, T. A., Nagasaki, T. and Gundersen, G. G., 1998. Rho guanosine triphosphatase mediates the selective stabilization of microtubules induced by lysophosphatidic acid. *J. Cell Biol.* **141**: 175–185.

Coquelle, F. M., Caspi, M., Cordelieres, F. P., Dompierre, J. P., *et al.*, 2002. LIS1, CLIP-170's key to the dynein/dynactin pathway. *Mol. Cell Biol.* **22**: 3089–3102.

Cross, R. A. and Carter, N. J., 2000. Molecular motors. *Curr. Biol.* **10**: R177–R179.

Cukierman, E., Pankov, R., Stevens, D. R. and Yamada, K. M., 2001. Taking cell-matrix adhesions to the third dimension. *Science* **294**: 1708–1712.

Danen, E. H., Sonneveld, P., Brakebusch, C., Fassler, R. and Sonnenberg, A., 2002. The fibronectin-binding integrins alpha5beta1 and alphavbeta3 differentially modulate RhoA-GTP loading, organization of cell matrix adhesions, and fibronectin fibrillogenesis. *J. Cell Biol.* **159**: 1071–1086.

Danowski, B. A., 1989. Fibroblast contractility and actin organization are stimulated by microtubule inhibitors. *J. Cell Sci.* **93**: 255–266.

Daub, H., Gevaert, K., Vandekerckhove, J., Sobel, A. and Hall, A., 2001. Rac/Cdc42 and p65PAK regulate the microtubule-destabilizing protein stathmin through phosphorylation at serine 16. *J. Biol. Chem.* **276**: 1677–1680.

Del Pozo, M. A., Kiosses, W. B., Alderson, N. B., Meller, N., *et al.*, 2002. Integrins regulate GTP-Rac localized effector interactions through dissociation of Rho-GDI. *Nat. Cell Biol.* **4**: 232–239.

Dembo, M. and Wang, Y. L., 1999. Stresses at the cell-to-substrate interface during locomotion of fibroblasts. *Biophys. J.* **76**: 2307–2316.

Elbaum, M., Chausovsky, A., Levy, E. T., Shtutman, M. and Bershadsky, A. D., 1999. Microtubule involvement in regulating cell contractility and adhesion-dependent signalling: a possible mechanism for polarization of cell motility. *Biochem. Soc. Symp.* **65**: 147–172.

Etienne-Manneville, S. and Hall, A., 2002. Rho GTPases in cell biology. *Nature* **420**: 629–635.

Euteneuer, U. and Schliwa, M., 1984. Persistent, directional motility of cells and cytoplasmic fragments in the absence of microtubules. *Nature* **310**: 58–61.

Evangelista, M., Pruyne, D., Amberg, D. C., Boone, C. and Bretscher, A., 2002. Formins direct Arp2/3-independent actin filament assembly to polarize cell growth in yeast. *Nat. Cell Biol.* **4**: 260–269.

Fuchs, E. and Karakesisoglou, I., 2001. Bridging cytoskeletal intersections. *Genes Dev.* **15**: 1–14.

Fukata, Y., Amano, M. and Kaibuchi, K., 2001. Rho-Rho-kinase pathway in smooth muscle contraction and cytoskeletal reorganization of non-muscle cells. *Trends Pharmacol. Sci.* **22**: 32–39.

Fukata, M., Watanabe, T., Noritake, J., Nakagawa, M., *et al.*, 2002. Rac1 and Cdc42 capture microtubules through IQGAP1 and CLIP-170. *Cell* **109**: 873–885.

Gail, M. H. and Boone, C. W., 1971. Effect of colcemid on fibroblast motility. *Exp. Cell Res.* **65**: 221–227.

Galbraith, C. G. and Sheetz, M. P., 1997. A micromachined device provides a new bend on fibroblast traction forces. *Proc. Natl. Acad. Sci. USA* **94**: 9114–9118.

Gardiner, E. M., Pestonjamasp, K. N., Bohl, B. P., Chamberlain, C., *et al.*, 2002. Spatial and temporal analysis of Rac activation during live neutrophil chemotaxis. *Curr. Biol.* **12**: 2029–2034.

Geiger, B. and Bershadsky, A., 2001. Assembly and mechanosensory function of focal contacts. *Curr. Opin. Cell Biol.* **13**: 584–592.

Geiger, B. and Bershadsky, A., 2002. Exploring the neighborhood: adhesion-coupled cell mechanosensors. *Cell* **110**: 139–142.

Geiger, B., Bershadsky, A., Pankov, R. and Yamada, K. M., 2001. Transmembrane extracellular matrix–cytoskeleton crosstalk. *Nat. Rev. Mol. Cell Biol.* **2**: 793–805.

Goldman, R. D., 1971. The role of three cytoplasmic fibers in BHK-21 cell motility. I. Microtubules and the effects of colchicine. *J. Cell Biol.* **51**: 752–762.

Gundersen, G. G., 2002. Evolutionary conservation of microtubule-capture mechanisms. *Nat. Rev. Mol. Cell Biol.* **3**: 296–304.

Harris, A. K., Wild, P. and Stopak, D., 1980. Silicone rubber substrata: a new wrinkle in the study of cell locomotion. *Science* **208**: 177–179.

Heald, R. and Nogales, E., 2002. Microtubule dynamics. *J. Cell Sci.* **115**: 3–4.

Helfman, D. M., Levy, E. T., Berthier, C., Shtutman, M., *et al.*, 1999. Caldesmon inhibits nonmuscle cell contractility and interferes with the formation of focal adhesions. *Mol. Biol. Cell* **10**: 3097–3112.

Holleran, E. A., Karki, S. and Holzbaur, E. L., 1998. The role of the dynactin complex in intracellular motility. *Int. Rev. Cytol.* **182**: 69–109.

Holleran, E. A., Ligon, L. A., Tokito, M., Stankewich, M. C., *et al.*, 2001. beta III spectrin binds to the Arp1 subunit of dynactin. *J. Biol. Chem.* **276**: 36 598–36 605.

Holly, S. P., Larson, M. K. and Parise, L. V., 2000. Multiple roles of integrins in cell motility. *Exp. Cell Res.* **261**: 69–74.

Ingber, D. E., 2002. Mechanical signaling and the cellular response to extracellular matrix in angiogenesis and cardiovascular physiology. *Circ. Res.* **91**: 877–887.

Ishizaki, T., Morishima, Y., Okamoto, M., Furuyashiki, T., *et al.*, 2001. Coordination of microtubules and the actin cytoskeleton by the Rho effector mDia1. *Nat. Cell Biol.* **3**: 8–14.

Itoh, R. E., Kurokawa, K., Ohba, Y., Yoshizaki, H., *et al.*, 2002. Activation of rac and cdc42 video imaged by fluorescent resonance energy transfer-based single-molecule probes in the membrane of living cells. *Mol. Cell Biol.* **22**: 6582–6591.

Ivanova, O. Y., Svitkina, T. M., Vasiliev, J. M. and Gelfand, I. M., 1980. Effect of colcemid on the distribution of pseudopodial activity in fibroblasts. Microtubule-independent stabilization of cell surface. *Exp. Cell Res.* **128**: 457–461.

Kaverina, I., Rottner, K. and Small, J. V., 1998. Targeting, capture, and stabilization of microtubules at early focal adhesions. *J. Cell Biol.* **142**: 181–190.

Kaverina, I., Krylyshkina, O. and Small, J. V., 1999. Microtubule targeting of substrate contacts promotes their relaxation and dissociation. *J. Cell Biol.* **146**: 1033–1044.

Kaverina, I., Krylyshkina, O., Gimona, M., Beningo, K., *et al.*, 2000. Enforced polarisation and locomotion of fibroblasts lacking microtubules. *Curr. Biol.* **10**: 739–742.

Kaverina, I., Krylyshkina, O., Beningo, K., Anderson, K., *et al.*, 2002. Tensile stress stimulates microtubule outgrowth in living cells. *J. Cell Sci.* **115**: 2283–2291.

Keating, T. J., Peloquin, J. G., Rodionov, V. I., Momcilovic, D. and Borisy, G. G., 1997. Microtubule release from the centrosome. *Proc. Natl. Acad. Sci. USA* **94**: 5078–5083.

Keller, H. U. and Niggli, V., 1993. Colchicine-induced stimulation of PMN motility related to cytoskeletal changes in actin, alpha-actinin, and myosin. *Cell Motil. Cytoskeleton* **25**: 10–18.

Khawaja, S., Gundersen, G. G. and Bulinski, J. C., 1988. Enhanced stability of microtubules enriched in detyrosinated tubulin is not a direct function of detyrosination level. *J. Cell Biol.* **106**: 141–149.

Kimura, K., Ito, M., Amano, M., Chihara, K., *et al.*, 1996. Regulation of myosin phosphatase by Rho and Rho-associated kinase (Rho-kinase). *Science* **273**: 245–248.

Kinoshita, K., Arnal, I., Desai, A., Drechsel, D. N. and Hyman, A. A., 2001. Reconstitution of physiological microtubule dynamics using purified components. *Science* **294**: 1340–1343.

Kirchner, J., Kam, Z., Tzur, G., Bershadsky, A. D. and Geiger, B., 2003. Live-cell monitoring of tyrosine phosphorylation in focal adhesions following microtubule disruption. *J. Cell Sci.* **116**: 975–986.

Kjoller, L. and Hall, A., 1999. Signaling to Rho GTPases. *Exp. Cell Res.* **253**: 166–179.

Kolodney, M. S. and Elson, E. L., 1995. Contraction due to microtubule disruption is associated with increased phosphorylation of myosin regulatory light chain. *Proc. Natl. Acad. Sci. USA* **92**: 10 252–10 256.

Komarova, Y. A., Akhmanova, A. S., Kojima, S., Galjart, N. and Borisy, G. G., 2002. Cytoplasmic linker proteins promote microtubule rescue *in vivo*. *J. Cell Biol.* **159**: 589–599.

Komarova, Y. A., Vorobjev, I. A. and Borisy, G. G., 2002. Life cycle of MTs: persistent growth in the cell interior, asymmetric transition frequencies and effects of the cell boundary. *J. Cell Sci.* **115**: 3527–3539.

Kraynov, V. S., Chamberlain, C., Bokoch, G. M., Schwartz, M. A., *et al.*, 2000. Localized Rac activation dynamics visualized in living cells. *Science* **290**: 333–337.

Lauffenburger, D. A. and Horwitz, A. F., 1996. Cell migration: a physically integrated molecular process. *Cell* **84**: 359–369.

Lee, J. and Jacobson, K., 1997. The composition and dynamics of cell–substratum adhesions in locomoting fish keratocytes. *J. Cell Sci.* **110**: 2833–2844.

Littlefield, R. and Fowler, V. M., 2002. A minor actin catastrophe. *Nat. Cell Biol.* **4**: E209–E211.

Lo, C. M., Wang, H. B., Dembo, M. and Wang, Y. L., 2000. Cell movement is guided by the rigidity of the substrate. *Biophys. J.* **79**: 144–152.

Machesky, L. M. and Insall, R. H., 1998. Scar1 and the related Wiskott–Aldrich syndrome protein, WASP, regulate the actin cytoskeleton through the Arp2/3 complex. *Curr. Biol.* **8**: 1347–1356.

McNally, F. J., 2001. Cytoskeleton: CLASPing the end to the edge. *Curr. Biol.* **11**: R477–R480.

Miao, H., Li, S., Hu, Y. L., Yuan, S., *et al.*, 2002. Differential regulation of Rho GTPases by beta1 and beta3 integrins: the role of an extracellular domain of integrin in intracellular signaling. *J. Cell Sci.* **115**: 2199–2206.

Mitchison, T. J., 1992. Compare and contrast actin filaments and microtubules. *Mol. Biol. Cell* **3**: 1309–1315.

Moritz, M. and Agard, D. A., 2001. Gamma-tubulin complexes and microtubule nucleation. *Curr. Opin. Struct. Biol.* **11**: 174–181.

Nakano, K., Imai, J., Arai, R., Toh, E. A., *et al.*, 2002. The small GTPase Rho3 and the diaphanous/formin For3 function in polarized cell growth in fission yeast. *J. Cell Sci.* **115**: 4629–4639.

Palazzo, A. F., Cook, T. A., Alberts, A. S. and Gundersen, G. G., 2001. mDia mediates Rho-regulated formation and orientation of stable microtubules. *Nat. Cell Biol.* **3**: 723–729.

Pelham, R. J., Jr. and Wang, Y., 1997. Cell locomotion and focal adhesions are regulated by substrate flexibility. *Proc. Natl. Acad. Sci. USA* **94**: 13 661–13 665.

Peterson, J. R. and Mitchison, T. J., 2002. Small molecules, big impact. A history of chemical inhibitors and the cytoskeleton. *Chem. Biol.* **9**: 1275–1285.

Pollard, T. D. and Beltzner, C. C., 2002. Structure and function of the Arp2/3 complex. *Curr. Opin. Struct. Biol.* **12**: 768–774.

Pollard, T. D. and Borisy, G. G., 2003. Cellular motility driven by assembly and disassembly of actin filaments. *Cell* **112**: 453–465.

Price, L. S., Leng, J., Schwartz, M. A. and Bokoch, G. M., 1998. Activation of Rac and Cdc42 by integrins mediates cell spreading. *Mol. Biol. Cell* **9**: 1863–1871.

Pruyne, D., Evangelista, M., Yang, C., Bi, E., *et al.*, 2002. Role of formins in actin assembly: nucleation and barbed-end association. *Science* **297**: 612–615.

Rehberg, M. and Graf, R., 2002. *Dictyostelium* EB1 is a genuine centrosomal component required for proper spindle formation. *Mol. Biol. Cell* **13**: 2301–2310.

Ren, X. D., Kiosses, W. B. and Schwartz, M. A., 1999. Regulation of the small GTP-binding protein Rho by cell adhesion and the cytoskeleton. *Embo J.* **18**: 578–585.

Ridley, A. J., 2001. Rho GTPases and cell migration. *J. Cell Sci.* **114**: 2713–2722.

Ridley, A. J. and Hall, A., 1992. The small GTP-binding protein rho regulates the assembly of focal adhesions and actin stress fibers in response to growth factors. *Cell* **70**: 389–399.

Riveline, D., Zamir, E., Balaban, N. Q., Schwarz, U. S., *et al.*, 2001. Focal contacts as mechanosensors: externally applied local mechanical force induces growth of focal contacts by an mDia1-dependent and ROCK-independent mechanism. *J. Cell Biol.* **153**: 1175–1186.

Rodionov, V. I. and Borisy, G. G., 1997. Microtubule treadmilling *in vivo*. *Science* **275**: 215–218.

Rodionov, V., Nadezhdina, E. and Borisy, G., 1999. Centrosomal control of microtubule dynamics. *Proc. Natl. Acad. Sci. USA* **96**: 115–120.

Rottner, K., Hall, A. and Small, J. V., 1999. Interplay between Rac and Rho in the control of substrate contact dynamics. *Curr. Biol.* **9**: 640–648.

Sagot, I., Klee, S. K. and Pellman, D., 2002. Yeast formins regulate cell polarity by controlling the assembly of actin cables. *Nat. Cell Biol.* **4**: 42–50.

Sagot, I., Rodal, A. A., Moseley, J., Goode, B. L. and Pellman, D., 2002. An actin nucleation mechanism mediated by Bni1 and profilin. *Nat. Cell Biol.* **4**: 626–631.

Schafer, D. A., Korshunova, Y. O., Schroer, T. A. and Cooper, J. A., 1994. Differential localization and sequence analysis of capping protein beta-subunit isoforms of vertebrates. *J. Cell Biol.* **127**: 453–465.

Schmidt, A. and Hall, A., 2002. Guanine nucleotide exchange factors for Rho GTPases: turning on the switch. *Genes Dev.* **16**: 1587–1609.

Schuyler, S. C. and Pellman, D., 2001. Microtubule 'plus-end-tracking proteins': The end is just the beginning. *Cell* **105**: 421–424.

Severson, A. F., Baillie, D. L. and Bowerman, B., 2002. A formin homology protein and a profilin are required for cytokinesis and Arp2/3-independent assembly of cortical microfilaments in *C. elegans*. *Curr. Biol.* **12**: 2066–2075.

Sharpless, K. E. and Harris, S. D., 2002. Functional characterization and localization of the *Aspergillus nidulans* formin SEPA. *Mol. Biol. Cell* **13**: 469–479.

Small, J. V., Geiger, B., Kaverina, I. and Bershadsky, A., 2002. Opinion: How do microtubules guide migrating cells? *Nat. Rev. Mol. Cell Biol.* **3**: 957–964.

Sroka, J., von Gunten, M., Dunn, G. A. and Keller, H. U., 2002. Phenotype modulation in non-adherent and adherent sublines of Walker carcinosarcoma cells: the role of cell–substratum contacts and microtubules in controlling cell shape, locomotion and cytoskeletal structure. *Int. J. Biochem. Cell Biol.* **34**: 882–899.

Suter, D. M., Errante, L. D., Belotserkovsky, V. and Forscher, P., 1998. The Ig superfamily cell adhesion molecule, apCAM, mediates growth cone steering by substrate–cytoskeletal coupling. *J. Cell Biol.* **141**: 227–240.

Tien, J., Nelson, C. M. and Chen, C. S., 2002. Fabrication of aligned microstructures with a single elastomeric stamp. *Proc. Natl. Acad. Sci. USA* **99**: 1758–1762.

Tirnauer, J. S., Grego, S., Salmon, E. D. and Mitchison, T. J., 2002. EB1-microtubule interactions in *Xenopus* egg extracts: role of EB1 in microtubule stabilization and mechanisms of targeting to microtubules. *Mol. Biol. Cell* **13**: 3614–3626.

Tominaga, T., Sahai, E., Chardin, P., McCormick, F., *et al.*, 2000. Diaphanous-related formins bridge Rho GTPase and Src tyrosine kinase signaling. *Mol. Cell* **5**: 13–25.

Tsuji, T., Ishizaki, T., Okamoto, M., Higashida, C., *et al.*, 2002. ROCK and mDia1 antagonize in Rho-dependent Rac activation in Swiss 3T3 fibroblasts. *J. Cell Biol.* **157**: 819–830.

Tsuruta, D., Gonzales, M., Hopkinson, S. B., Otey, C., *et al.*, 2002. Microfilament-dependent movement of the beta3 integrin subunit within focal contacts of endothelial cells. *Faseb J.* **16**: 866–868.

Turner, C. E. and Brown, M. C., 2001. Cell motility: ARNO and ARF6 at the cutting edge. *Curr. Biol.* **11**: R875–R877.

Tzima, E., Del Pozo, M. A., Kiosses, W. B., Mohamed, S. A., *et al.*, 2002. Activation of Rac1 by shear stress in endothelial cells mediates both cytoskeletal reorganization and effects on gene expression. *Embo J.* **21**: 6791–6800.

Vale, R. D. and Milligan, R. A., 2000. The way things move: looking under the hood of molecular motor proteins. *Science* **288**: 88–95.

Vasiliev, J. M., Gelfand, I. M., Domnina, L. V., Ivanova, O. Y., *et al.*, 1970. Effect of colcemid on the locomotory behaviour of fibroblasts. *J. Embryol. Exp. Morphol.* **24**: 625–640.

Verkhovsky, A. B., Svitkina, T. M. and Borisy, G. G., 1999. Self-polarization and directional motility of cytoplasm. *Curr. Biol.* **9**: 11–20.

Watanabe, N., Madaule, P., Reid, T., Ishizaki, T., *et al.*, 1997. p140mDia, a mammalian homolog of *Drosophila diaphanous*, is a target protein for Rho small GTPase and is a ligand for profilin. *Embo J.* **16**: 3044–3056.

Watanabe, N., Kato, T., Fujita, A., Ishizaki, T. and Narumiya, S., 1999. Cooperation between mDia1 and ROCK in Rho-induced actin reorganization. *Nat. Cell Biol.* **1**: 136–143.

Waterman-Storer, C. M., Karki, S. and Holzbaur, E. L., 1995. The p150Glued component of the dynactin complex binds to both microtubules and the actin-related protein centractin (Arp-1). *Proc. Natl. Acad. Sci. USA* **92**: 1634–1638.

Waterman-Storer, C. M. and Salmon, E. D., 1997. Microtubule dynamics: treadmilling comes around again. *Curr. Biol.* **7**: R369–R372.

Webster, D. R., Gundersen, G. G., Bulinski, J. C. and Borisy, G. G., 1987. Differential turnover of tyrosinated and detyrosinated microtubules. *Proc. Natl. Acad. Sci. USA* **84**: 9040–9044.

Wehrle-Haller, B. and Imhof, B., 2002. The inner lives of focal adhesions. *Trends Cell Biol.* **12**: 382–389.

Winter, D., Podtelejnikov, A. V., Mann, M. and Li, R., 1997. The complex containing actin-related proteins Arp2 and Arp3 is required for the motility and integrity of yeast actin patches. *Curr. Biol.* **7**: 519–529.

Wittmann, T. and Waterman-Storer, C. M., 2001. Cell motility: can Rho GTPases and microtubules point the way? *J. Cell Sci.* **114**: 3795–3803.

Zamir, E., Katz, M., Posen, Y., Erez, N., *et al.*, 2000. Dynamics and segregation of cell-matrix adhesions in cultured fibroblasts. *Nat. Cell Biol.* **2**: 191–196.

6

Initial Steps from Cell Migration to Cell–cell Adhesion

Jason S. Ehrlich, W. James Nelson
and **Marc D. H. Hansen**

High resolution time-lapse microscopy, quantitative image analysis and biochemical analysis of protein complexes reveal that cell–cell contact in MDCK epithelial cells coincides with a spatio-temporal reorganization of plasma membrane Rac1 and lamellipodia from non-contacting to contacting surfaces. Later stages of adhesion involve compaction of the contact and it appears that activation of actin-myosin contraction is required for this process and that the activity of this complex may be spatially constrained to the lateral edges of the contact. We propose a mechanism in which force vectors perpendicular to the contact initiate cell–cell adhesion by forcing membranes and cadherins into contact, but that subsequently, force vectors parallel to the contact further extend, strengthen and eventually compact the contact.

Introduction

Multicellular organisms comprise heterogeneous cell types that are organized during development into distinct patterns to form tissues and organs (Trinkaus, 1965). One of the most important primary processes involved in regulating the establishment and maintenance of these patterns is cell–cell adhesion. A central principle is that the interaction between cells within heterogeneous cell populations is based upon the specificity and extent of

Cell Motility. Edited by Anne Ridley, Michelle Peckham and Peter Clark.
 ISBN 0 470 84872 3

adhesion between cells; cells of one type aggregate together and sort-out from other cells (Steinberg, 1964; Takeichi, 1991). For the purpose of this review, we will focus on our recent studies to examine cell–cell interactions in mammalian simple epithelia and the question of how cell change from a migratory phenotype to a non-migratory, cell–cell adhesion phenotype. Much of our insight into mechanism has come from the study of live cells, and some of the phenomena were first observed over 30 years ago by Professor Michael Abercrombie including the formation of ruffling membranes and the dynamics of lamellipodia (Abercrombie *et al.*, 1970a,b).

Studies to understand the basis of cell–cell adhesion were underway at the turn of the 19th century and had a mechanical focus in which movements and forces underlying the properties of cell–cell interactions were measured (Trinkaus, 1965). When molecules involved in cell–cell adhesion were discovered in the 1980s (Steinberg, 1964; Gumbiner, 1996; Jamora and Fuchs, 2002), the focus shifted to understanding the molecular and biochemical nature of protein–protein interactions and assembly of protein complexes. A goal of the current work is to combine our new understanding of these protein complexes with the mechanics of cell–cell adhesion (Adams and Nelson, 1998).

Epithelial cell–cell adhesion complexes

While many cell types require cell–cell adhesion for function, it is in simple epithelial cells that the complexity of structural and functional organization is greatest. In addition to distinctive organizations of membrane proteins, cytoskeleton and organelles, simple epithelial cells have a higher-ordered organization involving cell–cell and cell–extracellular matrix contacts that orientates cells into a monolayer that separates different biological compartments in the body (Rodriguez-Boulan and Nelson, 1989). The cell surface bounded by these contacts (basal-lateral membrane) and facing the *inside* (serosa) of the organism is structurally and functionally distinct from the unbounded surface (apical membrane) facing a free space (lumen) that is usually continuous with the *outside* of the organism. Functional differences between these plasma membrane domains are required to regulate vectorial transport of ions and solutes across the epithelium, and abnormalities in epithelial organization, cell–cell adhesion and protein distributions are characteristic of many diseases (Rodriguez-Boulan and Nelson, 1989). Understanding how different membrane domains are organized in polarized epithelial cells requires knowledge of how cells adhere to each other and the extracellular matrix, and how the resulting bounded and unbounded cell surfaces are converted into different membrane domains by localized assembly and targeted delivery of specific proteins (Yeaman *et al.*, 1999).

Early electron microscopy of simple epithelia revealed both the complexity and order of cell–cell adhesion junctions (Farquhar and Palade, 1965). At the apex of the lateral membrane, the plasma membranes of adjacent cells are closely opposed at the tight junction. The tight junction acts both as a gate to regulate ion and solute flow in the paracellular pathway, and a fence to regulate lipid and protein diffusion between the apical and lateral membrane domains. Tight junction function is determined by the claudin family of tetraspanning membrane proteins, which are linked to the actin cytoskeleton by adapter proteins including ZO-1, -2 and -3 (Tsukita *et al.*, 2001). Immediately below the tight junction is located the adherens junction which maintains cell–cell attachment and through direct linkage to a circumferential bundle of actin filaments is also involved in regulating morphogenetic movements of the cells during processes requiring invagination of the monolayer in development and in wound-healing (Gumbiner, 1996). The cadherin superfamily of transmembrane Ca^{2+}-dependent cell–cell adhesion proteins appear to initiate adhesion and maintain the organization of the adherens junction (see below). Scattered down the lateral membrane below the adherens junction are desmosomes composed of desmosomal cadherins that are linked through specialized adapter proteins (desmoplakin, plakoglobin) to intermediate filaments to form a structural continuum across the epithelial monolayer (Green and Gaudry, 2000).

Molecular interactions and functions of classical cadherins

Initiation of cell–cell adhesion in simple epithelia appears to require a calcium-dependent mechanism involving members of the cadherin superfamily of proteins. Cadherins are single membrane spanning proteins with a divergent extracellular domain of five repeats and a conserved cytoplasmic domain (Gumbiner, 2000). Binding between extracellular domains, which requires Ca^{2+} for protein conformation (Pertz *et al.*, 1999), is thought to involve multiple *cis*-dimers of cadherin (Brieher *et al.*, 1996; Yap *et al.*, 1997) forming *trans*-oligomers between cadherins on opposing cell surfaces (Boggon *et al.*, 2002). Binding between cadherin extracellular domains is weak, but strong cell–cell adhesion develops during lateral clustering of cadherins by proteins that link cadherin to the actin cytoskeleton (Gumbiner, 2000); β-catenin binds to cadherin cytoplasmic domain and to α-catenin (Aberle *et al.*, 1994; Jou *et al.*, 1995), which is linked directly (Rimm *et al.*, 1995) or indirectly through vinculin/α-actinin (Nieset *et al.*, 1997; Hazan *et al.*, 1997) to the actin cytoskeleton. However, little is known about how these protein complexes assemble in cells following initial cell–cell contact events, how the cadherin/catenin complex binds and organizes the actin cytoskeleton, or how regulators of the actin cytoskeleton identified at cell–cell contacts [eg. Arp2/3 (Kovacs *et al.*, 2002), VASP (Vasioukhin *et al.*, 2000)] might link cadherin/catenin function with actin organization.

Examining E-cadherin distribution during cell–cell adhesion in live cells

Many studies have examined the subcellular distribution of cadherins, catenins and associated actin cytoskeleton proteins in static, fixed cells. However, dynamic cellular events, such as cell migration and cell–cell adhesion, are best examined by live cell imaging, as championed by Abercrombie and his contemporaries many decades previously (Abercrombie *et al.*, 1970a,b). Development of techniques to examine protein distributions at high resolution in single, living cells, in collaboration with Dr Stephen Smith of Stanford University, permitted investigation into the temporal and spatial regulation of assembly and function of the cadherin/catenin complex in MDCK cells. We developed time-lapse differential interference contrast (DIC) imaging to observe the development of cell–cell contacts (McNeill *et al.*, 1993). This was correlated with either quantitative retrospective immunocytochemistry (Adams *et al.*, 1996) or fluorescence microscopy of cells expressing GFP-tagged proteins (Adams *et al.*, 1998; Ehrlich *et al.*, 2002) in order to monitor dynamics of protein localization during cell–cell contact formation. Our initial analyses showed that E-cadherin, α-catenin and β-catenin, but not plakoglobin, co-assemble into Triton X-100 insoluble (TX-insoluble) structures at cell–cell contacts (Adams *et al.*, 1996). The assembly of the cadherin/catenin complex occurred with kinetics similar to that for strengthening of E-cadherin-mediated cell adhesion determined using a centrifugation force-based adhesion assay (Angres *et al.*, 1996). TX-insoluble E-cadherin, α-catenin and β-catenin co-localize along cell–cell contacts in spatially discrete micro-domains, which we term puncta, and the relative amounts of each protein in each punctum increased proportionally during cell–cell adhesion (Adams *et al.*, 1996). Subsequently, we imaged live cells expressing E-cadherinGFP (EcadGFP) and found that EcadGFP rapidly accumulated at cell–cell contacts and became concentrated in puncta identical to those we had found by retrospective immunohistochemistry (Adams *et al.*, 1998; see Figure 6.1A). Using photo-bleaching of EcadGFP, we showed directly that each punctum arose by coalescence of a highly mobile, diffuse pool of cell surface E-cadherin, and that E-cadherin rapidly became immobilized within each punctum (Adams *et al.*, 1998; see Figure 6.1B,C). These E-cadherin puncta were spatially coincident with membrane attachment sites for actin filaments branching off from circumferential actin cables that circumscribed each cell (Adams *et al.*, 1996, 1998; see Figure 6.1D). As the

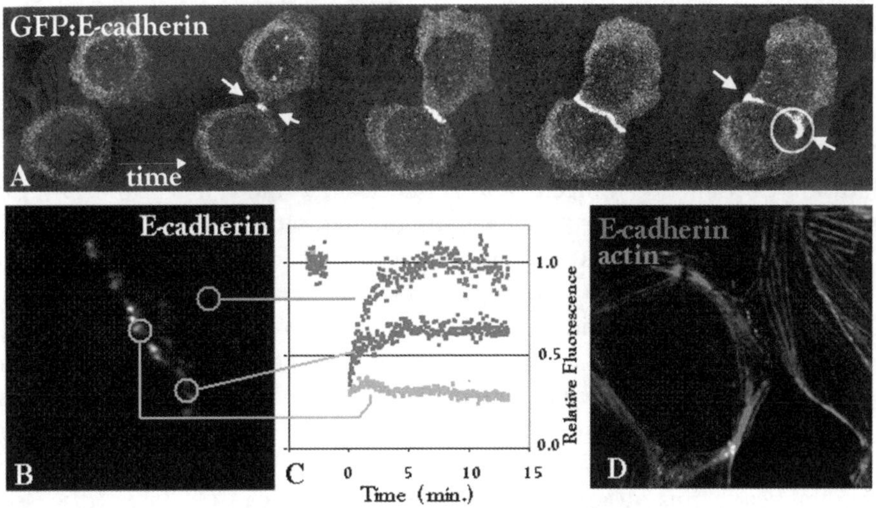

Figure 6.1 Distribution of E-cadherinGFP (EcadGFP) during initial stages of cell–cell adhesion in MDCK cells (for details of protein distributions and live cell imaging see Adams *et al.*, 1998)). (A) Still images from a time-lapse movie of EcadGFP distribution in live MDCK cells. Note rapid appearance of EcadGFP at initial sites of cell–cell adhesion, and the subsequent formation of individual puncta as the contact extends. (B) Higher magnification of a cell–cell contact showing distribution of EcadGFP in puncta along the contact. (C) Fluorescence recovery after photobleaching (FRAP) of EcadGFP in different regions of MDCK cells during cell–cell adhesion. Note that the EcadGFP is mobile in the plasma membrane outside the site of cell–cell contact (grey), less mobile at the newest sites of cell–cell contact (green) and immobile at the oldest sites of cell–cell contact (beige); for details, see Adams *et al.* (1998). (D) Double fluorescence microcopy of two MDCK cells showing the distribution of E-cadherin (green) and actin filaments (rhodamine-phalloidin); note close association of actin filaments coming from the cortical bundle with individual E-cadherin puncta. (A colour reproduction of this figure can be found in the colour plate section)

surface area of the contact increased, we found that new puncta were inserted at regularly spaced intervals along the contact, and bundles of actin filaments become organized perpendicularly to the membrane in association with each punctum (Adams *et al.*, 1998; see Figure 6.1D). Subsequently, the circumferential actin cables near cell–cell contact sites separated, and the resulting two ends of the cable swung outwards to the perimeter of the contact. Concomitantly, subsets of E-cadherin puncta were also swept to the margins of the contact where they further coalesced into large E-cadherin plaques (Adams *et al.*, 1998). The circumferential actin cable was embedded into each E-cadherin plaque at the contact margin, forming a continuous actin structure that circumscribed the entire cell couplet. At this stage, the two cells had achieved maximum contact, a process referred to as compaction.

Mechanistic insights into E-cadherin function during cell–cell adhesion

Cell–cell contact has been proposed to act as a trap for cadherin that is freely diffusing within the plasma membrane, as originally envisioned in the diffusion-trap model (Singer, 1976; McCloskey and Poo, 1986). In this model, E-cadherin at the contact site rapidly becomes immobile. Results with live cell imaging analysis strongly supports this model. Several roles for cadherin in this process can be considered. Trans interactions between adjacent cells could provide the initial mechanism for a diffusion trapping system. *Cis* interactions of cadherins, between cadherin molecules on the same cell, could provide a secondary mechanism to recruit and retain cadherins at contact sites. Besides these structural roles, signalling pathways initiated by cadherin engagement in *trans* could result in further accumulation of immobile cadherins at contact sites. Activation of Rho GTPases (discussed below) could direct changes in actin organization that provide additional binding sites for freely diffusive cadherins at cell–cell contacts. Diffusion and trapping of E-cadherin at the contact would increase the local concentration of cadherin, thereby resulting in strengthening of cell–cell adhesion. Thus, short-range diffusion trapping around immobilized E-cadherin may be important for clustering E-cadherin within the cell–cell contact.

The role of Rho family small GTPases and membrane dynamics in cell–cell adhesion

Our time-lapse imaging of live MDCK epithelial cells expressing a GFP fusion to E-cadherin demonstrates that contact between migratory cells is an opportunistic event to which migratory cells must respond rapidly in order to convert a transient contact into strong adhesion (Adams *et al.*, 1998). This description of adhesion dynamics has led us to inquire about the interplay between the cellular machinery involved in cell migration, the initiation of cell–cell adhesion, and the subsequent suppression of the migratory phenotype.

The Rho family of small GTPases has been shown to play an important role in cell migration and plasma membrane dynamics, and a role for these proteins in cell–cell adhesion has long been sought (Bishop and Hall, 2000). Prior studies from our laboratory indicate that lamellipodia extension, a Rac1-driven process (Nobes and Hall, 1995), is important in enlarging cell–cell contacts (McNeill *et al.*, 1990) as shown also in keratinocytes (Braga *et al.*, 1997, 1999). Recent reports demonstrate increased activity levels of these two Rho family members following cadherin-dependent cell–cell adhesion in MDCK cells (Kim *et al*., 2000; Nakagawa *et al*., 2001).

Attempts to understand the function of Rac1 and Cdc42 in cell–cell adhesion have led to varied and sometimes conflicting results (Vasioukhin *et al*., 2000; Braga *et al*., 1997, 1999, 2000). These are often attributed to poorly understood contextual variations in cell–cell adhesion between different experimental systems (Vasioukhin and Fuchs, 2001). Microinjection and over-expression approaches with mutant proteins have typically been used to evaluate GTPase functions, but the large number of potential GTPase binding partners complicates conclusions from these experiments. With over 45 such partners identified *in vitro* (Bishop and Hall, 2000), exogenously introduced, mutant Rho GTPases may bind partners that are not physiologically or contextually relevant, making it difficult to distinguish normal functions of GTPases from cellular adaptations to disrupted signalling networks. Finally, evaluations of adhesion between cells expressing different GTPase mutants have been static and descriptive, and have not exploited either dynamic cell imaging or quantitative functional assays.

Rac1-containing lamellipodia drive cell–cell contact formation between MDCK cells

As Rac1 function is classically linked to lamellipodia formation (Nobes and Hall, 1995; Ridley *et al*., 1992), we examined the subcellular distribution of a functional Rac1-GFP fusion protein (Subauste *et al*., 2000) using time-lapse, confocal imaging of live MDCK cells. In single, migratory cells, RacGFP localized diffusely throughout the cytoplasm (Ehrlich *et al*., 2002), consistent with the large cytosolic pool of Rac1 in these cells (Hansen and Nelson, 2001), and transiently at the tips of extending lamellipodia as predicted from previous static images of migrating cells (Ridley *et al*., 1992).

We found that initial, opportunistic contacts between cells occurred through exploratory lamellipodia that were randomly orientated on the cell perimeter. Concerted formation of cell–cell contacts coincided with repeated rounds of lamellipodia protrusions that were polarized towards the contacting membrane (Ehrlich *et al*., 2002; see also McNeill *et al*., 1993 and Adams *et al*., 1996). The concentration of lamellipodia at cell–cell contacts coincided with a substantial and constant accumulation of RacGFP along the contacting membranes. RacGFP accumulation was most pronounced at the ends of cell–cell contacts, and diminished at regions where cell–cell adhesion had stabilized.

The striking correlation between formation of lamellipodia and RacGFP localization implies that Rac1 is functionally active at these sites. Further, the concomitant decrease in lamellipodia activity at membranes lacking Rac1 indicates a tight spatial regulation of Rac1 localization and activity. That Rac1 localization and activity are tightly regulated was also evident in another observation that we made, namely that the relative frequency, number and perdurance of lamellipodia gradually increased at sites of cell–cell adhesion and concomitantly diminished elsewhere around the cell membrane (Ehrlich *et al.*, 2002). This change in localization of Rac1 and lamellipodia coincided with a change from a migratory to a non-migratory, cell–cell adherent phenotype.

Cell–cell contact induces changes in Rac1 complexes

Changes in RacGFP localization as cells transition from a migratory to a non-migratory, adherent state could reflect changes in molecular associations of endogenous Rac1 at the plasma membrane. To investigate this, isolated membranes from a heterogeneous population of migratory and adherent MDCK cells were extracted in buffer containing octyl-glucoside to solubilize membrane-bound Rac1; note that $<5\%$ of endogenous Rac1 is membrane-bound, but this population is tightly associated with membranes and can be removed only by extraction with detergents (Hansen and Nelson, 2001; Hansen *et al.*, 2002; see Figure 6.2). Solubilized proteins were separated by rate-zonal centrifugation in sucrose gradients. The Rac1 profile determined by Western blotting revealed three separate peaks of Rac1 distribution: a low molecular weight Rac1:PAK effector complex (Hansen and Nelson, 2001), and two higher molecular weight complexes of $\sim 11S$ (complex B); and $\sim 16S$ (complex C) (Hansen and Nelson, 2001; Hansen *et al.*, 2002; see Figure 6.2). Significantly, we found that Rac1 complex B is enriched in non-contacting, migratory MDCK cells and complex C is absent, while in contacted MDCK cells complex C is enriched and complex B is absent (the Rac1:RAK complex is present in both). This switch in molecular associations of endogenous Rac1 from complex B to complex C following cell–cell adhesion correlates with the redistribution of RacGFP at non-contacting surfaces to lamellipodia at cell–cell junctions and the concomitant change in lamellipodia activity (Hansen *et al.*, 2002; see Figure 6.2).

Effects of Rac1 mutant expression on endogenous Rac1 complexes and cell behaviour

Rac1 mutants are thought to exert biological activity by remaining in a specific nucleotide binding state (Bishop and Hall, 2000; Farnsworth and Feig, 1991; Stacey *et al.*, 1991). Dominant negative forms of Rac1 (T17N)

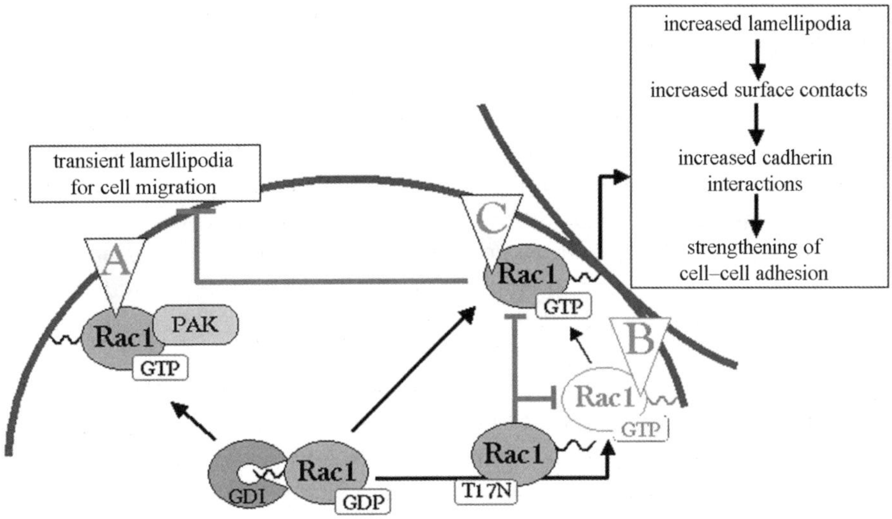

Figure 6.2 Description of Rac1 complexes in MDCK cells during cell–cell adhesion. (For details of protein distributions, live cell imaging and biochemical analysis of protein complexes, please see Ehrlich *et al.*, 2000, and Hansen *et al.*, 2000.) Greater than 95% of Rac1 is in a GDP bound and inactive form in the cytosol in a complex with Rho-GDI (Hansen and Nelson, 2001). We detected three Rac1 complexes in membranes of MDCK cells, termed A, B and C. Complex A comprises a Rac1:PAK complex (Hansen and Nelson, 2001). Complex B and C are high molecular weight complexes of unknown components. Complex B is enriched in single, migratory cells, but is almost absent in cells undergoing cell–cell adhesion (hence it is not coloured). Complex C is enriched during cell–cell adhesion, and we suggest that it localizes at cell–cell contacts (as shown for RacGFP, see Ehrlich *et al.*, 2002 and text). Localization of Rac1 complexes at cell–cell contacts would be expected to increase the number, frequency and perdurance of lamellipodia, which would lead to a cascade of events involving increased cell surface contacts between adhering cells, increased opportunity for cadherin–cadherin interactions and thereby strengthening of cell–cell adhesion. (A colour reproduction of this figure can be found in the colour plate section)

preferentially bind GDP and sequester machinery required to activate endogenous Rac1. Constitutively active Rac1 mutants (G12V and Q61L) have reduced GTPase activity and remain GTP bound for long periods, thereby generating a persistent signal through downstream effector proteins. Additionally, sequestration of GTPase activating proteins by active mutants slows inactivation of endogenous Rac1, leaving it in the active conformation.

We expressed RacT17NGFP in MDCK cells and found cell–cell contacts expanded very slowly compared with wild-type cells (Ehrlich *et al.*, 2002). Bursts of localized lamellipodia formation at sites of cell–cell contacts were markedly reduced in RacT17NGFP cells, and RacT17NGFP cells failed to restrict membrane protrusions to sites of cell–cell contact. In contrast to cells expressing RacT17NGFP, cells expressing constitutively active RacQ61LGFP exhibited an increased rate of lamellipodia formation, with numerous ruffles extending around the cell periphery, consistent with previous work demonstrating that activated Rac1 leads to an increase in lamellipodia formation in many cell types (Bishop and Hall, 2000). We found that broad lamellipodia extended over contacting cells and lateral expansion of cell–cell contacts occurred more quickly when RacQ16L was expressed.

To examine whether expression of Rac1 mutants affected endogenous Rac1 complexes A, B and C, we prepared membrane extracts from mutant Rac1 expressing cells and separated the proteins by rate-zonal centrifugation, and endogenous Rac1 and mutant Rac1 distributions were examined by Western blotting with a Rac1 antibody (Hansen and Nelson, 2001; Hansen *et al.*, 2002). In cells expressing Rac1T17N , endogenous Rac1 appeared as a single peak corresponding to complex A but was absent from complexes B and C. Rac1T17N sedimented as a single peak corresponding to the normal distribution of endogenous Rac1 complex B. In cells expressing Rac1G12V, endogenous Rac1 was detected in complexes A, B, and C, and Rac1G12V co-sedimented with endogenous Rac1 in peaks corresponding to complexes B and C.

Thus, in MDCK cells expressing mutant Rac1 there was a strong positive correlation between the presence of active Rac1 in complexes B and C, and the formation of lamellipodia-driven cell–cell contacts. In cells expressing Rac1T17N, endogenous Rac1 was excluded from complexes B and C; cell–cell adhesions formed tentatively and without lamellipodia. In cells expressing Rac1Q61L or Rac1G12V, complexes B and C contained constitutively active Rac1 and dynamic lamellipodia-driven adhesive contacts were formed rapidly (Hansen and Nelson, 2001; Hansen *et al.*, 2002).

Although live cell imaging revealed changes in the kinetics and degree of cell–cell contact formation in cells expressing Rac1T17N or Rac1G12V, we used a quantitative adhesion assay to measure directly the size of cell aggregates formed over time from a suspension of single cells, and the resistance of these aggregates to shearing force (Ehrlich *et al.*, 2002). Greater than 95% of control cells formed large clusters (>50 cells) within 3 to 6 h that were completely resistant to trituration. Rac1 T17N cells formed aggregates with kinetics similar to that of controls, but the aggregates were far less resistant to trituration than were control cells. Since Rac1 T17N blocks incorporation of endogenous Rac1 into complexes B and C, we conclude that Rac1 these complexes are required for the strengthening and compaction phase of cell–cell adhesion. Rac1 G12V cells displayed rapid kinetics of adhesion formation and aggregates became resistant to trituration earlier than control cells. That Rac1G12V incorporates into Rac1 complexes B and C is further evidence that Rac1 these complexes C are important for the strengthening and compaction phase of cell–cell adhesion.

Linking Rac1 complexes back to mechanisms of cell–cell adhesion

The identity of proteins in these Rac1 complexes, and hence the mechanism of their action on cell migration and cell–cell adhesion, remains unknown. Since activation of Rac1 increases upon cadherin engagement (Noren *et al.*, 2001), it could be physically localized to cell–cell contacts by binding specific proteins (see below), or activity of Rac1 could be controlled by local recruitment of guanine exchange factors (GEFs; reviewed in Hall, 1998; Evers *et al.*, 2000; Fukata and Kaibuchi, 2001). Note that many GEFs localize to membranes through a plekstrin homology domain that binds phospholipid products of PI 3-kinase (Hurley and Meyer, 2001; Martin, 2001), and that PI 3-kinase is also activated by cadherin engagement (Nakagawa *et al.*, 2001) and may bind E-cadherin directly (Nakagawa *et al.*, 2001; Espada *et al.*, 1999; Pece *et al.*, 1999; Woodfield *et al.*, 2001). Indeed, we found that a fusion protein containing the PH domain of AKT and GFP [a marker for $PI(3,4,5)P_3$ and $PI(3,4)P_2$] is rapidly and precisely localized to the edges of the expanding MDCK cell–cell contacts where Rac1 and lamellipodia are active (Ehrlich *et al.*, 2002). Inhibitors of PI-3 kinase have been shown to prevent activation of Rac1 and Cdc42 in response to cadherin-based adhesion. However, inhibition of PI 3-kinase activity does not appear to block cell–cell adhesion (Nakagawa *et al.*, 2001; Ehrlich *et al.*, 2002) suggesting that cell–cell adhesion involves more than one mechanism working in parallel.

The discrepancy in these results may be explained by the effect of PI3 kinase inhibition on Rac1 complexes. Contacting MDCK cells treated with PI-3 kinase inhibitors assemble Rac1 complexes A and B, but not complex C. This indicates that complex C is dispensable for cell–cell contact formation. If active Rac1 is displaced from complexes B and C, as in cells expressing dominant negative Rac1, however, formation of cell–cell contacts is disrupted. We conclude that active Rac1 in complexes B and C perform an identical function, that of initiating the protrusion of lamellipodia at the plasma membrane. Coincidence of Rac1GFP and lamellae in live cell imaging studies supports this conclusion. The observation by Ehrlich *et al.* (2002) that dominant negative Rac1 does not accumulate at nascent cell–cell contacts, while constitutively active Rac1 does, indicates that Rac1 complexes B and C might have distinct diffusive properties in the plane of the membrane. Thus, the transition of Rac1 from complex B to complex C might reflect a molecular mechanism of orientating Rac1 activity to sites of cell–cell adhesion. Given the results with inhibitors of PI-3 kinase, such a mechanism appears to be mediated by the generation of specific phospholipids. Future studies that identify the link between Rac1 localization and phospholipid generation are ongoing.

Several Rac1 GEFs contain pleckstrin homology motifs that bind PI 3-kinase targets and may serve to localize or activate GEFs at sites of PI 3-kinase activity. One example is Tiam1, a Rac1-specific GEF which localizes to cell–cell junctions and is sensitive to PI 3-kinase inhibitors (Hordijk *et al.*, 1997; Sander *et al.*, 1998). Another example is Vav, a GEF reported to activate Rac1 downstream of PI 3-kinase (Gringhuis *et al.*, 1998). The Vav2 isoform binds p120 catenin, possibly localizing the GEF to cell–cell contacts through the interaction of p120 with E-cadherin (Noren *et al.*, 2000). IQGAP, a putative binding partner of Rac1, also associates with developing cell–cell adhesions (Nakagawa *et al.*, 2001; Fukata *et al.*, 1999; Kuroda *et al.*, 1998, 1999). IQGAP is thought to bind β-catenin and block its interaction with α-catenin, preventing linkage of E-cadherin to actin. Active Rac1 could compete with β-catenin for IQGAP, releasing β-catenin to bind α-catenin and thus linking cadherin to the cytoskeleton to form strong adhesions.

Another mechanism by which Rac1 and Cdc42 could affect cell–cell contacts is by regulating actin structure and dynamics. Of particular interest is the Cdc42-WASp-Arp2/3 pathway, which has been particularly well defined (Higgs and Pollard, 2001; Prehoda *et al.*, 2000; Rohatgi *et al.*, 1999). Cdc42 and PIP2 bind WASp coordinately, resulting in a conformational change in WASp (Higgs and Pollard, 2001), and allowing WASp to bind the Arp2/3 complex via its VCA domain (Hufner *et al.*, 2001). Normally weak Arp2/3 activity in initiating actin polymerization is dramatically increased by WASp binding (Higgs and Pollard, 2001; Prehoda *et al.*, 2000; Machesky and Insall, 1998). Rac1-mediated activation of Arp2/3 in lamellipodia is thought to occur in a similar mechanism involving WAVE/Scar proteins and IRSp53 (Miki *et al.*, 2000). Recently, Arp2/3 binding and regulation by Abp1 in response to Rac1 activity has been reported (Goode *et al.*, 2001; Kessels *et al.*, 2000). Thus, structural diversity generated by Arp2/3 activation may arise by orchestrating reactions between different binding proteins (Higgs and Pollard, 2001).

Acknowledgements

J.S.E. was supported by the Medical Scientist Training Program (grant 5T32GM07365 from the National Institute of General Medical Sciences), M.D.H.H. was supported by the Smith Fellowship in the Stanford Graduate Fellowship Program and the Cancer Biology Graduate Program, and work was also supported by NIH grant GM35527 to W.J.N.

References

Abercrombie, M., Heaysman, J. E. and Pegrum, S. M., 1970a. The locomotion of fibroblasts in culture. I. Movements of the leading edge. *Exptl. Cell Res.* **59**: 393–398.

Abercrombie, M., Heaysman, J. E. and Pegrum, S. M., 1970b. The locomotion of fibroblasts in culture. II. Ruffling. *Exp. Cell Res.* **60**: 437–444.

Aberle, H. Schwartz, H. and Kermler, R., 1994. Assembly of the cadherin–catenin complex *in vitro* with recombinant proteins. *J. Cell Sci.* **107**: 3655–3663.

Adams, C. and Nelson, W. J., 1998. Cytomechanics of cadherin-mediated cell–cell adhesion. *Cur. Opin. Cell Biol.* **10**: 572–577.

Adams, C. A., Nelson, W. J. and Smith, S. J., 1996. Quantitative, single-cell analysis of cadherin–catenin–actin reorganization during development of cell–cell adhesion. *J. Cell Biol.* **135**: 1899–1911.

Adams, C. A., Chen, Y.-T., Smith, S. J. and Nelson, W. J., 1998. Mechanisms of epithelial cell–cell adhesion and cell compaction revealed by high resolution tracking of E-cadherin/GFP. *J. Cell Biol.* **142**: 1105–1119.

Angres, B. A., Barth, A. I. M. and Nelson, W. J., 1996. Mechanism for transition from initial to stable cell–cell adhesion: kinetic analysis of E-cadherin-mediated adhesion using a quantitative adhesion assay. *J. Cell Biol.* **134**: 549–558.

Bishop, A. L. and Hall, A., 2000. Rho GTPases and their effector proteins. *Biochem. J.* **348**: 241–255.

Boggon, T. J., Murray, J., Chappuis-Flament, S., Wong, E., *et al.*, 2002. C-cadherin ectodomain structure and implications for cell adhesion mechanisms. *Science*, **296**: 1308–1313.

Braga, V. M., Machesky, L. M., Hall, A. and Hotchin, N. A., 1997. The small GTPases Rho and Rac are required for the establishment of cadherin-dependent cell–cell contacts. *J. Cell Biol.* **137**: 1421–1431.

Braga, V. M., Del Maschio, A., Machesky, L. and Dejana, E., 1999. Regulation of cadherin function by Rho and Rac: modulation by junction maturation and cellular context. *Mol. Biol. Cell* **10**: 9–22.

Braga, V. M., Betson, M., Li, X. and Lamarche-Vane, N., 2000. Activation of the small GTPase Rac is sufficient to disrupt cadherin-dependent cell–cell adhesion in normal human keratinocytes. *Mol. Biol. Cell* **11**: 3703–3710.

Brieher, W. M., Yap, A. S. and Gumbiner, B. M., 1996. Lateral dimerization is required for the homophilic binding activity of C-cadherin. *J. Cell Biol.* **135**: 487–496.

Ehrlich, J. S., Hansen, M. D. H. and Nelson. W. J., 2002. Spatio-temporal regulation of Rac1 localization and lamellipodia dynamics during epithelial cell–cell adhesion. *Dev. Cell* **3**: 259–270.

Espada, J., Perez-Moreno, M., Braga, V. M., Rodriguez-Viciana, P. and Cano, A., 1999. H-Ras activation promotes cytoplasmic accumulation and phosphoinositide 3-OH kinase association of beta-catenin in epidermal keratinocytes. *J. Cell Biol.* **146**: 967–980.

Evers, E. E., Zondag, G. C., Malliri, A., Price, L. S., *et al.*, 2000. Rho family proteins in cell adhesion and cell migration. *Eur. J. Cancer* **36**: 1269–1274.

Farnsworth, C. L. and Feig, L. A., 1991. Dominant inhibitory mutations in the Mg(2+)-binding site of RasH prevent its activation by GTP. *Mol. Cell Biol.* **11**: 4822–4829.

Farquhar, M. G. and Palade, G. E., 1965. Cell junctions in amphibian skin. *J. Cell Biol.* **26**: 263–291.

Fukata, M., Kuroda, S., Nakagawa, M., Kawajiri, A., *et al.*, 1999. Cdc42 and Rac1 regulate the interaction of IQGAP1 with beta-catenin. *J. Biol. Chem.* **274**: 26 044–26 050.

Fukata, M. and Kaibuchi, K., 2001. Rho-family GTPases in cadherin-mediated cell–cell adhesion. *Nature Rev. Mol. Cell Biol.* **2**: 887–897.

Goode, B. L., Rodal, A. A., Barnes, G. and Drubin, D. G., 2001. Activation of the Arp2/3 complex by the actin filament binding protein Abp1p. *J. Cell Biol.* **153**: 627–634.

Green, K. J. and Gaudry, C. A., 2000. Are desmosomes more than tethers for intermediate filaments? *Nat. Rev. Mol. Cell Biol.* **1**: 208–216.

Gringhuis, S. I., de Leij, L. F., Coffer, P. J. and Vellenga, E., 1998. Signaling through CD5 activates a pathway involving phosphatidylinositol 3-kinase, Vav, and Rac1 in human mature T lymphocytes. *Mol. Cell Biol.* **18**: 1725–1735.

Gumbiner, B. M., 1996. Cell adhesion: the molecular basis of tissue architecture and morphogenesis. *Cell* **84**: 345–357.

Gumbiner, B. M., 2000. Regulation of cadherin adhesive activity. *J. Cell Biol.* **148**: 399–403.

Hall, A., 1998. Rho GTPases and the actin cytoskeleton. *Science* **279**: 509–514.

Hansen, M. D. and Nelson, W. J., 2001. Serum-activated assembly and membrane translocation of an endogenous Rac1: effector complex. *Curr. Biol.* **11**: 56–60.

Hansen, M. D. H., Ehrlich, J. S. and Nelson, W. J., 2002. Molecular mechanism for orienting membrane and actin dynamics to nascent epithelial cell–cell contacts. *J. Biol. Chem.* **277**: 45 371–45 376.

Hazan, R., Kang, L., Row, S. P., Borgen, I. and Rimm, D. L., 1997. Vinculin is associated with the E-cadherin adhesion complex. *J. Biol. Chem.* **272**: 32 448–32 453.

Higgs, H. N. and Pollard. T. D., 2001. Regulation of actin filament network formation through ARP2/3 complex: activation by a diverse array of proteins. *Annu. Rev. Biochem.* **70**: 649–676.

Hordijk, P. L., ten Klooster, J. P., van der Kammen, R. A., Michiels, F., *et al.*, 1997. Inhibition of invasion of epithelial cells by Tiam1-Rac signaling. *Science* **278**: 1464–1466.

Hufner, K., Higgs, H. N., Pollard, T. D., Jacobi, C., *et al.*, 2001. The verprolin-like central (vc) region of Wiskott-Aldrich syndrome protein induces Arp2/3 complex-dependent actin nucleation. *J. Biol. Chem.* **276**: 35 761–35 767.

Hurley, J. H. and Meyer, T., 2001. Subcellular targeting by membrane lipids. *Curr. Opin. Cell Biol.* **13**: 146–152.

Jamora, C. and Fuchs, E., 2002. Intercellular adhesion, signaling and the cytoskeleton. *Nature Cell Biol.* **4**: 101–108.

Jou, T.-S., Stewart D. B., Stappert, J., Nelson, W. J. and Marrs, J. A., 1995. Genetic and biochemical dissection of protein linkages in the cadherin-catenin complex. *Proc. Natl. Acad. Sci. USA* **92**: 5067–5071.

Kessels, M. M., Engqvist-Goldstein, A. E. and Drubin, D. G., 2000. Association of mouse actin-binding protein 1 (mAbp1/SH3P7), an Src kinase target, with dynamic regions of the cortical actin cytoskeleton in response to Rac1 activation. *Mol. Biol. Cell* **11**: 393–412.

Kim, S. H., Li, Z. and Sacks, D. B., 2000. E-cadherin-mediated cell–cell attachment activates Cdc42. *J. Biol. Chem.* **275**: 36 999–37 005.

Kovacs, E. M., Goodwin, M., Ali, R. G., Paterson A. D. and Yap, A. S., 2002. Cadherin-directed actin assembly. E-cadherin physically associates with the arp2/3 complex to direct actin assembly in nascent adhesive contacts. *Curr. Biol.* **12**: 379–382.

Kuroda, S., Fukata, M., Nakagawa, M., Fujii, K., *et al.*, 1998. Role of IQGAP1, a target of the small GTPases Cdc42 and Rac1, in regulation of E-cadherin-mediated cell–cell adhesion. *Science* **281**: 832–835.

Kuroda, S., Fukata, M., Nakagawa, M. and Kaibuchi, K., 1999. Cdc42, Rac1, and their effector IQGAP1 as molecular switches for cadherin-mediated cell–cell adhesion. *Biochem. Biophys. Res. Commun.* **262**: 1–6.

Machesky, L. M. and Insall, R. H., 1998. Scar1 and the related Wiskott-Aldrich syndrome protein, WASP, regulate the actin cytoskeleton through the Arp2/3 complex. *Curr. Biol.* **8**: 1347–1356.

Martin, T. F. J., 2001. PI(4,5)P_2 regulation of surface membrane traffic. *Curr. Opin. Cell Biol.* **13**: 493–499.

McCloskey, M. S. and Poo, M.-M., 1986. Contact-induced redistribution of specific membrane components: local accumulation and development of adhesion. *J. Cell Biol.* **102**: 2185–2196.

McNeill, H., Ozawa, M., Kemler, R. and Nelson, W. J., 1990. Novel function of the cell adhesion molecule uvomorulin as an inducer of cell surface polarity. *Cell* **62**: 309–316.

McNeill, H, Ryan, T. A., Smith, S. J. and Nelson, W. J., 1993. Spatial and temporal dissection of immediate and early events following cadherin-mediated epithelial cell adhesion. *J. Cell Biol.* **120**: 1217–1227.

Miki, H., Yamaguchi, H., Suetsugu, S. and Takenawa, T., 2000. IRSp53 is an essential intermediate between Rac1 and WAVE in the regulation of membrane ruffling. *Nature* **408**: 732–735.

Nakagawa, M., Fukata, M., Yamaga, M., Itoh, N. and Kaibuchi, K., 2001. Recruitment and activation of Rac1 by the formation of E-cadherin-mediated cell–cell adhesion sites. *J. Cell Sci.* **114**: 1829–1838.

Nieset, E., Redfield, A. R., Jin, F., Knudsen, K. A., *et al.*, 1997. Characterization of the interactions of alpha-catenin with alpha-actinin and beta-catenin/plakoglobin. *J. Cell Sci.* **110**: 1013–1022.

Nobes, C. D. and Hall, A., 1995. Rho, Rac, and Cdc42 GTPases regulate the assembly of multimolecular focal complexes associated with actin stress fibers, lamellipodia, and filopodia. *Cell* **81**: 53–62.

Noren, N. K., Liu, B. P., Burridge, K. and Kreft, B., 2000. p120 catenin regulates the actin cytoskeleton via Rho family GTPases. *J. Cell Biol.* **150**: 567–580.

Noren, N. K., Niessen, C. M., Gumbiner, B. M. and Burridge, K., 2001. Cadherin engagement regulates Rho GTPases. *J. Biol. Chem.* **16**: 33 305–33 308.

Pece, S., Chiariello, M., Murga, C. and Gutkind, J. S., 1999. Activation of the protein kinase Akt/PKB by the formation of E-cadherin-mediated cell–cell junctions. Evidence for the association of phosphatidylinositol 3-kinase with the E-cadherin adhesion complex. *J. Biol. Chem.* **274**: 19 347–19 351.

Pertz, O., Bozic, D., Koch, A. W., Fauser, C., *et al.*, 1999. A new crystal structure, Ca^{2+} dependence and mutational analysis reveal molecular details of E-cadherin homoassociation. *EMBO J.* **18**: 1738–1749.

Prehoda, K. E., Scott, J. A., Mullins, R. D. and Lim., W. A., 2000. Integration of multiple signals through cooperative regulation of the N-WASP-Arp2/3 complex. *Science* **290**: 801–806.

Ridley, A. J., Paterson, H. F., Johnston, C. L., Diekmann, D. and Hall, A., 1992. The small GTP-binding protein rac regulates growth factor-induced membrane ruffling. *Cell* **70**: 401–410.

Rimm, D. L., Koslov, E., Kebriaei, P., Cianci, C. and Morrow, J., 1995. α_1(E)-catenin is an actin-binding and -bundling protein mediating the attachment of F-actin to the membrane adhesion complex. *Proc. Natl. Acad. Sci. USA* **92**: 8813–8817.

Rodriguez-Boulan, E. and Nelson, W. J., 1989. Morphogenesis of the polarized epithelial cell phenotype. *Science* **245**: 718–725.

Rohatgi, R., Ma, L., Miki, H., Lopez, M., *et al.*, 1999. The interaction between N-WASP and the Arp2/3 complex links Cdc42-dependent signals to actin assembly. *Cell* **97**: 221–231.

Sander, E. E., van Delft, S., ten Klooster, J. P., Reid, T., *et al.*, 1998. Matrix-dependent Tiam1/Rac signaling in epithelial cells promotes either cell–cell adhesion or cell migration and is regulated by phosphatidylinositol 3-kinase. *J. Cell Biol.* **143**: 1385–1398.

Singer, S. J., 1976. The fluid mosaic model of membrane structure. Some applications to ligand–receptor and cell–cell interactions. In: *Surface Membrane Receptors* (Bradshaw, R. A., Frazier, W. A., Merrell, R. E. C., Gottlieb, D. I., Hogue-Angeletti, R. A., eds). New York, Plenum Press, pp. 1–24.

Stacey, D. W., Feig, L. A. and Gibbs, J. B., 1991. Dominant inhibitory Ras mutants selectively inhibit the activity of either cellular or oncogenic Ras. *Mol. Cell Biol.* **11**: 4053–4064.

Steinberg, M. S., 1964. The problem of adhesive selectivity in cellular interactions. In: *Cellular Membranes in Development* (M. Locke, Ed.). New York, Academic Press, pp. 321–366.

Subauste, M. C., Von Herrath, M., Benard, V., Chamberlain, C. E., *et al.*, 2000. Rho family proteins modulate rapid apoptosis induced by cytotoxic T lymphocytes and Fas. *J. Biol. Chem.* **275**: 9725–9733.

Takeichi, M., 1991. Cadherin cell adhesion receptors as a morphogenetic regulator. *Science* **251**: 1451–1455.

Trinkaus, J. P., 1965. Mechanisms of morphogenetic movements. In: *Organogenesis* (DeHaan, R. L. and Ursprung, H., eds). New York, Holt, Rinehart and Winston, pp. 55–104.

Tsukita, S., Furuse, M. and Itoh, M., 2001. Multifunctional strands in tight junctions. *Nat. Rev. Mol. Cell Biol.* **2**: 285–293.

Vasioukhin, V., Bauer, C., Yin, M. and Fuchs, E., 2000. Directed actin polymerization is the driving force for epithelial cell–cell adhesion. *Cell* **100**: 209–219.

Vasioukhin, V. and Fuchs, E., 2001. Actin dynamics and cell–cell adhesion in epithelia. *Curr. Opin. Cell Biol.* **13**: 76–84.

Woodfield, R. J., Hodgkin, M. N., Akhtar, N., Morse, M. A., *et al.*, 2001. The p85 subunit of phosphoinositide 3-kinase is associated with beta-catenin in the cadherin-based adhesion complex. *Biochem. J.* **360**: 335–344.

Yap, A. S., Brieher, W. M., Pruschy, M. and Gumbiner, B. M., 1997. Lateral clustering of the adhesive ectodomain: a fundamental determinant of cadherin function. *Curr. Biol.* **7**: 308–315.

Yeaman, C., Grindstaff, K. K. and Nelson, W. J., 1999. New perspectives on mechanisms involved in generating epithelial cell polarity. *Physiol. Rev.* **79**: 73–98.

7

Using Bioprobes to follow Protein Dynamics in Living Cells

Mark R. Holt, Daniel Y. H. Soong, James Monypenny, Ian M. Dobbie, Daniel Zicha and **Graham A. Dunn**

Green fluorescent protein (GFP) from the jellyfish *Aequorea victoria*, its designer variants and new analogues from other marine organisms, are revolutionizing our understanding of many cell biological processes such as cell signalling, membrane- and cytoskeletal dynamics. Contributing to this progress are parallel developments in fluorescence microscopy enabling these bioprobes to reveal information about molecules that goes far beyond their mere localization, including data on protein–ligand interactions, post-translational modification, conformational state and molecular translocation. We discuss the application of bioprobes in the following techniques: fluorescence resonance energy transfer (FRET), fluorescence lifetime imaging (FLIM), total internal reflection microscopy (TIRF), fluorescence speckle microscopy (FSM) and the recently developed fluorescence localization after photobleaching (FLAP). These enable the imaging of many aspects of molecular dynamics directly in the living cell, thus complementing advances made with genetic and biochemical methods and extending our understanding of biological hierarchies from the molecular to the cellular level and beyond.

The green fluorescent protein (GFP) of the jellyfish *Aequorea victoria* has literally transformed the way we see cells. Extensive engineering of the native molecule has resulted in a versatile and rapidly expanding arsenal of multicoloured probes. Meanwhile, new living fluorophores are being

Cell Motility. Edited by Anne Ridley, Michelle Peckham and Peter Clark.
 ISBN 0 470 84872 3

discovered in other marine coelenterates and reef corals have provided a particularly rich hunting ground. The research impact of these living probes derives chiefly from the ability of cells of widely differing species to express them as fully functional fluorescent proteins fused to proteins under study. This feature has sparked the development of new methods, not only for locating specific proteins in living cells but also for investigating their interactions and dynamics. The focus of this review is to describe how these methods are being used to study molecular translocation, post-translational modifications and changes in protein activity and conformation. The techniques discussed include fluorescence (or Förster) resonance energy transfer (FRET), fluorescence lifetime imaging microscopy (FLIM), (frustrated) total internal reflection fluorescence microscopy (TIRF), fluorescence speckle microscopy (FSM), and the more recently developed fluorescence localization after photobleaching (FLAP), which will be discussed in most detail. Particular reference will be made to the use of these methods in studies on cytoskeletal dynamics and cellular signalling.

Fluorescence resonance energy transfer (FRET)

FRET is a quantum mechanical process that has been used as a biochemical tool to study protein structure, conformational change and protein–ligand interactions. Excitation of a donor fluorophore can give rise to non-radiative transfer of the absorbed energy to an adjacent acceptor fluorophore provided that the emission spectrum of the donor overlaps sufficiently with the excitation spectrum of the acceptor. This is not a simple process of emission and reabsorption but direct energy transfer dependent on coupling of the respective dipole moments. Since the probability of transfer is proportional to orientation and inversely proportional to the sixth power of the distance ($1/R^6$) between the donor and acceptor, FRET can be used to deduce the respective positional or rotational contexts of the two molecules. The positional constraints limit efficient FRET to distances of less than 10 nm and therefore sufficiently close to infer interaction. An example of FRET is shown in Figure 7.1A, which shows the biochemically well characterized interaction of paxillin with vinculin occurring in the focal adhesions of live fibroblasts (Ballestrem and Geiger, unpublished).

Two different approaches to obtain FRET using GFP and variants have been followed. The first uses cells expressing chimaeric GFP that are then microinjected with protein, e.g., a specific primary antibody, or peptide, conjugated with an acceptor dye, usually the red AlexaFluor-546. The second approach relies on multicoloured variants of GFP. A suitable GFP-variant pair for FRET, and by far the most widely used (Pollok and Heim, 1999), is the combination of cyan and yellow fluorescent proteins (CFP and YFP). CFP

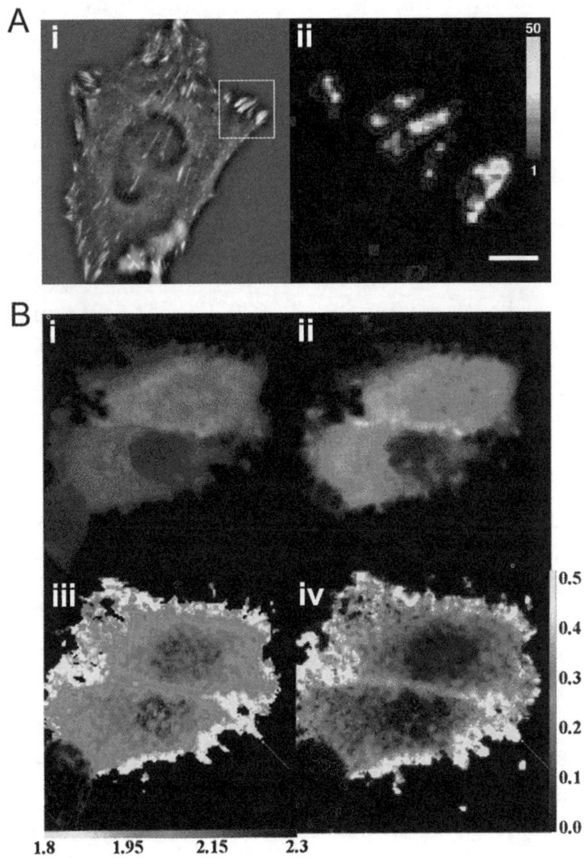

Figure 7.1 Examples of FRET and FLIM. (A) Ballestrem, C., and Geiger, B., unpublished material: FRET demonstrates interaction of paxillin and vinculin in live cells. NIH3T3 cells were co-transfected with cDNA encoding CFP-paxillin and YFP-vinculin. Image shows (i) FRET in focal adhesions at the cell periphery and (ii) a spectrum colour scale representation of the region highlighted in (i). Red corresponds to high FRET efficiency, blue corresponds to low FRET efficiency. FRET values were calculated as previously described (*J. Cell Sci.* **108**, 1051–1062. 1995). Bar, 2 μm. (B) Edme *et al.*, unpublished material: Spatially variant interaction between ezrin and PKCα V1V3 shown by FRET in multi-photon FLIM measurements. MCF-7 cells were dually transfected with both a GFP-PKCα (aa1-337) and a VSVG-tagged ezrin construct. After 15 h, cells were fixed in 4% PFA for 15 min at room temperature. Cells were permeabilized with 0.1% Triton-X100 and then stained with a Cy3-labelled anti-VSVG antibody. The GFP-intensity images were acquired and fluorescence GFP-lifetimes in cells were determined by multi-photon FLIM. The fluorescent images of the donor (GFP-PKCα aa1-337) (i) and the acceptor (anti-VSVG-Cy3) (ii) are shown. Both ezrin-positive cells exhibit a shortening of average GFP-fluorescence lifetimes in their peripheries from 2.3 ns to 1.8 ns (iii). The population map of interacting species is shown in (iv). (A colour reproduction of this figure can be found in the colour plate section)

acts as the donor molecule and YFP as the acceptor, the coupling between them being particularly efficient. Recent publications have detailed the development of elegant probes for following cell signalling based on these two approaches. Specific examples are probes for visualizing activation of Rho-family GTPases and tyrosine phosphorylation.

Activation of Rho-family GTPases involves conversion from a GDP-bound form to a GTP-bound one. Activated GTPases exert effects through binding to specific effector proteins (for a historical account of Rho family signalling see Ridley, 2001). These interactions have been visualized using FRET to reveal spatial and temporal aspects of Rac GTPase activation in growth factor-stimulated cells (Kraynov *et al.*, 2000), in neutrophils undergoing chemotaxis (Gardiner *et al.*, 2002), and in integrin-stimulated cells (Del Pozo *et al.*, 2002). In all cases, the donor is GFP-Rac, whereas the acceptor is the effector domain of PAK (p21-binding domain or PBD) conjugated with AlexaFluor-546. The FRET signal is thought to be proportional to the amount of GTP bound to Rac. Chemoattractant, applied by the micropipette method, resulted in a rapid activation, as judged by an increased FRET signal (Gardiner *et al.*, 2002). These data suggested that, as expected, Rac was involved in lamellipodial extension. Surprisingly, it appeared that Rac was also activated in the retracting tail, suggesting that Rac played a role in tail retraction.

Ting *et al.* (2001) have used multicoloured GFP variants to measure tyrosine phosphorylation in live cells. Here, reporters consisted of the contiguous fusion of CFP, a phosphotyrosine-binding domain, a consensus substrate for the kinase under observation, and YFP. Phosphorylation of the substrate by a specific tyrosine kinase enabled an intramolecular association with the phosphotyrosine-binding domain. This resulted in the close apposition of, and FRET between, the CFP and YFP modules. The FRET signal is thought to be proportional to the extent of tyrosine phosphorylation. One such reporter was used to provide evidence for PDGF-dependent activation of the tyrosine kinase Abl in the regulation of cell morphology (membrane ruffling) with a possible link to paxillin via the adaptor protein Crk.

Fluorescence lifetime imaging microscopy (FLIM)

A number of techniques have been developed that permit the analysis of FRET interactions such as monitoring sensitized emission by the acceptor fluorophore (as described above), monitoring donor fluorescence intensity following acceptor photobleaching, and the fluorescence lifetime imaging of the donor fluorophore. Techniques that observe the sensitized emission of the acceptor molecule upon donor excitation are not independent of fluorophore concentrations and are prone to complications due to channel cross-talk. Dequenching of donor fluorescence through the photobleaching of the acceptor population provides a more quantitative approach to FRET analysis as pre-bleach acceptor intensities can be compared with post-bleach values. Therefore, where possible, it is desirable to combine these two techniques as each provides an independent evaluation of the FRET process. An excellent review and comparison of these and other techniques is provided elsewhere (Wouters *et al.*, 2001).

FLIM provides one method that enables a quantitative measure of FRET. The fluorescence lifetime of a fluorophore is defined as the average duration the molecule remains in the excited state following photon absorption (Bastiaens and Squire, 1999). The lifetime of a donor fluorophore is a useful parameter to measure because it is sensitive to local environmental changes such as FRET interactions whilst insensitive to fluorophore concentration, a factor that can complicate measurements based on absolute intensities. Two techniques have been developed that enable fluorescence lifetimes to be measured: frequency domain and time domain FLIM. In frequency domain FLIM the continuous hi-frequency modulation of the excitation intensity is used to modulate the intensity of fluorescence emission. The frequency of intensity modulation is the same for both excitation and emission wavelengths but differences exist between their phase and relative modulation depth. These differences are determined by the fluorescence lifetime of the fluorophore and can be used as independent parameters for its derivation. By modulating the sensitivity/gain of the detector at the same frequency as the light intensity modulation, phase-sensitive imaging is enabled allowing the differences in phase and modulation depth to be calculated (Bastiaens and Squire, 1999; Squire and Bastiaens, 1999). Alternatively, the time domain technique uses pulsed excitation of the sample. The pulsed excitation of a population of a single species of fluorophore results in fluorescence emission that decays exponentially with time. The lifetime of the fluorophore is derived from this decay curve and is defined as $1/e$, which is the time taken for the intensity of fluorescence emission to decay to approximately 37% of its initial value (Bastiaens and Squire, 1999; Benny Lee *et al.*, 2001).

One of the most powerful features of FRET/FLIM analysis is that it enables biochemical processes to be imaged in individual living cells. Any spatial constraints that affect the subcellular localization of a protein should apply and therefore the technique is ideal for the study of spatio-temporal signalling relationships. Furthermore, single cell imaging provides the opportunity to study the signalling heterogeneity that can exist within a population of cells, something that is impossible to achieve using conventional biochemical techniques.

The use of FLIM in the analysis of ErbB1 receptor activation dynamics provides an elegant example of how microscopy-based analysis of biochemical interactions has revealed a spatio-temporal pattern of signalling events at the single-cell level (Verveer *et al.*, 2000). FLIM-based measurements of FRET interactions between ErbB1-GFP donor and a Cy3-conjugated phosphotyrosine-specific acceptor antibody were used to measure the activation state of the receptor. The antibody only recognized the active phosphorylated form of the receptor and thus only the active ErbB1-GFP population could undergo FRET. The study revealed rapid lateral propagation of receptor activation within the plasma membrane following only localized stimulation with EGF-coated beads. This finding suggests that a relatively small and localized pool of ligand-occupied receptor can initiate a transactivation cascade that ultimately leads to the activation of the remaining ligand-free receptor population within the plasma membrane.

A similar approach has been used to image protein kinase Cα (PKCα) activation upon TPA treatment (Ng *et al.*, 1999). In this study a Cy3.5-labelled acceptor antibody specific only to the active, phosphorylated form of the protein was microinjected into living cells and used as a dynamic indicator of GFP-PKCα activity. An example of the data produced from FLIM experiments can be found in Figure 7.1B. Here, GFP-PKCα (V1V3 domains only) and a Cy3-labelled anti-VSVG antibody (to locate VSVG-tagged ezrin) showed that interaction between PKCα and ezrin only occurs at the cell periphery (Edme *et al.*, unpublished).

More recently, FLIM has been used to reveal interactions between the CD44 hyaluronan receptor and the actin binding protein ezrin in a PKC-dependent manner (Legg *et al.*, 2002). Activation of PKC with phorbol ester resulted in the loss of a basal level of FRET between GFP-CD44 and ezrin-VSVG indirectly labelled with Cy3-conjugated anti-VSVG antibody. The authors concluded that PKC-mediated disruption of CD44-ezrin interaction is necessary for directional cell motility.

Total internal reflection fluorescence (TIRF)/ evanescent wave microscopy

TIRF, also known as evanescent wave microscopy is not a new technique but recent developments have increased its range of application. While TIRF has traditionally been used for studying cell surface features, such as focal adhesions or endo/exocytosis, it is also useful for the study of single molecule fluorescence.

When light travelling in a high refractive index medium (such as a glass coverslip) strikes an interface with a material of lower refractive index (such as water or culture medium) above a critical angle of incidence, the result is known as total internal reflection. However, some energy penetrates the distal medium as an evanescent wave that can excite fluorescent molecules. The light intensity penetrating into the medium falls off exponentially with distance from the coverslip. The exact depth of penetration depends on the angle of incidence, the refractive indices of the two media and the wavelength of the light. In practice, the propagation distance into the distal medium (the distance over which the intensity falls by $1/e$) can be less than 100 nm (for a more detailed description see Toomre and Manstein, 2001). The emitted fluorescent signal can be detected either from below through the same objective used to launch the evanescent wave or from above using an immersion objective.

The distinct advantage of TIRF is the very thin optical sectioning achieved (<100 nm), which is significantly better than on confocal systems (>250 nm). However, TIRF is not a replacement for confocal or other fluorescence imaging techniques since it is not able to penetrate into a sample and 3D imaging is not possible. In fact, TIRF is often combined with conventional wide-field epi-fluorescence microscopy in order to relate surface effects to internal cellular structures (Merrifield *et al.*, 2000).

The very thin optical sectioning is critical to both applications of TIRF. For viewing ventral cell surface processes, the exponential decay of illumination intensity restricts fluorescent emission to the cell surface. For single molecule fluorescence studies, the same limited illumination depth dramatically reduces background fluorescence, increasing the signal-to-noise ratio and allowing discrimination of single fluorescent molecules.

Two elegant examples of TIRF applications that show its versatility involve imaging the invagination of clathrin-coated pits and measuring the oligomerization of E-cadherin on the free cell surface. In the first example a fusion protein, clathrin-dsRed (a red-coloured fluorescent protein), is imaged at the ventral membrane using TIRF. The associated proteins, dynamin and actin, are also imaged using either TIRF or conventional epi-fluorescence enabling the study of the dynamics of the invagination process (Merrifield *et al.*, 2002 and Figure 7.2A and B). In the second example E-cadherin-GFP fusion proteins were introduced into cell lines lacking E-cadherin. The expressed fusion protein was detected in the ventral cell membrane by TIRF and the intensity of fluorescent spots was measured. It was found that the intensity was quantized, one quantum being the intensity due to a single GFP molecule. The single molecule detection sensitivity was further demonstrated by single step photobleaching from the single quantum intensity level. The intensity of individual spots was then used to assess the oligomerization state of the E-cadherin complexes showing that E-cadherin oligomerizes before binding to its partner in cell–cell junctions (Iino *et al.*, 2001).

Finally, Justin Molloy's group (unpublished data) have used TIRF to follow single molecules of GFP fused to the PH domain of myosin X. This

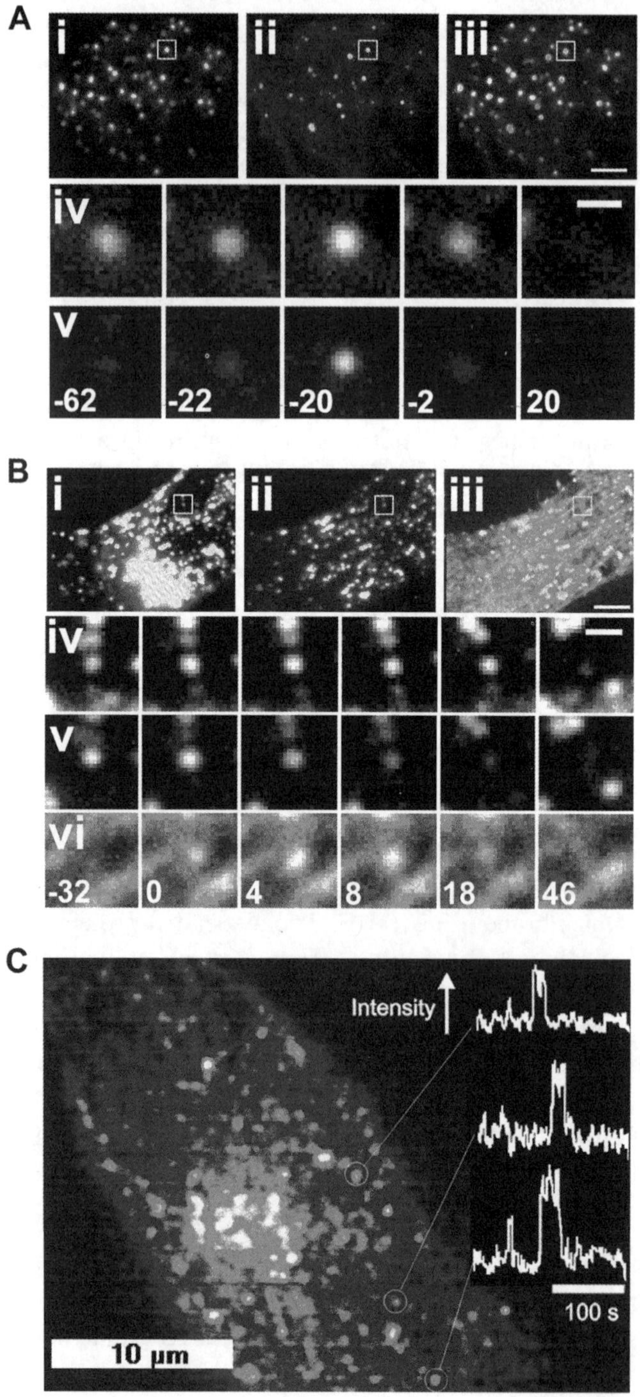
A
i
ii
iii
iv
v
-62
-22
-20
-2
20
B
i
ii
iii
iv
v
vi
-32
0
4
8
18
46
C
Intensity
100 s
10 µm

particular PH domain binds specifically to phosphatidylinositol (4,5)-bisphosphate (PIP_2). Surprisingly, the spatial localization of PIP_2 in mouse myoblasts appeared to be more stable than expected (Figure 7.2C).

Fluorescence speckle microscopy (FSM)

Conventional fluorescence microscopy works on the basis that a greater fluorescence signal results in a less noisy image with a higher contrast. While this is true when looking at stationary structures or fixed cells, it is not always an advantage when imaging a moving structure or dynamic process. When studying cytoskeletal proteins, for instance, it is not possible to see the internal dynamics of a particular structure if it is uniformly labelled due to a high density of fluorescent molecules. FSM was developed (Waterman-Storer *et al.*, 1998) in response to a phenomenon observed in an earlier study of microtubule retrograde flow (Waterman-Storer and Salmon, 1997). Using FSM, the underlying dynamic organization of labelled proteins can be observed. Instead of a static homogeneously stained structure, moving speckles are seen, indicative of molecular translocation. FSM relies on the incorporation of such small amounts of fluorophore-conjugated protein into the biopolymers of the cytoskeleton that the fluorescence can only be seen when chance causes the incorporation of enough labelled molecules in one small area to register a charge on a CCD camera (Waterman-Storer and Salmon, 1998). The principle is that only relatively immobile molecules will show as speckles since freely diffusing molecules move too fast to register during a long exposure time ($\sim$0.5–2 s). Fluorophores can be conjugated chemically, e.g., AlexaFluor or Cy dyes, or genetically, i.e., GFP and variants.

Figure 7.2 (*opposite*) Observation of plasma membrane events using TIRF microscopy. (A) Merrifield, C. *et al.*, unpublished material: A Swiss 3T3 fibroblast expressing (i) clathrin-DsRed, (ii) dynamin 1-GFP, (iii) and as they appear in an overlay. The lower panels (iv, v) show a time-resolved montage of an example event (i–ii, white box in upper panel) showing the transient recruitment of dynamin 1-GFP to a clathrin-coated pit. Numerals refer to time in seconds relative to the recorded event (t = 0). Scale bars (iii) 5 μm, (iv) 1 μm. (B) Merrifield, C. *et al.*, unpublished material: A Swiss 3T3 fibroblast stably expressing GFP-β-actin and transiently expressing clathrin-DsRed: (i) clathrin-DsRed, taken with epifluorescent illumination, (ii) clathrin-DsRed, taken with TIRF illumination, (iii) overlay of clathrin-DsRed and GFP-β-actin taken with TIRF illumination. The lower panels (iv, v, vi) show a time resolved montage of the coated pit indicated in (i–iii, white square). Numerals refer to time in seconds relative to the recorded event (t = 0). Scale bars (iii) 5 μm, (iv) 1 μm. (C) Mashanov, G. and Molloy, J., unpublished material: A mouse myoblast transiently transfected with eGFP fused to the pleckstrin homology domain of Myosin-X. This image of single fluorophores on the cell membrane was obtained by time-lapse TIRF microscopy. Note the intensity plots revealing the lifetime of these spots. Scale bars for space and time are 10 μm and 100 s respectively. (A colour reproduction of this figure can be found in the colour plate section)

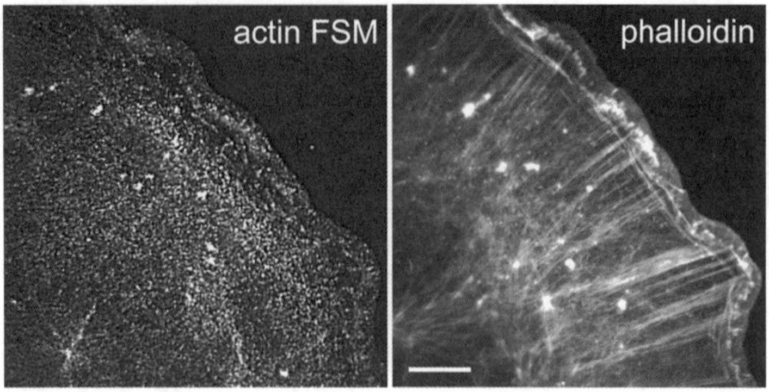

Figure 7.3 Fluorescence speckle microscopy with labelled actin. A newt lung epithelial cell that was injected with a small amount of X-rhodamine-labelled actin and imaged with a high magnification, high resolution digital epi-fluorescence microscope system. In the FSM image, the actin in the lamella and lamellipodium appears as a relatively even distribution of fluorescent speckles, whose motion, appearance and disappearance can be tracked over time to reveal molecular dynamics of the actin cytoskeleton. The image on the right shows the same cell after fixation and staining with alexa-488 phalloidin to reveal in greater detail the structural organization of the actin. Scale bar is 10 μm

The amount of labelled protein used is generally between 0.1–1% of total cellular protein (Waterman-Storer and Salmon, 1999). Two incidental advantages of this technique are, therefore, the low quantities of biological material required and the lack of overexpression artefacts. An example of speckles of fluorescently-labelled actin observed in epithelial cells is shown in Figure 7.3 (Waterman-Storer, unpublished data).

When viewing time-lapse images from FSM, the internal dynamics of structures such as microtubules, actin stress fibres and the dendritic actin network in the lamella become apparent (Waterman-Storer and Salmon, 1997; Waterman-Storer *et al.*, 1999). Speckles can be seen to make up structures and move in unison in specific identifiable directions. These images can be used to calculate retrograde flow velocity of filaments and microtubules in different parts of the cell.

Recent technological advances have allowed progression of the FSM technique to a level where single fluorophores, and therefore single molecules of protein, can be detected (Watanabe and Mitchison, 2002). This eliminated the previous requirement for fluorophores to aggregate although signal integration during long exposures still ensures that only molecules that are relatively immobilized, for example by polymerization, can give rise to speckles. This is an additional advantage of the single-molecule method since the lifetime of a speckle is a direct indication of the time that the molecule remains polymerized or otherwise anchored to a quasi-stable structure. In the pilot study, single-molecule speckle fluorescence has revealed valuable information about the turnover of actin in dendritic lamellipodial networks and thus increased our knowledge of cytoskeletal dynamics at the leading edge of motile cells.

Fluorescence localization after photobleaching (FLAP)

Conventional fluorescence microscopy can reveal non-steady-state processes such as the distribution of a molecular species within a cell. In the previous two sections we have seen how, by detecting single molecules or small groups of molecules, we can extend our knowledge of dynamics to reveal the flow of molecules through a structure in the steady state. Many fundamental cell biological processes, such as the turnover of actin during cell motility, have a large steady-state component. For example, a smoothly gliding cell such as a fish keratocyte would reveal little internal redistribution of actin by conventional fluorescence methods.

Although single-molecule methods can reveal some of the underlying steady-state processes, they can only track the fate of relatively immobilized molecules. A more general approach requires the ability to label specific molecules at a given locality and subsequently follow their fate whether they are anchored, polymerized or freely diffusing. Traditionally, this has been done by photoactivation of fluorescence (PAF), which requires the construction of caged fluorescent probes that can be locally activated to fluoresce by uncaging them using ultraviolet light. Unfortunately, the PAF approach is not applicable to biofluorescent proteins expressed by the cells but GFP variants have been produced recently that can be directly photoactivated in mammalian cells without the requirement for harmful UV light (Patterson and Lippincott-Schwartz, 2002; Chudakov *et al.*, 2003). In one particular GFP variant, a threonine at position 203 has been mutated to a histidine residue, a change that results in poor emission when excited at 488 nm. However, following intense irradiation with 413 nm light, GFP emission increases 100-fold. This is due to conversion of non-fluorescent phenolic acid into the fluorescent anionic derivative, phenolate. This results in photo-activated molecules that can be tracked by conventional means. The recently developed FLAP technique also enables effective photolabelling of specific molecules using standard fluorescent probes (Dunn *et al.*, 2002). The FLAP method retains important advantages of the photoactivatable GFP – that it can be used with fluorescent fusion proteins expressed by the cells and that no harmful UV radiation is required – but it has the additional advantage that ratiometry between labelled and unlabelled molecules of the same species is possible (Zicha *et al.*, 2003).

In FLAP, the conventional photobleaching process is used to label a specific pool of fluorescent molecules. While fluorescence recovery after photo-bleaching (FRAP) has long been used to study the repopulation of a bleached region by unbleached molecules, it cannot follow the fate of the bleached molecules themselves. In contrast, FLAP does allow the bleached molecules to be tracked. This is facilitated by comparison of the distribution of remaining unbleached molecules with that of another reference fluorophore. For example, to visualize the steady-state dynamics of actin turnover, YFP-actin and CFP-actin fusion proteins were co-expressed in the same cell (Dunn *et al.*, 2002). The YFP-actin was then bleached in a predefined region of the cell. The CFP was not photobleached and acted as a reference for the YFP-actin, since statistically the two molecules would have co-localized under the observed resolution. Subtraction of the YFP signal from the CFP signal resulted in a difference image reflecting the position of the bleached YFP molecules. When this approach was used to observe actin dynamics in the membrane ruffles at the leading edge of a cell, labelled actin was seen rapidly to pervade the ruffle system (Dunn *et al.*, 2002). In contrast, actin was found to be extremely stable in the perinuclear region of the same cell. The approach has also been used to observe actin dynamics in actin bundles or stress fibres and in the cytokinetic ring formed during the later stages of mitosis. In both cases these actin bundle structures were surprisingly unstable and showed rapid interchange of actin with surrounding structures – a process that we believe to occur by rapid depolymerization and repolymerization.

FLAP: Image acquisition and image processing

In experiments that analyse actin dynamics, transformed rat fibroblasts (a gift from Dr Pavel Vesely) are microinjected with DNA constructs encoding yellow and cyan N-terminal fluorescent proteins fused to β-actin (the β-actin cDNA was a gift from Dr John Copeland) (Figure 7.4A). Vectors encoding pECFP-C1 and pEYFP-C1 were obtained from ClonTech. Cells are visualized using a Zeiss LSM 510 laser scanning confocal microscope and a 63 × Plan-Apochromat NA 1.4 PH 3 oil objective. Temperature is stabilized using a 37°C temperature-controlled hood, since minor changes can cause focus drifts that result in artefacts due to chromatic aberration of the objective. The CFP and YFP channels use the 458 nm and 514 nm argon lines respectively for excitation. Simultaneous phase contrast is collected with the CFP channel, whilst simultaneous interference reflection is collected with the YFP channel to enable visualization of focal adhesions (Figure 7.4B and 7.4C). The two emission channels are split by a 545 nm dichroic mirror followed by a bandpass 475–525 nm filter (CFP) and a longpass 530 nm filter (YFP). The pinholes are set to give 3 μm depth of focus. Each image line is scanned as rapidly as possible, first by the 514 nm laser line with the YFP channel detector and interference reflection detector active followed by the 458 nm laser line with the CFP detector and the phase contrast detector active.

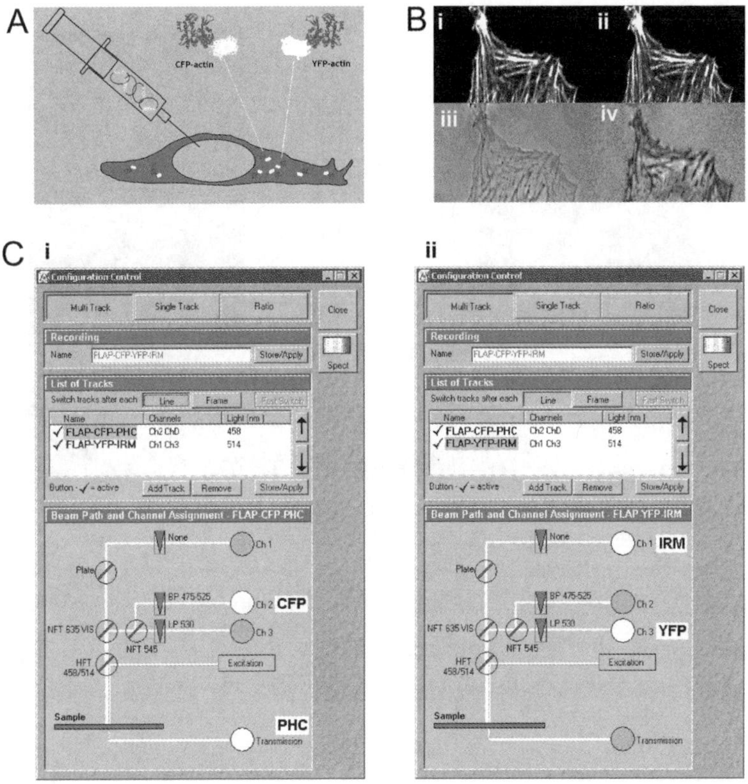

Figure 7.4 Basic FLAP methodology using CFP and YFP-conjugated proteins. (A) Cartoon depicting nuclear microinjection of plasmids encoding CFP- and YFP-conjugated actin, which are expressed in the cytoplasm. (B) A collection of images of the same fibroblast showing (i) CFP-actin localization, (ii) YFP-actin localization, (iii) phase contrast, (iv) interference reflection microscopy (IRM). Note the co-localization of CFP- and YFP-actin in stress fibres and how these terminate in the IRM dark patches that represent focal adhesions. (C) Screen capture image from the Zeiss LSM 510 software showing filters and mirror set-ups for collecting (i) CFP emission and phase contrast images, and (ii) YFP emission and interference reflection images

After a suitable cell has been found, the image is rotated to fit into a rectangular region of interest that defines the scan area. If a rectangular strip is to be bleached across the cell, then the strip is defined horizontally in order to reduce the number of scan lines in the bleach region and hence the bleaching time. The gain for each channel is then adjusted so that the CFP and YFP fluorescent intensities are matched without saturation and the offsets are set so that backgrounds have zero intensity. The time-lapse mode is then used to collect an image sequence. At least two frames are collected before bleaching. Bleaching consists of 50 scans using the 514 nm laser line at maximum power for 2–3 s. Time-lapse image recording is resumed

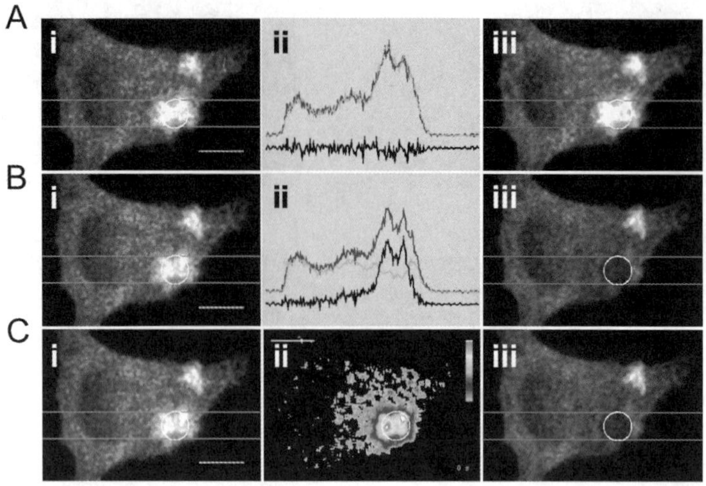

Figure 7.5 A typical FLAP experiment following actin dynamics. (A) A pre-bleach fibroblast showing (i) CFP-actin expression, (iii) YFP-actin expression and (ii) integrated intensity traces (cyan line is CFP, yellow line is YFP) through the cell encompassing the proposed bleach region. The red line corresponds to the difference in intensities of the YFP and CFP. Since this is a pre-bleach image the difference signal will be zero with noise artefacts superimposed over it. (B) The same cell after bleaching (the bleach region is defined by the white circle). Note the red difference signal now registering a significant difference in integrated CFP and YFP intensities. Descriptions for parts (i–iii) are as in (A) in this legend. (C) The same as (B) except that (ii) the integrated difference signal has been replaced with a pseudocolour representation (Hall LUT) of the difference signal for the whole cell. Red regions correspond to a high FLAP signal (large difference between CFP and YFP) whereas blue regions correspond to low FLAP or difference signals. (The reader is referred to the colour montage.) (A colour reproduction of this figure can be found in the colour plate section)

immediately afterwards. Data from a typical FLAP experiment are shown in Figure 7.5 (from Dunn *et al.*, 2002).

Simple image processing is required to produce the FLAP signal. Using Mathematica 4.2, the pre-bleach CFP and YFP images are smoothed using a 3×3 pixel median filter followed by calculation of the offsets required to bring both image backgrounds to a mean intensity of zero. User-defined algorithms are then applied to calculate a variable that when applied to one image produces a precise match of the integrated intensities of the two images. To calculate the FLAP image, the YFP image is simply subtracted from the CFP reference image (Figure 7.5). In the case of the pre-bleach images, the FLAP image should have a uniform intensity of zero, although in practice the images are noisy when viewed with a pseudocolour look-up table (Hall LUT). This noise can be reduced satisfactorily with the minimum of thresholding. Non-uniformity of the pre-bleach images, particularly around the cell edges, indicates a problem often due to chromatic aberration of the objective when images are captured far away from the optic axis. Factors defined for offsets and thresholding in the pre-bleach images are applied to all images in the time-lapse sequence. For long sequences, estimates for the rate of fade of the images are required. These can be obtained by integration of the CFP intensity over the whole cell and application of a second factor to both images to compensate for this. We have found it advantageous to match the rates of fade of YFP and CFP by careful adjustment of the laser excitation intensities.

FLAP: Some points to consider

Besides chromatic aberration, two possible causes of non-uniformity of prebleach FLAP images are (1) the two fluorophores are not co-localized and (2) FRET is occurring. We have never experienced the first case but there is often evidence for FRET in dense regions of F-actin. As discussed previously, CFP and YFP constitute an effective FRET pair (Pollok and Heim, 1999) and one effect of FRET occurring is to reduce the donor (CFP) signal when exciting the CFP channel. The CFP detector does not register the increased YFP emission caused by FRET and no FRET occurs when the YFP channel is excited. The net result is that the FLAP image shows less intensity in regions of dense actin though the effect is seldom more than one or two grey levels and can be ignored. Moreover, the effect is reduced in regions where the YFP has been photobleached and so the strength of the FLAP signal is hardly affected by FRET. It is likely that the FRET efficiency is always going to be low in the system described above since the statistical probability of two labelled actin molecules being within 10 nm of one another is not high.

Theoretically, FLAP could be simplified if the actin molecule carried both fluorophores, one on the N-terminus and the other on the C-terminus. In this case, co-localization of CFP and YFP would be perfect, not a statistical probability, eliminating probabilistic noise in the FLAP signal. Early indications suggest that the additional noise due to statistical co-localization is small compared to instrumental noise.

Significant levels of FRET would be expected in this simplified FLAP since the N- and C-termini of actin are adjacent to one another and this could act to enhance the FLAP signal. CFP emission would be reduced except in cases where CFP was fused to the same molecule as a bleached YFP (remember that the unbleached YFP signal would be unaltered). This would increase the FLAP signal. However, complications arise since incorporation of G-actin into F-actin results in a marked conformational change. This would be expected to lead to repositioning of the CFP-YFP pair with respect to one another, a consequent change in FRET efficiency and therefore an altered difference image. The FLAP signal would thus represent two variables: number of bleached molecules and polymerization state of the actin. The combination of variables would make the difference image very difficult to interpret.

We believe that FLAP and derivative methods will be extremely useful in observing and analysing the complex spatio-temporal dynamics of the cytoskeleton during such processes as cell adhesion, cell motility and cell signalling.

Concluding remarks

The techniques based on bio-fluorescent probes presented here are being used by an expanding number of scientists in biological research. The techniques provide important information on dynamic localization with high sensitivity covering steady state processes and specific protein–protein interactions. This information complements advances in molecular genetics and biochemistry and generates data essential for computer simulation models whichever concept they may be based on: stochastic, systems biology, differential equations or non-linear thermodynamics. These models are having an increasing impact on our understanding of biological systems at different levels: molecular, supra-molecular and cellular.

Acknowledgements

We would like to thank the following authors for allowing us to reproduce some of their fine work (some of it unpublished): Christophe Ballestrem and Benny Geiger (FRET), Natacha Edme, Simon Ameer-Beg, Boris Vojnovic and Tony Ng (FLIM), Christien Merrifield and co-workers (TIRF), Gregory Mashanov and Justin Molloy (TIRF), and Clare Waterman-Storer (FSM). FLAP experiments were performed with the help of Colin Gray (Cancer Research UK), Peter Jordan (Cancer Research UK) and Paul Fraylich (MRC). Supported by a MRC programme grant (G.A.D.).

References

Bastiaens, P. I. and Squire, A., 1999. Fluorescence lifetime imaging microscopy: spatial resolution of biochemical processes in the cell. *Trends Cell Biol.* **9**: 48–52.

Benny Lee, K. C., Siegel, J., Webb, S. E. D., Lévêque-Fort, S., *et al.*, 2001. Application of the stretched exponential function to fluorescence lifetime imaging. *Biophys. J.* **81**: 1265–1274.

Chudakov D. M., Belousov V. V., Zaraisky A. G., Novoselov V. V., *et al.*, 2003. Kindling fluorescent proteins for precise in vivo photolabeling. *Nature Biotechnol.* **21**: 191–194.

Del Pozo M. A., Kiosses W. B., Alderson N. B., Meller N., *et al.*, 2002. Integrins regulate GTP-Rac localized effector interactions through dissociation of Rho-GDI. *Nature Cell Biol.* **4**: 232–239.

Dunn, G. A., Dobbie, I. M., Monypenny, J., Holt, M. R. and Zicha, D., 2002. Fluorescence localisation after photobleaching (FLAP): a new method for studying protein dynamics in living cells. *J. Microscopy* **205**: 109–112.

Gardiner E. M., Pestonjamasp K. N., Bohl B. P., Chamberlain C., *et al.*, 2002. Spatial and temporal analysis of Rac activation during live neutrophil chemotaxis. *Curr. Biol.* **12**: 2029–2034.

Iino, R., Koyama, I. and Kusumi, A., 2001. Single molecule imaging of green fluorescent proteins in living cells: E-cadherin forms oligomers on the free cell surface. *Biophys. J.* **80**: 2667–2677.

Kraynov V. S., Chamberlain C., Bokoch G. M., Schwartz M. A., *et al.*, 2000. Localized Rac activation dynamics visualized in living cells. *Science* **290**: 333–337.

Legg, J. W., Lewis, C. A., Parsons, M., Ng, T. and Isacke, C. M., 2002. A novel PKC-regulated mechanism controls CD44-ezrin association and directional cell motility. *Nature Cell Biol.* **4**: 399–407.

Merrifield C. J., Feldman, M. E., Wan, L. and Almers, W., 2002. Imaging actin and dynamin recruitment during invagination of single clathrin-coated pits. *Nature Cell Biol.* **9**: 691–698.

Ng, T., Squire, A., Hansra, G., Bornancin, F., *et al.*, 1999. Imaging protein kinase Cα activation in cells. *Science* **283**: 2085–2089.

Patterson G. H. and Lippincott-Schwartz, J., 2002. A photoactivatable GFP for selective photolabeling of proteins and cells. *Science* **297**: 1873–1877.

Pollok, B. A. and Heim, R., 1999. Using GFP in FRET-based applications. *Trends Cell Biol.* **9**: 57–60.

Ridley A. J., 2001. Rho family proteins: coordinating cell responses. *Trends Cell Biol.* **11**: 471–477.

Squire, A. and Bastiaens, P. I., 1999. Three dimensional image restoration in fluorescence lifetime imaging microscopy. *J. Microscopy* **193**: 36–49.

Ting A. Y., Kain K. H., Klemke R. L. and Tsien R. Y., 2001. Genetically encoded fluorescent reporters of protein tyrosine kinase activities in living cells. *Proc. Natl. Acad. Sci. USA* **98**: 15003–15008.

Toomre, D. and Manstein, D. J., 2001. Lighting up the cell surface with evanescent wave microscopy. *Trends Cell Biol.* **11**: 298–303.

Verveer P. J., Wouters F. S., Reynolds A. R. and Bastiaens P. I., 2000. Quantitative imaging of lateral ErbB1 receptor signal propagation in the plasma membrane. *Science* **290**: 1567–1570.

Watanabe, N. and Mitchison, T. J., 2002. Single-molecule speckle analysis of actin filament turnover in lamellipodia. *Science* **295**: 1083–1086.

Waterman-Storer, C., Desai, A. and Salmon, E. D., 1999. Fluorescent speckle microscopy of spindle microtubule assembly and motility in living cells. *Methods Cell Biol.* **61**: 155–173.

Waterman-Storer, C. M. and Salmon, E. D., 1997. Actomyosin-based retrograde flow of microtubules in the lamella of migrating epithelial cells influences microtubule dynamic instability and turnover and is associated with microtubule breakage and treadmilling. *J. Cell Biol.* **139**: 417–434.

Waterman-Storer, C. M. and Salmon, E. D., 1998. How microtubules get fluorescent speckles. *Biophys. J.* **75**: 2059–2069.

Waterman-Storer, C. M. and Salmon, E. D., 1999. Fluorescent speckle microscopy of microtubules: how low can you go? *Faseb J.* **13**: S225–S230.

Waterman-Storer, C. M., Desai, A., Bulinski, J. C. and Salmon, E. D., 1998. Fluorescent speckle microscopy, a method to visualize the dynamics of protein assemblies in living cells. *Curr. Biol.* **8**: 1227–1230.

Wouters, F. S., Verveer, P. J. and Bastiaens, P. I., 2001. Imaging biochemistry inside cells. *Trends Cell Biol.* **11**: 203–211.

Zicha, D., Dobbie, I. M., Holt, M. R., Monypenny, J., *et al.*, 2003. Rapid actin transport during cell protrusion. *Science* **300**: 142–145.

8

Actin Filament Assembly: The Search for a Barbed End

Craig F. Stovold, Stewart J. Sharp and
Laura M. Machesky

Protrusion of the leading edge of a motile cell is thought to be largely driven by actin assembly near the plasma membrane. Cells initially extend filopodia, which are long thin processes composed of bundled actin or lamellipodia, which are thin sheets containing highly branched arrays of actin filaments. It is becoming clear that cells use a variety of mechanisms to initiate the assembly of actin-based structures, centring around the need for the creation of free actin filament ends for rapid polymerization. The Arp2/3 complex is a main source of regulation for the availability of free barbed ends in cells and is thought to nucleate branched actin filaments under the regulation of the Wiskott–Aldrich Syndrome Family (WASp family) proteins. Additionally, cells seem to be able to use other proteins, such as the formins in yeast, to nucleate unbranched filaments, leading to the possibility that formation of specific actin structures could be mediated by different sets of nucleating proteins. Finally, the Ena/VASP proteins modulate the balance between branching and growth of filaments by forming a complex that competes with capping protein at the free growing end of the filament. Together, a variety of actin modulating proteins are now thought to be responsible for the complexity of actin-based protrusion and we are starting to understand how they work together and in parallel to provide diversity in different cells under various conditions.

Cell Motility. Edited by Anne Ridley, Michelle Peckham and Peter Clark.
 ISBN 0 470 84872 3

Coordinated actin filament assembly and disassembly are required for cell motility, integrity and shape changes. Actin filaments have an inherent polarity, conferred by the asymmetry of the monomer, and the two ends of the filament are commonly termed the barbed (fast growing) and pointed (slower growing) ends. These terms arise due to the appearance of actin filaments decorated with the head of the motor protein myosin-II in electron micrographs. Due to the nucleotide (ATP) bound to actin monomers, and the orientation of barbed ends towards the plasma membrane, the barbed end is able to undergo rapid growth that drives pseudopod protrusion (reviewed by Pollard *et al.*, 2000). Interestingly, cellular concentrations of monomeric actin greater than 100 μM are not uncommon, e.g., 160 μM in the amoeba *Dictyostelium* and 230 μM in human platelets (Hartwig, 1992). This is surprising, as pure actin has a critical concentration of only 0.1–1 μM at its barbed end; clearly the cell has evolved a system to control the availability of actin monomers for polymerization. Two monomer-sequestering proteins have been identified, which are abundant enough in most cell types to sequester the majority of the monomeric actin: thymosin-β4 and profilin (Carlsson *et al.*, 1977; Cassimeris *et al.*, 1992; Safer *et al.*, 1990).

As well as controlling the availability of actin monomers, cells also regulate actin filament growth by using capping proteins that bind with nanomolar affinity to the barbed ends of a filament. If free, barbed ends would rapidly elongate, depleting the pool of actin monomers. Gelsolin and capping protein are two barbed end capping proteins; the later with a barbed end affinity of only 0.1 nM is able to cap the majority of barbed ends within a cell (Schafer *et al.*, 1996). It is clear therefore that as cells have developed binding proteins able to prevent 'nucleation' of unpolymerized actin and capping proteins that protect barbed ends, that the creation of free barbed ends is limiting for actin assembly in cells. Currently there are three proposed mechanisms for the generation of free barbed ends in cells.

Perhaps the most obvious process of generating free barbed ends for elongation is that of uncapping. Indeed it has been shown that both capping protein and gelsolin can be removed from filaments on interaction with polyphosphoinositides (Schafer *et al.*, 1996). Gelsolin is removed from filaments in activated human platelets in response to the production of phosphatidylinositol(4,5)bisphosphate (PI(4,5)P_2), promoting a burst of actin assembly (Hartwig *et al.*, 1995).

Secondly, the severing of established actin filaments also provides a source of free barbed ends. Again the best example of this comes from platelets whose activation results in an increase in intracellular Ca^{2+} levels which in turn activates gelsolin to sever actin filaments resulting in actin polymerization (Hartwig *et al.*, 1995). Cofilin is another severing protein that provides free barbed ends in response to signals such as epidermal growth factor (EGF) which induce lamellipodial protrusion (Ichetovkin *et al.*, 2002).

Finally, *de novo* actin polymerization from a 'nucleus' of three associated actin monomers could provide new filament templates (Pollard *et al.*, 2000). The formation of actin trimers is kinetically unfavourable though, and would need a factor to accelerate and stabilize it. For many years nucleating factors were postulated but never found. However, in 1994 the discovery of a multiprotein complex, the Arp2/3 complex, and subsequent finding that this complex promoted actin-based motility of *Listeria monocytogenes* bacteria, has had a profound effect on the understanding of *de novo* actin polymerization (Machesky *et al.*, 1994; Welch *et al.*, 1997a). Indeed it is now apparent that the Arp2/3 complex is a major 'nucleator' of new actin filaments.

The Arp2/3 complex

Together, the actin-related proteins Arp2 and Arp3 and five unique proteins p41-Arc/ARPC1 (Arp complex/accepted human genome nomenclature = ARPC), p34-Arc/ARPC2, p21-Arc/ARPC3, p20-Arc/ARPC4 and p16-Arc/ARPC5 form the Arp2/3 complex. Each subunit has homologues throughout eukaryotes, including *Drosophila melanogaster*, *Caenorhabditis elegans*, *Acanthamoeba castellanii*, *Saccharomyces cerevisiae* and *Homo sapiens*. The structures of Arp2 and Arp3 greatly resemble actin, with the exception that their central clefts are spread open approximately 15° further and that some of their surface loops differ (Robinson *et al.*, 2001). The other five subunits are novel and do not share any significant homology with previously identified proteins. However, ARPC1 contains seven WD (tryptophan and aspartate) repeat motifs that fold to give a β-propeller conformation characteristic of protein–protein interaction domains (Robinson *et al.*, 2001; Westerberg *et al.*, 2001).

In mammals, there appear to be isoforms of several of the Arp2/3 complex subunits. Humans express two genes for ARPC1, p41-Arc and SOP2Hs. SOP2Hs functionally complements loss of the 41-kDa subunit (SOP2), in the fission yeast *Schizosaccharomyces pombe* as does p41-Arc (Balasubramanian *et al.*, 1996). Recent research also indicates the expression of two forms of Arp3 (Arp3 and Arp3β) and ARPC5 (p16A and p16B) (Jay *et al.*, 2000; Millard *et al.*, 2003). As yet the functional relevance and tissue expression pattern of most of these different isoforms is unknown, but some examples are emerging. In mouse, p16A and p16B show different expression patterns, with p16A being highly abundant in haematopoietic cells and p16B being more abundant in brain (Millard *et al.*, 2003). It also appears based on peptide sequences, that the Arp2/3 complex from human platelets contains p41-Arc but not SOP2Hs and p16A but not p16B (Machesky *et al.*, 1997).

The first evidence that the Arp2/3 complex was able to stimulate de novo actin polymerization *in vivo* came in 1997 through studies with the motile intracellular pathogen *L. monocytogenes* (Welch *et al.*, 1997b). In these experiments Arp2/3 complex isolated from human platelets is sufficient for the induction of actin polymerization and motility at the cell surface of the bacterium. Since then the Arp2/3 complex has been shown to both nucleate the formation of new actin filaments, crosslink existing filaments and cap the pointed end with high affinity, leading to the proposition of the now widely accepted dendritic nucleation model of actin organization in lamellipodium protrusion (Mullins *et al.*, 1998) (see also Chapter 1).

The many roles of the Arp2/3 complex *in vivo* are rapidly being uncovered. For example the Arp2/3 complex has been shown to be necessary for phagocytosis in mammals and the amoeba *Dictyostelium discoideum* (Insall *et al.*, 2001; May *et al.*, 2000), endocytosis in yeast (Warren *et al.*, 2002) and macropinocytosis (Insall *et al.*, 2001). The Arp2/3 complex also appears to be involved in cell polarity establishment and migration in fibroblast monolayers in a wound-healing model (Magdalena *et al.*, 2003). Studies utilizing loss of function mutations in the *Drosophila* homologues of p41-Arc and Arp3 also reveal a requirement for the Arp2/3 complex for ring-canal expansion during oogenesis but not for the formation of parallel actin bundles in the cytoplasm of nurse cells (Hudson and Cooley, 2002). In *Drosophila*, the Arp2/3 complex is also required for embryogenesis, as pseudocleavage furrow assembly, necessary for proper embryonic cell division, is disrupted when a mutation in the p41-Arc subunit is introduced (Stevenson *et al.*, 2002). Arp2/3 complex-dependent actin polymerization is also necessary for actin polymerization in platelets, leading to the formation of filopodia and lamellipodia (Li *et al.*, 2002).

Purified Arp2/3 complex nucleates actin filaments with free barbed ends, but this activity is very weak (Mullins *et al.*, 1998). However, several proteins activate the complex at sites of new actin polymerization. Of these, the most well-understood are members of the WASp family, which sport domains that interact with signalling molecules (such as the small GTPases Rac and Cdc42, G-proteins and tyrosine kinases), actin monomers and the Arp2/3 complex (Machesky and Insall, 1998).

WASp family proteins

The WASp family consists of five proteins in mammals: WASp, N-WASp and Scar1-3 (also known as WAVE1-3). WASp is expressed only in haematopoietic cells and when mutated results in Wiskott–Aldrich syndrome (WAS), a rare X-linked disease, in which sufferers have severely impaired immune function (Derry *et al.*, 1994). N-WASp is more widely expressed, as are the Scars. Scar was originally identified in *Dictyostelium* in a screen for suppressors of a cAMP receptor mutation (Bear *et al.*, 1998).

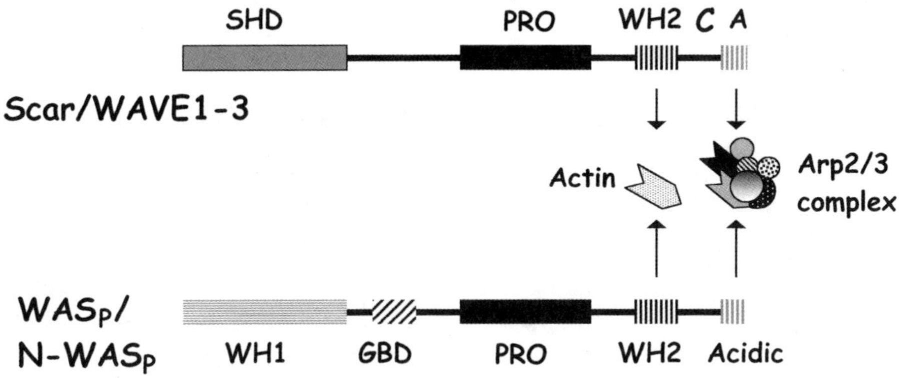

Figure 8.1 The WASp family proteins and their interactions. SHD = Scar homology domain, PRO = proline-rich region, WH2 = WASp homology 2 actin monomer-binding domain, C = central basic domain, A = Arp2/3 complex-binding domain, WH1 = WASp homology 1, GBD = GTPase binding domain

All members of the WASp family show considerable conservation, particularly in their C-terminal domain containing three common motifs (Figure 8.1). A central proline rich region (PRO) that can bind to SH3 domain-containing proteins such as the adapter protein Nck and thus indirectly associate with receptor tyrosine kinases (Rivero-Lezcano *et al.*, 1995); a WH2 (WASp homology 2) domain that binds monomeric actin, a central basic patch of amino acids that appears to be important both for actin and Arp2/3 complex binding (C) and at the extreme C-terminus a region of net negative charge A (Acidic) that binds to the Arp2/3 complex (Machesky and Insall, 1998). Both WASp and N-WASp also contain a GTPase binding domain (GBD) that binds to Cdc42 (Aspenstrom *et al.*, 1996). The N-terminal regions of WASp and N-WASp also interact with PIP_2 *in vitro* (Miki *et al.*, 1996) (see also Chapter 9).

The WASp-Arp2/3 pathway

WASp and Scar1 bind to the Arp2/3 complex via the p21-Arc subunit and likely other subunits as well. Over-expression of C-terminal fragments of Scar1 and WASp results in both delocalization of the Arp2/3 complex and loss of lamellipodia and actin spots (Machesky and Insall, 1998). Cdc42 and PIP_2 as well as adapter proteins (Nck and Grb2) also activate WASp and N-WASp. Thus the discovery of these connections provided an insight for a complete signalling pathway from external stimuli to controlled actin polymerization: the WASp-Arp2/3 pathway (Figure 8.2).

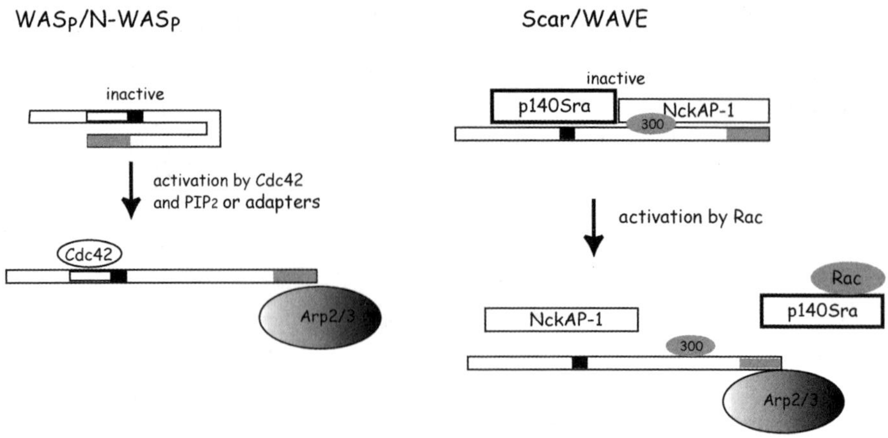

Figure 8.2 Activation of WASp/N-WASp and Scar/WAVE proteins WASp and N-WASp are folded such that the tail (grey) interacts with a Cdc42 GTPase binding domain (GBD, white box). Activation occurs when GTP-bound Cdc42 and possibly also membrane polyphosphoinositides interact with the GBD and a nearby basic patch (black) and open it up to cause activation and interaction with the Arp2/3 complex (gradient shading). Scars/WAVES lack a GBD and do not show this autoinhibition *in vitro*. Instead, bovine brain Scars have been reported to bind to cytoplasmic inhibitors, including the Rac-binding protein p140-Sra and the Nck-interacting protein Nck-AP1. Scar1 also binds to the small protein HSPC300, but the consequence of this interaction has not been reported. The model for Scar1 activation in response to signalling is that activated Rac GTPase binds to p140-Sra and causes it and Nck-AP to fall off Scar1, leaving the active Scar1-HSPC300 complex to interact with and activate the Arp2/3 complex (Eden *et al.*, 2002). This model is attractive but has not yet been tested *in vivo*

But how do these proteins activate Arp2/3 complex and stimulate actin polymerization? At present it is believed that WASp family proteins bind to actin monomers, which subsequently are passed on to the Arp2/3 complex. The two Arp subunits interact with the pointed end of the actin monomer thus allowing polymerization in the direction of the free barbed end – with Arp2 and Arp3 forming the first two subunits of the filament (Mullins *et al.*, 1997). Recent biochemical evidence supports this hypothesis, as recombinant Arp2/3 complex that lacks Arp3 is unable to nucleate actin filaments (Gournier *et al.*, 2001). The recently published crystal structure of the Arp2/3 complex may explain why activation by WASp proteins is required for efficient activation of Arp2/3 mediated actin polymerization (Dayel *et al.*, 2001). Although the two Arps lie in an approximate head-to-tail fashion (like subunits of an actin filament) they are rotated 180° relative to each other (Robinson *et al.*, 2001). It is believed WASp family proteins bind the Arp2/3 complex promoting a conformational change resulting in a translation of Arp2 relative to Arp3. Such a translation would position the Arp subunits relative to each other in an identical way to that observed in actin filaments thus providing the necessary actin dimer to trigger polymerization (Robinson *et al.*, 2001).

According to the dendritic nucleation model, WASp proteins activate Arp2/3 complex to initiate the formation of 70° actin branches at the leading edge of migrating cells. Currently, models predict that the Arp2/3 complex binds to the mother filament and is simultaneously activated by WASp/Scar proteins at the plasma membrane. The complex presumably then becomes incorporated into the branch point with the two Arps forming the first subunits of the branch (Volkmann *et al.*, 2001). Exactly which subunits contact the filament is still subject to debate. While the crystal structure of the complex suggests p34-Arc and p40-Arc are the leading candidates (Robinson *et al.*, 2001), recent work using recombinant Arp2/3 complex reveals that p40-Arc is not required and that the anchoring subunits may be p34-Arc and p20-Arc (Gournier *et al.*, 2001).

So far we have discussed how WASp family proteins may activate the Arp2/3 complex, but WASp family proteins are also subject to tight regulation. Since there are five WASp-related proteins in mammals, the different mechanisms of activation might provide specificity for the context and type of cellular process (such as lamellipodia versus phagosomes, for example). WASp and N-WASp are apparently regulated by autoinhibition (Figure 8.2, Miki, H. *et al.*, 1998). In their native states, WASp and N-WASp are held in an auto-inhibited confirmation in which the WA region is prevented from activating/binding the Arp2/3 complex by interaction with its own C-terminal region (Figure 8.2). This inactive confirmation is released on binding of the GBD with activated Cdc42 and perhaps PIP_2, allowing activation of the Arp2/3 complex and new filament nucleation (Rohatgi *et al.*, 2000). WASp and N-WASp can also bind to the SH3 domains of adaptor proteins such as Grb2 via their proline-rich sequences providing yet another potential route to activation (Carlier *et al.*, 2000). Another adapter protein WISH (WASp interacting SH3 protein) also binds both WASp and N-WASp via its SH3 domain leading to increased N-WASp activation of Arp2/3 complex-mediated actin polymerization (Fukuoka *et al.*, 2001).

While the mechanisms of WASp/N-WASp activation have begun to be understood activation of Scar proteins have, until this year, remained a mystery. Previous reports have suggested that they can be regulated by the GTPase Rac (Miki *et al.*, 1998b). However as Scars lack the GBD found in WASp/N-WASp the mechanism was unclear. Now Eden and co-workers (Eden *et al.*, 2002) have shown that Scar1/WAVE1 exists as a heteropentameric complex consisting of the proteins PIR121, Nap124, Abi2 and HSPC300 (Figure 8.2) which is inactive *in vitro*. Excitingly, Rac1 can relieve this trans-inhibition by binding to the PIR-21-Nap124-Abi2 sub-complex resulting in release of the Scar1-HSPC300 complex that is able to stimulate actin polymerization. The adapter protein Nck can also induce this dissociation suggesting an additional pathway for Scar activation (Eden *et al.*, 2002). The relative contributions of Nck and Rac1 in modifying Scar activity are however unknown.

Thus, while it is apparent that signals from many cellular receptors converge on the common WASp-Arp2/3 pathway, the many roles of WASp proteins in the cell are only beginning to be understood. N-WASp has been localized to actin comets that propel intracellular vesicles (Rozelle *et al.*, 2000) and in N-WASp-defective cells, actin comets are not observed (Benesch *et al.*, 2002). The adapter proteins Nck, Grb2 and WIP have also been implicated in the recruitment of N-WASp to the vesicle surface (Benesch *et al.*, 2002) and the large GTPase dynamin, important in triggering the pinching off nascent vesicles from both the Golgi apparatus and plasma membrane, is co-localized to actin in comet structures (Orth *et al.*, 2002). N-WASp has also been identified as a downstream effector of Cdc42 important in regulating the retrograde transport of proteins from the Golgi apparatus to the endoplasmic reticulum (Luna *et al.*, 2002). The physiological significance of actin in such vesicle trafficking is not understood but these results provide mounting evidence that there is a direct link between actin and membrane dynamics.

Cdc42-activated N-WASp is also implicated in filopodium formation (Miki *et al.*, 1998a) although more recent studies using N-WASp-defective cell lines have shown that Cdc42-mediated filopodium formation is unaffected and thus may provide evidence of an alternative mechanism leading to their formation (Snapper *et al.*, 2001).

It is well documented that WASp is required for proper development of leukocytes and the immune system (Thrasher *et al.*, 2000). Recently, however, studies in *Drosophila* have revealed an interesting requirement of the WASp homologue Wsp in the control of cell fate decisions mediated by Notch. Ben-Yaacov and colleagues show that flies with mutant Wsp are unable to undergo normal sensory organ development resulting in increased neuronal differentiation at the expense of other cell types (Ben-Yaacov *et al.*, 2001). As Notch signalling is believed to involve receptor-mediated endocytosis the authors postulate that WASp may function to link endocytosis of Notch with the underlying actin cytoskeleton.

Compared with WASp/N-WASp, the cellular roles of the Scar proteins have been more elusive. In the first studies of their kind, Zallen and co-workers have examined the role of the *Drosophila* homologue of Scar and the Arp2/3 complex during embryonic development and found both are required for proper cell shape and motility (Zallen *et al.*, 2002). Interestingly, WASp was not required for these events providing evidence that the Arp2/3 complex governs distinct cellular events dependent on either Scar or WASp activation.

Arp2/3 complex-independent actin nucleation

Whilst the WASp-Arp2/3 pathway of actin filament nucleation is now becoming fairly well documented, the possibility remains for actin filament nucleation by alternative mechanisms. It is suggested that although the Arp2/3 complex is responsible for nucleation of actin in the formation of branched microfilaments (Mullins *et al.*, 1998), a separate mechanism may exist for the production of unbranched filaments such as those found in stress fibres. Current research has identified additional proteins capable of nucleating unbranched actin polymerization from barbed ends.

The *L. monocytogenes* protein ActA has been known to be involved in the intracellular motility of the bacterium for a long time, recruiting actin-binding proteins to enable actin tail formation (Loisel *et al.*, 1999). Recently Fradelizi *et al.* (2001) showed that ActA, and the related human protein zyxin, which localizes to stress fibres and focal contacts, are capable of nucleating actin polymerization in cell-free and permeabilized cell systems. Actin assembles on ActA-coated beads as well as on ActA and zyxin targeted to mitochondria. This actin polymerization is dependent upon VASP in the case of zyxin, but independent of the Arp2/3 complex in both cases. Currently no mechanisms for control of this potential actin nucleation system have been described.

Two groups have recently implicated the budding yeast formin Bni1 as an actin nucleator (Pruyne *et al.*, 2002; Sagot *et al.*, 2002b). This formin, previously identified as being required for polarized arrays of actin cables in budding yeast (Evangelista *et al.*, 2002; Sagot *et al.*, 2002a), is thought to be involved in the formation of long unbranched actin filaments which are organized into bundles. A purified constitutively active C-terminal construct of Bni1 containing the formin homology 1 (FH1) and 2 (FH2) domains (Figure 8.3) causes a concentration-dependent increase in the rate of *in vitro* actin filament assembly. Removal or mutation of the FH2 domain but not the FH1 domain inhibits this apparent nucleating activity (Sagot *et al.*, 2002b), and addition of profilin shortens the lag time of polymerization. The effect of profilin is dependent upon its capacity to bind to both actin and the polyproline-rich FH1 domain of Bni1 (Sagot *et al.*, 2002b).

Several formins are Rho GTPase effectors, believed to be activated by the relief of an intramolecular interaction between a GTPase binding domain (GBD) and an auto-regulatory C-terminal domain named DAD (Dia autoregulatory domain) upon binding of an activated GTPase (Figure 8.3) (Alberts, 2001). Specifically, the GTPases Rho1p and Cdc42p, interact with Bni1 at the N-terminal GBD. This causes the formin to adopt an open or active conformation in a similar manner to WASp (Evangelista *et al.*, 1997; Kohno *et al.*, 1996). Activation of formins allows interaction with a variety of binding partners at either the FH1 or FH2 domains to facilitate actin filament nucleation.

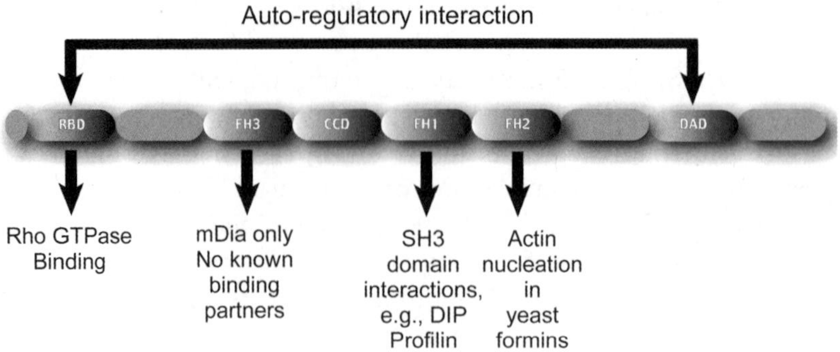

Figure 8.3 The predicted domain structure of mDia (mammalian diaphanous). RBD = Rho-binding domain; FH1 = poly-proline rich formin homology domain 1, which binds to SH3 domains; FH2 = formin homology domain 2, which shows actin nucleation activity in yeast formins; FH3 = formin homology domain 3; CCD = coiled-coil domain; DAD = Dia auto-regulatory domain, which interacts with the RBD to form the inactive conformation of the protein

The FH1 domain contains a proline-rich module that interacts with several Src Homology 3 (SH3) domain-containing proteins (Krebs *et al.*, 2001), linking formins to tyrosine kinase signalling pathway proteins. Satoh and Tominaga show the FH1 domain to be involved in the interaction between mDia1 and Dia-interacting protein (DIP), a ubiquitously expressed protein which is involved in stress fibre formation and to affect focal adhesion turnover through activation of Src (Satoh and Tominaga, 2001).

Evidence for Arp2/3 complex-independent nucleation focuses on the formation of actin cable structures and unbranched actin filaments. This does largely suit the idea that stress fibres and lamellipodia are formed by different nucleation mechanisms, but it is difficult to exclude the concept that the Arp2/3 complex may be recruited to sites of unbranched filament formation and inhibited from forming 70° branches or that branches are short-lived and may be remodelled into other types of structures.

The role of Ena/VASP proteins

The architecture of actin arrays appears to be regulated by the Ena/VASP proteins, possibly providing some of the diversity of structures observed in lamellipodia, filopodia and other actin-rich protrusions. The Ena/VASP proteins are a family that share two domains termed Ena/VASP homology domains (EVH1 and EVH2). The family consists of several members; vasodilator stimulated phosphoprotein (VASP) originally identified as a substrate for cyclic nucleotide-dependent protein kinases, *Drosophila* enabled (Ena) and its mammalian homologue, Mena and the Ena/VASP-like protein

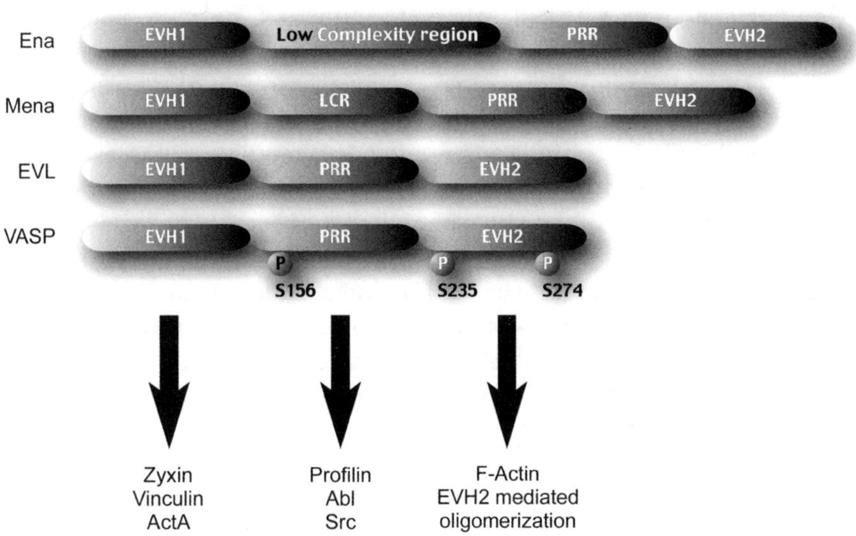

Figure 8.4 The predicted domain structure of Ena (*Drosophila* enabled), Mena (Mammalian enabled), EVL (Ena/VASP-like protein) and VASP (vasodilator stimulated phosphoprotein) showing phosporylation sites on VASP at serines 156, 235 and 274. EVH1 = Ena-VASP homology 1 domain, which binds to ligands such as vinculin, zyxin and ActA from *L. monocytogenes* via proline-rich repeats. EVH2 = Ena-VASP homology 2 domain, which binds to filamentous actin and also mediates oligomerization of the Ena/VASP proteins. PRR = proline rich region, which binds to profilin and to Src and Abl kinases. LCR = Low complexity region

EVL. Ena/VASP family members are largely located in focal adhesions, cell–cell adherens junctions, tips of stress fibres and the leading edges of filopodia and lamellipodia, all generally regarded as being sites of active actin polymerization.

Ena/VASP proteins largely consist of several modular domains (Figure 8.4), contributing to interactions with many proteins. Principally this is through the central polyproline-rich region, involved in interactions with SH3 domain-containing proteins. An important ligand for the proline-rich region is profilin (Reinhard *et al.*, 1995), an actin monomer-sequestering protein that facilitates increased actin filament dynamics by increasing the rate of ADP/ATP exchange on actin monomers and sequestering them in a soluble pool (Wolven *et al.*, 2000). Mena has been identified as a binding partner for IRS-p53 (Insulin receptor substrate of 53 kDa). Relief of an auto-inhibitory interaction between the N- and C-terminals of IRS-p53 by binding to activated Cdc42 leads to formation of excessive filopodia which is thought to involve Mena activity (Krugmann *et al.*, 2001) (see also Chapter 9).

The C-terminal EVH2 domain of Ena/VASP proteins is involved in both homo- and hetero-oligomerization between members of the family (Bachmann *et al.*, 1999), and in VASP has been identified as an actin-binding motif (Huttelmaier *et al.*, 1999). The actin-binding activity of the EVH2 domain is dependent upon the phosphorylation state of the protein at sites flanking the proline-rich domain (Harbeck *et al.*, 2000). Phosphorylation is necessary for normal Ena/VASP function during random cell motility, but does not affect the cellular location of the proteins (Loureiro *et al.*, 2002). Harbeck *et al.*, 2000) showed that VASP phosphorylation negatively regulates actin polymerization through inhibition of the interaction between the VASP EVH2 domain and F-actin. The EVH1 domain located at the N-terminus contributes to interactions with several focal adhesion proteins, such as vinculin (Huttelmaier *et al.*, 1998) and zyxin (Drees *et al.*, 2000), serving to recruit Ena/VASP proteins to focal adhesions, as well as the zyxin-related *Listeria* protein ActA (Niebuhr *et al.*, 1997).

VASP has an important role in actin polymerization during intracellular motility of *Listeria*, being required for increased rates of bacterial movement in reconstituted cell-free systems (Geese *et al.*, 2002; Laurent *et al.*, 1999; Skoble *et al.*, 2001). This suggests that VASP facilitates an increased rate of actin polymerization. VASP over-expression, however, negatively regulates fibroblast motility in a dose-dependent manner (Bear *et al.*, 2000) suggesting that the simple correlation between filament assembly and cell movement observed with Listeria does not hold in mammalian cells (Machesky, 2002).

Recently, Bear and colleagues have deepened our understanding of the cellular role of Ena/VASP proteins to resolve this apparent controversy (Bear *et al.*, 2002). They show that Ena/VASP proteins compete in an antagonistic manner with capping proteins to promote filament elongation. VASP over-expression reduces branched filament network formation as caused by Arp2/3 complex-dependent nucleation to affect production of lamellipodia (Bear *et al.*, 2002). This observation explains both the activity of VASP to allow increased *Listeria* motility and its ability to slow cell motility when over-expressed, which could be due to inhibiting formation of the actin network structure required for membrane ruffling whilst promoting the formation of longer filaments.

The function of Ena/VASP localization in focal adhesions and the tips of stress fibres is not yet entirely clear, although several hypotheses are possible. The proteins may have a molecular linker function through their ability to bind to both F-actin and to anchor fibres to focal adhesion complexes through interaction with vinculin and zyxin. As well as anchoring stress fibres it is feasible that Ena/VASP proteins have a role in stress fibre formation through promotion of filament elongation and bundling.

Conclusions

Clearly, many proteins work together to regulate actin dynamics in cells and we have only touched on a few of these. Uncapping and severing of filaments are important ways to induce actin assembly in response to signals. However, in the past five years, we have gained an understanding of how the WASp-Arp2/3 pathway, together with Ena/VASP proteins, can mediate actin assembly in dynamic cell structures. Recent evidence indicates that there are other ways to nucleate filament networks independently of the Arp2/3 complex, such as through the formins. As time progresses, we will hopefully develop a deeper understanding of how the different systems for generating actin filaments work together and thus a better idea of how the extremely complex processes of cell movement and shape changes are regulated. No doubt the next five years will also bring some new surprises and additional players to the field – we can only look forward to a clearer, if more detailed, picture of how actin dynamics allows cells to move and adapt to their environment.

References

Alberts, A. S., 2001. Identification of a carboxyl-terminal diaphanous-related formin homology protein autoregulatory domain. *J. Biol. Chem.* **276**: 2824–2830.

Aspenstrom, P., Lindberg, U. and Hall, A., 1996. Two GTPases, Cdc42 and Rac, bind directly to a protein implicated in the immunodeficiency disorder Wiskott–Aldrich syndrome. *Curr. Biol.* **6**: 70–75.

Bachmann, C., Fischer, L., Walter, U. and Reinhard, M., 1999. The EVH2 domain of the vasodilator-stimulated phosphoprotein mediates tetramerization, F-actin binding, and actin bundle formation. *J. Biol. Chem.* **274**: 23 549–23 557.

Balasubramanian, M. K., Feoktistova, A., McCollum, D. and Gould, K. L., 1996. Fission yeast Sop2p: a novel and evolutionarily conserved protein that interacts with Arp3p and modulates profilin function. *Embo J.* **15**: 6426–6437.

Bear, J. E., Loureiro, J. J., Libova, I., Fassler, R., *et al.*, 2000. Negative regulation of fibroblast motility by Ena/VASP proteins. *Cell* **101**: 717–728.

Bear, J. E., Rawls, J. F. and Saxe, C. L. 3rd., 1998. SCAR, a WASP-related protein, isolated as a suppressor of receptor defects in late *Dictyostelium* development. *J. Cell Biol.* **142**: 1325–1335.

Bear, J. E., Svitkina, T. M., Krause, M., Schafer, D. A., *et al.*, 2002. Antagonism between Ena/VASP proteins and actin filament capping regulates fibroblast motility. *Cell* **109**: 509–521.

Ben-Yaacov, S., Le Borgne, R., Abramson, I., Schweisguth, F. and Schejter, E. D., 2001. Wasp, the Drosophila Wiskott-Aldrich syndrome gene homologue, is required for cell fate decisions mediated by Notch signaling. *J. Cell Biol.* **152**: 1–13.

Benesch, S., Lommel, S., Steffen, A., Stradal, T. E., *et al.*, 2002. Phosphatidylinositol 4,5-biphosphate (PIP2)-induced vesicle movement depends on N-WASP and involves Nck, WIP, and Grb2. *J. Biol. Chem.* **277**: 37 771–37 776.

Carlier, M. F., Nioche, P., Broutin-L'Hermite, I., Boujemaa, R., *et al.*, 2000. GRB2 links signaling to actin assembly by enhancing interaction of neural Wiskott–Aldrich syndrome protein (N-WASp) with actin-related protein (ARP2/3) complex. *J. Biol. Chem.* **275**: 21 946–21 952.

Carlsson, L., Nystrom, L. E., Sundkvist, I., Markey, F. and Lindberg, U., 1977. Actin polymerizability is influenced by profilin, a low molecular weight protein in non-muscle cells. *J. Mol. Biol.* **115**: 465–483.

Cassimeris, L., Safer, D., Nachmias, V. T. and Zigmond, S. H., 1992. Thymosin beta 4 sequesters the majority of G-actin in resting human polymorphonuclear leukocytes. *J. Cell Biol.* **119**: 1261–1270.

Dayel, M. J., Holleran, E. A. and Mullins, R. D., 2001. Arp2/3 complex requires hydrolyzable ATP for nucleation of new actin filaments. *Proc. Natl. Acad. Sci. USA* **98**: 14 871–14 876.

Derry, J. M., Ochs, H. D. and Francke, U., 1994. Isolation of a novel gene mutated in Wiskott–Aldrich syndrome. *Cell* **79**: 635–644.

Drees, B., Friederich, E., Fradelizi, J., Louvard, D., *et al.*, 2000. Characterization of the interaction between zyxin and members of the Ena/vasodilator-stimulated phosphoprotein family of proteins. *J. Biol. Chem.* **275**: 22 503–22 511.

Eden, S., Rohatgi, R., Podtelejnikov, A. V., Mann, M. and Kirschner, M. W., 2002. Mechanism of regulation of WAVE1-induced actin nucleation by Rac1 and Nck. *Nature* **418**: 790–793.

Evangelista, M., Blundell, K., Longtine, M. S., Chow, C. J., *et al.*, 1997. Bni1p, a yeast formin linking cdc42p and the actin cytoskeleton during polarized morphogenesis. *Science* **276**: 118–122.

Evangelista, M., Pruyne, D., Amberg, D. C., Boone, C. and Bretscher, A., 2002. Formins direct Arp2/3-independent actin filament assembly to polarize cell growth in yeast. *Nat. Cell Biol.* **4**: 32–41.

Fradelizi, J., Noireaux, V., Plastino, J., Menichi, B., *et al.*, 2001. ActA and human zyxin harbour Arp2/3-independent actin-polymerization activity. *Nat. Cell Biol.* **3**: 699–707.

Fukuoka, M., Suetsugu, S., Miki, H., Fukami, K., *et al.*, 2001. A novel neural Wiskott–Aldrich syndrome protein (N-WASP) binding protein, WISH, induces Arp2/3 complex activation independent of Cdc42. *J. Cell Biol.* **152**: 471–482.

Geese, M., Loureiro, J. J., Bear, J. E., Wehland, J., *et al.*, 2002. Contribution of Ena/VASP proteins to intracellular motility of *Listeria* requires phosphorylation and proline-rich core but not F-actin binding or multimerization. *Mol. Biol. Cell* **13**: 2383–2396.

Gournier, H., Goley, E. D., Niederstrasser, H., Trinh, T. and Welch, M. D., 2001. Reconstitution of human Arp2/3 complex reveals critical roles of individual subunits in complex structure and activity. *Mol. Cell* **8**: 1041–1052.

Harbeck, B., Huttelmaier, S., Schluter, K., Jockusch, B. M. and Illenberger, S., 2000. Phosphorylation of the vasodilator-stimulated phosphoprotein regulates its interaction with actin. *J. Biol. Chem.* **275**: 30 817–30 825.

Hartwig, J. H., 1992. Mechanisms of actin rearrangements mediating platelet activation. *J. Cell Biol.* **118**: 1421–1442.

Hartwig, J. H., Bokoch, G. M., Carpenter, C. L., Janmey, P. A., *et al.*, 1995. Thrombin receptor ligation and activated Rac uncap actin filament barbed ends through phosphoinositide synthesis in permeabilized human platelets. *Cell* **82**: 643–653.

Hudson, A. M. and Cooley, L., 2002. A subset of dynamic actin rearrangements in Drosophila requires the Arp2/3 complex. *J. Cell Biol.* **156**: 677–687.

Huttelmaier, S., Mayboroda, O., Harbeck, B., Jarchau, T., *et al.*, 1998. The interaction of the cell-contact proteins VASP and vinculin is regulated by phosphatidylinositol-4,5-bisphosphate. *Curr. Biol.* **8**: 479–488.

Huttelmaier, S., Harbeck, B., Steffens, O., Messerschmidt, T., *et al.*, 1999. Characterization of the actin binding properties of the vasodilator-stimulated phosphoprotein VASP. *FEBS Lett.* **451**: 68–74.

Ichetovkin, I., Grant, W. and Condeelis, J., 2002. Cofilin produces newly polymerized actin filaments that are preferred for dendritic nucleation by the Arp2/3 complex. *Curr. Biol.* **12**: 79–84.

Insall, R., Muller-Taubenberger, A., Machesky, L., Kohler, J., *et al.*, 2001. Dynamics of the *Dictyostelium* Arp2/3 complex in endocytosis, cytokinesis, and chemotaxis. *Cell Motil. Cytoskeleton* **50**: 115–128.

Jay, P., Berge-Lefranc, J. L., Massacrier, A., Roessler, E., *et al.*, 2000. ARP3beta, the gene encoding a new human actin-related protein, is alternatively spliced and predominantly expressed in brain neuronal cells. *Eur. J. Biochem.* **267**: 2921–2928.

Kohno, H., Tanaka, K., Mino, A., Umikawa, M., *et al.*, 1996. Bni1p implicated in cytoskeletal control is a putative target of Rho1p small GTP binding protein in *Saccharomyces cerevisiae. Embo J.* **15**: 6060–6068.

Krebs, A., Rothkegel, M., Klar, M. and Jockusch, B. M., 2001. Characterization of functional domains of mDia1, a link between the small GTPase Rho and the actin cytoskeleton. *J. Cell Sci.* **114**: 3663–3672.

Krugmann, S., Jordens, I., Gevaert, K., Driessens, M., 2001. Cdc42 induces filopodia by promoting the formation of an IRSp53:Mena complex. *Curr. Biol.* **11**: 1645–1655.

Laurent, V., Loisel, T. P., Harbeck, B., Wehman, A., *et al.*, 1999. Role of proteins of the Ena/VASP family in actin-based motility of *Listeria monocytogenes. J. Cell Biol.* **144**: 1245–1258.

Li, Z., Kim, E. S. and Bearer, E. L., 2002. Arp2/3 complex is required for actin polymerization during platelet shape change. *Blood* **99**: 4466–4474.

Loisel, T. P., Boujemaa, R., Pantaloni, D. and Carlier, M. F., 1999. Reconstitution of actin-based motility of *Listeria* and *Shigella* using pure proteins. *Nature* **401**: 613–616.

Loureiro, J. J., Rubinson, D. A., Bear, J. E., Baltus, G. A., *et al.*, 2002. Critical roles of phosphorylation and actin binding motifs, but not the central proline-rich region, for Ena/vasodilator-stimulated phosphoprotein (VASP) function during cell migration. *Mol. Biol. Cell.* **13**: 2533–2546.

Luna, A., Matas, O. B., Martinez-Menarguez, J. A., Mato, E., *et al.*, 2002. Regulation of protein transport from the Golgi complex to the endoplasmic reticulum by CDC42 and N-WASP. *Mol. Biol. Cell* **13**: 866–879.

Machesky, L. M., 2002. Sharks' teeth and dunes. *Nature* **417**: 494–497.

Machesky, L. M. and Insall, R. H., 1998. Scar1 and the related Wiskott–Aldrich syndrome protein, WASP, regulate the actin cytoskeleton through the Arp2/3 complex. *Curr. Biol.* **8**: 1347–1356.

Machesky, L. M., Atkinson, S. J., Ampe, C., Vandekerckhove, J. and Pollard, T. D., 1994. Purification of a cortical complex containing two unconventional actins from *Acanthamoeba* by affinity chromatography on profilin-agarose. *J. Cell Biol.* **127**: 107–115.

Machesky, L. M., Reeves, E., Wientjes, F., Mattheyse, F. J., *et al.*, 1997. Mammalian actin-related protein 2/3 complex localizes to regions of lamellipodial protrusion and is composed of evolutionarily conserved proteins. *Biochem J.* **328**: 105–112.

Magdalena, J., Millard, T. H., Etienne-Manneville, S., Launay, S., Warwick, H. K. and Machesky, L. M., 2003. Involvement of the Arp2/3 Complex and Scar2 in Golgi polarity in scratch wound models. *Mol. Biol. Cell* **14**(2): 670–684.

May, R. C., Caron, E., Hall, A. and Machesky, L. M., 2000. Involvement of the Arp2/3 complex in phagocytosis mediated by FcgammaR or CR3. *Nat. Cell Biol.* **2**: 246–248.

Miki, H., Miura, K. and Takenawa, T., 1996. N-WASP, a novel actin-depolymerizing protein, regulates the cortical cytoskeletal rearrangement in a PIP2-dependent manner downstream of tyrosine kinases. *Embo J.* **15**: 5326–5335.

Miki, H., Sasaki, T., Takai, Y. and Takenawa, T., 1998a. Induction of filopodium formation by a WASP-related actin-depolymerizing protein N-WASP. *Nature* **391**: 93–96.

Miki, H., Suetsugu, S. and Takenawa, T., 1998b. WAVE, a novel WASP-family protein involved in actin reorganization induced by Rac. *Embo J.* **17**: 6932–6941.

Millard, T. H., Behrendt, B., Launay, S., Fütterer, K. and Machesky, L. M., 2003. Identification and characterisation of a novel human isoform of Arp2/3 complex subunit p16-Arc/ARPC5. *Cell Motil. Cytoskel.* **54**(1): 81–90.

Mullins, R. D., Stafford, W. F. and Pollard, T. D., 1997. Structure, subunit topology, and actin-binding activity of the Arp2/3 complex from *Acanthamoeba*. *J. Cell Biol.* **136**: 331–343.

Mullins, R. D., Heuser, J. A. and Pollard, T. D., 1998. The interaction of Arp2/3 complex with actin: nucleation, high affinity pointed end capping, and formation of branching networks of filaments. *Proc. Natl. Acad. Sci. USA* **95**: 6181–6186.

Niebuhr, K., Ebel, F., Frank, R., Reinhard, M., *et al.*, 1997. A novel proline-rich motif present in ActA of *Listeria monocytogenes* and cytoskeletal proteins is the ligand for the EVH1 domain, a protein module present in the Ena/VASP family. *Embo J.* **16**: 5433–5444.

Orth, J. D., Krueger, E. W., Cao, H. and McNiven, M. A., 2002. The large GTPase dynamin regulates actin comet formation and movement in living cells. *Proc. Natl. Acad. Sci. USA* **99**: 167–172.

Pollard, T. D., Blanchoin, L. and Mullins, R. D., 2000. Molecular mechanisms controlling actin filament dynamics in nonmuscle cells. *Annu. Rev. Biophys. Biomol. Struct.* **29**: 545–576.

Pruyne, D., Evangelista, M., Yang, C., Bi, E., *et al.*, 2002. Role of formins in actin assembly: nucleation and barbed-end association. *Science* **297**: 612–615.

Reinhard, M., Giehl, K., Abel, K., Haffner, C., *et al.*, 1995. The proline-rich focal adhesion and microfilament protein VASP is a ligand for profilins. *Embo J.* **14**: 1583–1589.

Rivero-Lezcano, O. M., Marcilla, A., Sameshima, J. H. and Robbins, K. C., 1995. Wiskott–Aldrich syndrome protein physically associates with Nck through Src homology 3 domains. *Mol. Cell Biol.* **15**: 5725–5731.

Robinson, R. C., Turbedsky, K., Kaiser, D. A., Marchand, J. B., *et al.*, 2001. Crystal structure of Arp2/3 complex. *Science* **294**: 1679–1684.

Rohatgi, R., Ho, H. Y. and Kirschner, M. W., 2000. Mechanism of N-WASP activation by CDC42 and phosphatidylinositol 4, 5-bisphosphate. *J. Cell Biol.* **150**: 1299–1310.

Rozelle, A. L., Machesky, L. M., Yamamoto, M., Driessens, M. H., *et al.*, 2000. Phosphatidylinositol 4,5-bisphosphate induces actin-based movement of raft-enriched vesicles through WASP-Arp2/3. *Curr. Biol.* **10**: 311–320.

Safer, D., Golla, R. and Nachmias, V. T., 1990. Isolation of a 5-kilodalton actin-sequestering peptide from human blood platelets. *Proc. Natl. Acad. Sci. USA* **87**: 2536–2540.

Sagot, I., Klee, S. K. and Pellman, D., 2002a. Yeast formins regulate cell polarity by controlling the assembly of actin cables. *Nat. Cell Biol.* **4**: 42–50.

Sagot, I., Rodal, A. A., Moseley, J., Goode, B. L. and Pellman, D., 2002b. An actin nucleation mechanism mediated by Bni1 and Profilin. *Nat. Cell Biol.* **4**: 626–631.

Satoh, S. and Tominaga, T., 2001. mDia-interacting protein acts downstream of Rho-mDia and modifies Src activation and stress fiber formation. *J. Biol. Chem.* **276**: 39 290–39 294.

Schafer, D. A., Jennings, P. B. and Cooper, J. A., 1996. Dynamics of capping protein and actin assembly in vitro: uncapping barbed ends by polyphosphoinositides. *J. Cell Biol.* **135**: 169–179.

Skoble, J., Auerbuch, V., Goley, E. D., Welch, M. D. and Portnoy, D. A., 2001. Pivotal role of VASP in Arp2/3 complex-mediated actin nucleation, actin branch-formation, and *Listeria monocytogenes* motility. *J. Cell Biol.* **155**: 89–100.

Snapper, S. B., Takeshima, F., Anton, I., Liu, C. H., *et al.*, 2001. N-WASP deficiency reveals distinct pathways for cell surface projections and microbial actin-based motility. *Nat. Cell Biol.* **3**: 897–904.

Stevenson, V., Hudson, A., Cooley, L. and Theurkauf, W. E., 2002. Arp2/3-dependent psuedocleavage furrow assembly in syncytial *Drosophila* embryos. *Curr. Biol.* **12**: 705–711.

Thrasher, A. J., Burns, S., Lorenzi, R. and Jones, G. E., 2000. The Wiskott–Aldrich syndrome: disordered actin dynamics in haematopoietic cells. *Immunol. Rev.* **178**: 118–128.

Volkmann, N., Amann, K. J., Stoilova-McPhie, S., Egile, C., *et al.*, 2001. Structure of Arp2/3 complex in its activated state and in actin filament branch junctions. *Science* **293**: 2456–2459.

Warren, D. T., Andrews, P. D., Gourlay, C. W. and Ayscough, K. R., 2002. Sla1p couples the yeast endocytic machinery to proteins regulating actin dynamics. *J. Cell Sci.* **115**: 1703–1715.

Welch, M. D., DePace, A. H., Verma, S., Iwamatsu, A. and Mitchison, T. J., 1997a. The human Arp2/3 complex is composed of evolutionarily conserved subunits and is localized to cellular regions of dynamic actin filament assembly. *J. Cell Biol.* **138**: 375–384.

Welch, M. D., Iwamatsu, A. and Mitchison, T. J., 1997b. Actin polymerization is induced by Arp2/3 protein complex at the surface of *Listeria monocytogenes*. *Nature* **385**: 265–269.

Westerberg, L., Greicius, G., Snapper, S. B., Aspenstrom, P. and Severinson, E., 2001. Cdc42, Rac1, and the Wiskott–Aldrich syndrome protein are involved in the cytoskeletal regulation of B lymphocytes. *Blood* **98**: 1086–1094.

Wolven, A. K., Belmont, L. D., Mahoney, N. M., Almo, S. C. and Drubin, D. G., 2000. *In vivo* importance of actin nucleotide exchange catalyzed by profilin. *J. Cell Biol.* **150**: 895–904.

Zallen, J. A., Cohen, Y., Hudson, A. M., Cooley, L., *et al.*, 2002. SCAR is a primary regulator of Arp2/3-dependent morphological events in *Drosophila*. *J. Cell Biol.* **156**(4): 689–701.

9

Role of WASp Family Proteins in Cytoskeletal Reorganization and Cell Motility

Tadaomi Takenawa and **Shiro Suetsugu**

WASp and WAVE family proteins act downstream of Cdc42 and Rac, to induce formation of filopodia and lamellipodia, respectively. All WASp family proteins have VCA (verprolin-like, cofilin-like and acidic region) at the C-terminus. V region binds to monomer actin and CA region binds to Arp2/3 complex, resulting in actin polymerization at the leading edge of cells. This WASp/WAVE-Arp2/3 complex-mediated actin polymerization is essential for generating movement force. Thus, WASp family proteins play important roles in phenomena accompanied with cell movement such as podosome formation and epithelial tubulogenesis.

Introduction

Directed cell movement or chemotaxis is essential for a wide variety of cellular functions, including wound healing, immunoactivity, angiogenesis, axon guidance and embryonic development. Eukaryotic cells of diverse origins share many common mechanisms for sensing and responding to chemoattractant gradients, which occurs in two steps: cell polarity formation and leading edge formation. When cells encounter a chemoattractant gradient,

Cell Motility. Edited by Anne Ridley, Michelle Peckham and Peter Clark.
 ISBN 0 470 84872 3

they respond with local activation and amplification of signals at the site facing the gradient (cell polarity formation). These signals cause localized polymerization of actin and formation of membrane protrusions such as filopodia and lamellipodia (leading edge formation). These structures are thought to generate the force for movement toward the gradient. Thus, it has been of interest to understand how these structures are formed in response to extracellular chemoattractants. Recent studies clarified the signalling pathways that link extracellular stimuli to actin polymerization at the leading edge. One signalling cascade is governed by the small GTPases of the Rho family. Among them, RhoA, Rac and Cdc42 play a critical role in the formation and organization of cortical actin networks (Bishop and Hall, 2000; Hall, 1998). RhoA controls assembly of stress fibres and focal adhesions. Rac regulates formation of lamellipodia, and Cdc42 controls extension of filopodia and microspikes. During cell movement, Rac and Cdc42 stimulate formation of protrusions at the leading edges of cells, and RhoA induces retraction of the tail ends of cells. This coordinated reorganization of actin filaments permits cells to move toward a target. Though Rac and Cdc42 are essential for cell movement, the downstream molecules involved directly in actin filament reorganization were unknown until the WASp (Wiskott–Aldrich Syndrome protein) family proteins were identified. These proteins have garnered a great deal of attention as the links between the small GTPases and the actin cytoskeleton. The WASp family proteins are classified into two structural groups (Figure 9.1): WASps, which comprise WASp and N-WASp (Derry *et al.*, 1994; Miki *et al.*, 1996; Fukuoka *et al.*, 1997), and WAVEs (also called Scar), which comprise WAVE1, 2 and 3 (Miki *et al.*, 1998; Suetsugu *et al.*, 1999).

WASp and WAVE family proteins

WASp was first identified as the cause of Wiskott–Aldrich Syndrome (WAS), which is characterized by eczema, bleeding and recurrent infections (Derry *et al.*, 1994). This protein is expressed exclusively in hematopoietic cells. Lymphocytes from WAS patients show cytoskeletal abnormalities with a reduction in the number of cell surface microvilli (Kenney *et al.*, 1986; Molina *et al.*, 1992). WASp binds Cdc42 through its GBD/CRIB (GTPase-binding domain/Cdc42 and Rac interactive binding) motif, and overexpression of WASp induces formation of actin clusters, suggesting a role in actin polymerization (Symons *et al.*, 1996). N-WASp was isolated as an Ash/Grb2 SH3 domain-binding protein. This protein, in contrast to WASp, is ubiquitously expressed in a variety of tissues, though it was named neural (N)-WASp because of its high expression in neural tissues. N-WASp shows approximately 50% amino-acid homology with WASp, and it contains several

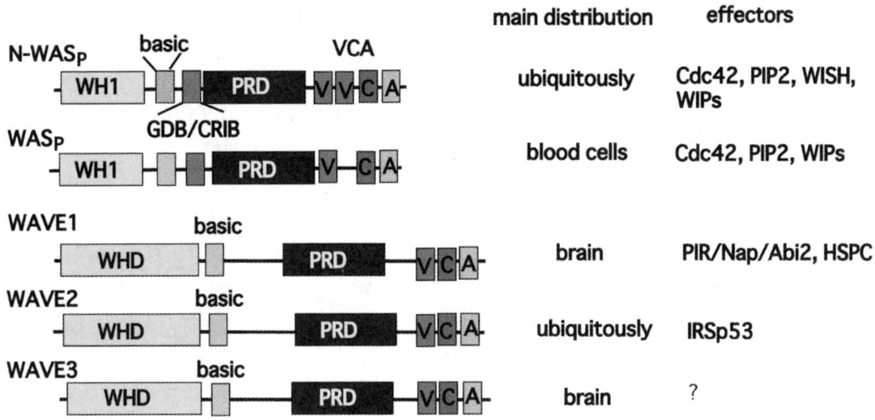

Figure 9.1 WASp and WAVE family proteins. The WASp family comprises N-WASp and WASp, and the WAVE family comprises, WAVE 1, 2, and 3. These proteins all have VCA region at the C-terminus through which these proteins activate the Arp2/3 complex, leading to actin nucleation and polymerization

multifunctional domains including a WH1/EVH1 (WASp homology/Ena VASP homology) domain, IQ motif, basic region, GBD/CRIB motif, proline-rich region, verprolin-homology region (V), cofilin-like region (C), and acidic region (A) (Figure 9.1). Available data suggest that WASp and N-WASp are downstream targets of Cdc42, leading to filopodium formation (Miki *et al.*, 1996; Symons *et al.*, 1996). WAVE1 was identified as a novel protein with a V domain, following the identification of WAVE2 and WAVE3 (Miki *et al.*, 1998; Suetsugu *et al.*, 1999). At the same time, a *Dictyostelium* homologue of WAVE was identified and named Scar (Bear *et al.*, 1998). WAVE1, 2 and 3 are highly homologous to each other, and all contain a WAVE/Scar homology domain (WHD) at the N-terminus, a basic region, a proline-rich region, and a VCA region at C-terminus (Figure 9.1). Therefore, the C-terminal regions of WAVEs are highly homologous to that of N-WASp. However, WAVEs do not have a Cdc42-binding site or GBD/CRIB motif, suggesting that WAVEs are regulated differently from WASps.

WASp and WAVE activate the Arp2/3 complex through the VCA region

In resting cells, the barbed ends of actin filaments are protected by capping proteins such as gelsolin superfamily proteins and capping protein, CapZ, to prevent spontaneous, unregulated actin polymerization. Therefore, to induce rapid actin polymerization at the leading edge in response to stimuli, generation of new barbed ends is necessary. It is thought that there are three mechanisms for generation of new barbed ends: uncapping of pre-existing filaments, severing of pre-existing filaments, and *de novo* nucleation of actin core (Suetsugu *et al.*, 2002; Condeelis, 2001; Wear *et al.*, 2000). Among these mechanisms, *de novo* nucleation appears to play the most important role in leading edge formation (see also chapter by Condeelis). The recent progress in our understanding of the process of actin network formation began with the discovery of the Arp2/3 complex, a complex of seven proteins that is the only known nucleator of new actin filaments. However, the Arp2/3 complex alone can not enhance actin polymerization, suggesting that some other factor is necessary for Arp2/3 complex activation.

In studies of the regulation of the Arp2/3 complex, pathogenic bacteria such as *Shigella* and *Listeria* have played important roles. These bacteria are known to form actin comets and move freely within infected cells. Actin comet structures resemble those at the leading edge, which consist of filopodia and lamellipodia (Cameron *et al.*, 2001). The ActA surface protein from *Listeria* was found to bind directly to the Arp2/3 complex, leading to its activation (Welch *et al.*, 1997). Surprisingly, the amino acid sequence of the ActA protein is homologous to the VCA region of WASp family proteins. The VCA region alone can activate the Arp2/3 complex, enhancing actin polymerization, suggesting that the VCA region is the minimum essential region for Arp2/3 complex activation. The V region is a monomeric actin-binding region, and the CA region is an Arp2/3 complex-binding region (Miki and Takenawa, 1998; Rohatgi *et al.*, 1999). This appears to be the basis for the activation of the actin nucleating activity of the Arp2/3 complex by the VCA region. Of the WASp family proteins, only N-WASp has two V motifs in the VCA region. However, the VCA regions of all WASp family proteins can activate the actin-nucleating activity of the Arp2/3 complex and induce rapid polymerization of actin (Yamaguchi *et al.*, 2000). The N-WASp VCA region is the most potent activator of Arp2/3 complex-induced-actin polymerization among WASp family proteins. We previously reported that the tandem V motifs are responsible for the strong activity of the N-WASp VCA region (Yamaguchi *et al.*, 2000). However, another group reported that a three-amino-acid stretch in the acidic region is important for the strong activity of N-WASp VCA and that the extra V motif has little effect on activity (Zalevsky *et al.*, 2000). There is a difference in the Arp2/3 complex-mediated actin polymerization activity between the full-length WAVE1 and the free WAVE1 VCA fragment. Full-length WAVE1 protein induces actin polymerization more effectively than does the free WAVE1 VCA fragment (Yamaguchi *et al.*, 2002), suggesting that regions other than the VCA play important roles in the activity of VCA. Thus, we added linker proteins to the WAVE1 VCA region. When fused to GST, maltose binding protein (MBP) and WAVE1 proline-rich domain, N-WASp VCA and WAVE1 VCA mutants containing two V motifs showed stronger activities than wild-type WAVE1 VCA with one V motif, indicating the importance of two V motifs in strong VCA activity. In contrast, WAVE1 VCA fragment tagged with six histidine (His) residues showed markedly reduced activity in comparison with GST-fused VCA, whereas His-tagged N-WASp VCA showed similar activity. Furthermore, an additional V motif failed to enhance activity of the His-tagged form of WAVE1 VCA. Thus, free WAVE1 VCA may assume an unfavourable conformation for activation of the Arp2/3 complex, suggesting that there are structural differences between WAVE1 and N-WASp VCA in addition to the number of V motifs.

Mechanism of activation of N-WASp, WAVE1, and WAVE2

Although the VCA region of N-WASp is the strongest activator among the VCA regions of WASp and WAVE family members, full-length N-WASp did not cause remarkable activation of the Arp2/3 complex, suggesting that the N-WASp molecule assumes a tertiary structure that masks the VCA region. Indeed, the VCA region of N-WASp and WASp was found to be masked by intramolecular interactions (Rohatgi *et al.*, 1999; Miki *et al.*, 1998; Higgs and Pollard, 2000; Suetsugu *et al.*, 2001). This auto-inhibition is released when signalling molecules bind to N-WASp. Cdc42 and phosphatidylinositol (4,5)-bisphosphate (PIP_2) bind to N-WASp through the GDB/CRIB motif and WH1 and basic regions respectively and activate the Arp2/3 complex to a level similar to that of the VCA region. Furthermore, proteins with SH3 domains such as WISH (Fukuoka *et al.*, 2001), Ash/Grb2 (Carlier *et al.*, 2000), Nck (Rohatgi *et al.*, 2001), and profilin (Yang *et al.*, 2000; Suetsugu *et al.*, 1998), bind to the proline-rich region of N-WASp and completely or partially release the auto-inhibition, thereby activating N-WASp. Taken together, these data indicate that molecules that bind to N-WASp or WASp activate them, although the potencies of these binding molecules are different.

WAVE1, 2 and 3 also have VCA regions similar to those of WASp and N-WASp (Suetsugu *et al.*, 1999). Thus, a similar mechanism of activation is likely for WAVEs. However, it was unclear how WAVEs are activated because WAVE proteins are already active when purified. WAVE1 is localized at membrane ruffles, and a dominant negative WAVE1, which has the verprolin-homology region (V region) deleted, inhibited Rac-induced formation of membrane ruffles, suggesting that WAVEs might function downstream of Rac (Miki *et al.*, 1998). Yeast two-hybrid system analyses yielded an adapter protein, IRSp53, that associates with WAVE2 (Miki *et al.*, 2000). IRSp53 binds to the proline-rich region of WAVE2 through the SH3 domain present in its C-terminus. IRSp53 also associates with activated Rac through a Rac-binding (RCB) domain in the N-terminus, linking the Rac signal to Arp2/3 complex-mediated actin polymerization and formation of lamellipodia (Miki

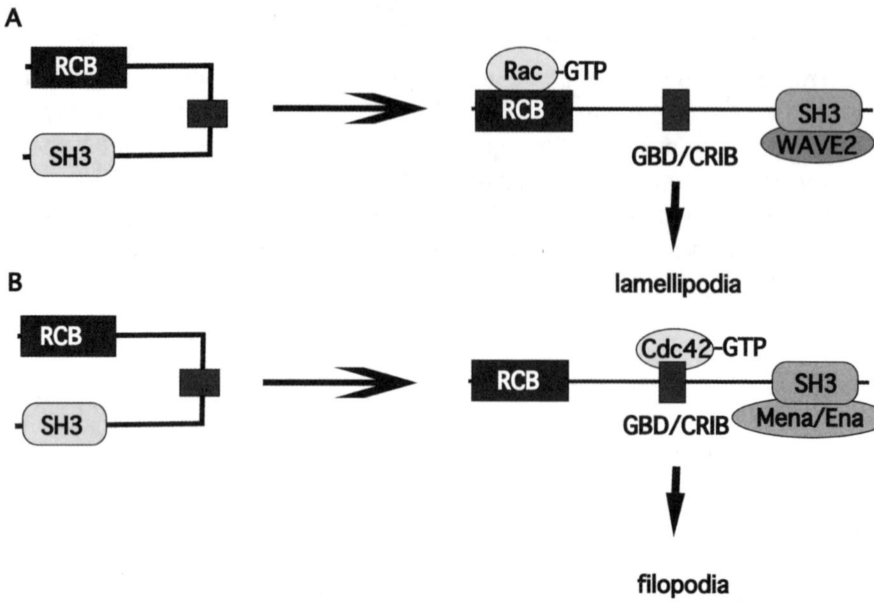

Figure 9.2 IRSp53 transmits signals from both Rac and Cdc42 for lamellipodia and filopodia formation. IRSp53 has an SH3 domain at C-terminus, a centrally located GBD/CRIB motif and an RCB (Rac binding site) at the N-terminus. IRSp53 binds to WAVE2 through its SH3 domain and to Rac through its RCB domain (upper panel), leading to lamellipodium formation. IRSp53 also binds to Cdc42 through its GDB/CRIB motif and associates with Mena/Ena through its C-terminal SH3 domain (lower panel), leading to filopodium formation

et al., 2000; Miki and Takenawa, 2002) (Figure 9.2). Curiously, IRSp53 was found to bind predominantly to WAVE2, and binding to WAVE1 and 3 was very weak. This pathway presumably plays a significant role in several kinds of cells because dominant-negative (ΔSH3) IRSp53 inhibits Rac-induced formation of lamellipodia (Miki *et al.*, 2000). However, it was recently reported that IRSp53 has a CRIB domain at its centre, to which Cdc42 can bind (Krugmann *et al.*, 2001; Govind *et al.*, 2001). Addition of Cdc42 to cells expressing IRSp53 induces formation of filopodia but not lamellipodia (Figure 9.2), suggesting that IRSp53 alone does not directly influence the shape of actin structures but instead transmits upstream signals to downstream proteins involved in actin assembly. In this case, Rac or Cdc42 rather than IRSp53 may determine the shape of actin filaments.

Eden *et al.* (2002) recently proposed a new mechanism of activation of WAVE1 (Figure 9.3). They isolated WAVE1 protein complexes from bovine brain and identified the protein components by mass spectrometry. This complex contained WAVE1, PIR121 (p53-inducible messenger RNA),

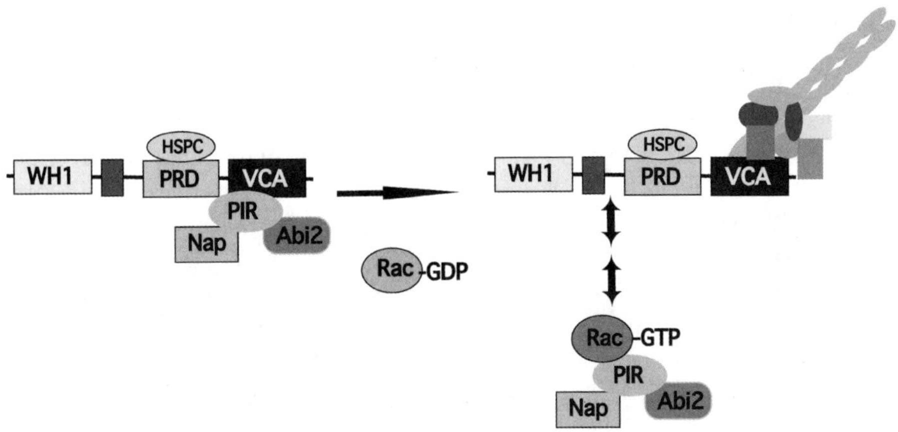

Figure 9.3 Mechanism activation of WAVE1. Under resting conditions, WAVE1 is inactive in a complex with Nap125, PIR121, Abi2, and HSPC300. When cells are stimulated, active Rac binds Nap-PIR-Abi2, and causes them to dissociate from WAVE1-HSPC, resulting in the activation of WAVE1

Nap125 (NCK-associated protein), Abi2 and HSPC300. In this complex, activity of WAVE1 is suppressed. However, active Rac or Nck caused dissociation of WAVE1 and HSCP300 from the complex. This dissociation releases WAVE1, which is thought to be constitutively active by itself, resulting in actin nucleation through Arp2/3 complex activation. However, it remains unclear whether this activation mechanism is valid for other WAVEs or if WAVE1 and WAVE2 are regulated differently.

N-WASp is involved in podosome formation and tubulogenesis

Podosomes occur specifically in monocyte-derived haematopoietic cells including macrophages and osteoclasts. Podosomes contain cell adhesion integrin receptors and several actin-regulating proteins such as cortactin, talin and vinculin that link integrins to the actin cytoskeleton. Podosome formation is also observed in cells transformed with the v-src oncogene. Podosome formation allows transformed cells to invade into extracellular matrix. N-WASp was found to localize at podosomes but WAVEs are not present in this area. Expression of a dominant-negative N-WASp, ΔV N-WASp, disrupted formation of podosomes, suggesting that N-WASp is crucial for podosome formation (Mizutani *et al.*, 2002). However, it is unclear how N-WASp localizes at podosomes. Because cortactin and N-WASp co-localize at podosomes, it is possible that cortactin recruits N-WASp through its SH3 domain. Indeed, the SH3 domain of cortactin binds to N-WASp. In addition, immunoprecipitation experiments revealed that endogenous N-WASp forms complexes with cortactin (Mizutani *et al.*, 2002; Weaver *et al.*, 2000). During tumour metastasis, degradation of the extracellular matrix (ECM) is considered a key process for malignant cells to escape the primary tumour and to invade other tissues. It was shown that matrix metalloproteases (MMPs), which degrade ECM, are localized at the leading edges of migrating cells. MMPs were also found to be concentrated at podosomes. Interestingly, MMP localization at podosomes was dependent on N-WASp. Wild-type N-WASp expression in 3Y1-src cells induced formation of larger podosomes in comparison with controls and localization of high levels of MMPs in podosomes. In contrast, expression of a dominant-negative N-WASp inhibited formation of podosomes and concentration of MMP in podosomes (Mizutani *et al.*, 2002). These results suggest the importance of N-WASp in podosome formation and MMP localization, which are processes underlying the invasive phenotype of 3Y1-src cells.

In multicellular organisms, movement of cells is essential in the normal development of many organs and tissues. Epithelial tubulogenesis, which is necessary for proper development of organs, including the kidneys, lungs, and mammary glands, requires complex cell rearrangements involving cell–cell adhesion, cell polarity, migration and invasion. The organization of the actin cytoskeleton is fundamental to these processes. When cultured in collagen gels, MDCK cells form spherical cysts with fluid-filled lumens. In the presence of hepatocyte growth factor (HGF), MDCK cysts form branching tubules, and branching tube formation is suppressed in MDCK cells expressing a dominant-negative N-WASp (Yamaguchi *et al.*, 2002). During cyst formation, cells in the cyst wall produce small extensions containing actin filaments into the surrounding gel matrix. N-WASp accumulates in these extensions of the cyst walls and in the tips of the extending tubules (Yamaguchi *et al.*, 2002). However, inhibition of N-WASp function by expression of a dominant-negative N-WASp did not affect cell–cell adhesion, cell polarity, or scattering in response to HGF. In contrast, directed migration toward HGF through membrane filters was inhibited, though the inhibitory effect of the dominant-negative N-WASp was relatively small compared with its effect on tubulogenesis. When cultured on collagen gel, MDCK cells grow as a monolayer confined to the surface of the gel. In the presence of HGF, MDCK cells invade the underlying collagen gel and form foci in the gel. MDCK cells expressing a dominant-negative N-WASp showed impairment of the invasive phenotype. In this case, N-WASp was also concentrated at the cell projections that extended into the collagen gel. Strong F-actin signals were also observed at these projections and were clearly co-localized with those of N-WASp (Yamaguchi *et al.*, 2002). These data indicate that N-WASp regulates HGF-induced cell migration and invasion, which are required for epithelial tubulogenesis.

Conclusions

WASp and WAVE family proteins act downstream of Cdc42 and Rac to induce formation of filopodia and lamellipodia, respectively. These structures are necessary for generating movement force at the leading edge of the cell. However, WASp and WAVE family proteins may not be critically involved in random migration of cells. Formation of invasive projections such as podosomes, and invasion into a collagen gel appear to be N-WASp-dependent processes. These data indicate that N-WASp assembles actin filaments and consequently concentrates MMPs at the tips of cellular projections for movement. WAVEs then assemble actin filaments to a mesh-like structure that generates moving force to invade through the ECM.

References

Bear, J. E., Rawls, J. F. and Saxe III, C. L., 1998. SCAR, a WASP-related protein, isolated as a suppressor of receptor defects in late *Dictyostelium* development. *J. Cell Biol.* **142**: 1325–1335.

Bishop, A. L. and Hall, A., 2000. Rho GTPases and their effector proteins. *Biochem. J.* **348**(2): 241–255.

Cameron, L. A., Svitkina, T. M., Vignjevic, D., Theriot, J. A. and Borisy, G. G., 2001. Dendritic organization of actin comet tails. *Curr. Biol.* **11**, 130–135.

Carlier, M.-F., Nioche, P., Broutin-L'Hermite, I., Boujemaa, R., *et al.*, 2000. Grb2 links signaling to actin assembly by enhancing interaction of neural Wiskott–Aldrich syndrome protein (N-WASP) with actin-related protein (Arp2/3) complex. *J. Biol. Chem.* **275**: 21946–21952.

Condeelis, J., 2001. How is actin polymerization nucleated in vivo? *Trends Cell Biol.* **11**: 288–293.

Derry, J. M., Ochs, H. D. and Francke, U., 1994. Isolation of a novel gene mutated in Wiskott-Aldrich syndrome. *Cell* **78**: 635–644.

Eden, S., Rohatgi, R., Podtelejnikov, A. V., Mann, M. and Kirschner, M. W., 2002. Mechanism of regulation of WAVE1-induced actin nucleation by Rac1 and Nck. *Nature* **418**: 790–793.

Fukuoka, M., Miki, H. and Takenawa, T., 1997. Identification of N-WASP homologues in human and rat brain. *Gene* **196**: 43–48.

Fukuoka, M., Suetsugu, S., Miki, H., Fukami, K., *et al.*, 2001. A novel neural Wiskott–Aldrich Syndrome Protein (N-WASP) binding protein, WISH, induces Arp2/3 complex activation independent of Cdc42. *J. Cell Biol.* **152**: 471–482.

Govind, S., Kozma, R., Monfries, C., Lim, L. and Ahmed, S., 2001. Cdc42Hs facilitates cytoskeletal reorganization and neurite outgrowth by localizing the 58-kD insulin receptor substrate to filamentous actin. *J. Cell Biol.* **152**: 579–594.

Hall, A., 1998. Rho GTPase and the actin cytoskeleton. *Science* **279**: 509–514.

Higgs, H. N. and Pollard, T. D., 2000. Activation by Cdc42 and PIP(2) of Wiskott–Aldrich syndrome protein (WASp) stimulates actin nucleation by Arp2/3 complex. *J. Cell Biol.* **150**: 1311–1320.

Kenney, D., Cairns, L., Remold-O'Donnell, E., Peterson, J., *et al.*, 1986. Morphological abnormalities in the lymphocytes of patients with the Wiskott–Aldrich syndrome. *Blood* **68**: 1329–1332.

Krugmann, S., Jordens, I., Gevaert, K., Driessens, M., *et al.*, 2001. Cdc42 induces filopodia by promoting the formation of an IRSp53:Mena complex. *Curr. Biol.* **11**: 1645–1655.

Miki, H. and Takenawa, T., 1998. Direct binding of the verprolin-homology domain in N-WASP to actin is essential for cytoskeletal reorganization. *Biochem. Biophys. Res. Commun.* **243**: 73–78.

Miki, H. and Takenawa, T., 2002. WAVE2 serves a functional partner of IRSp53 by regulating its interaction with Rac. *Biochem. Biophys. Res. Commun.* **293**: 93–99.

Miki, H., Miura, K. and Takenawa, T., 1996. N-WASP, a novel actin-depolymerizing protein, regulates the cortical cytoskeletal rearrangement in a PIP2-dependent manner downstream of tyrosine kinases. *EMBO J.* **15**: 5326–5335.

Miki, H., Sasaki, T., Takai, Y. and Takenawa, T., 1998a. Induction of filopodium formation by WASP-related actin-depolymerizing protein N-WASP. *Nature* **391**: 93–96.

Miki, H., Suetsugu, S. and Takenawa, T., 1998b. WAVE, a novel WASP-family protein involved in actin reorganization induced by Rac. *EMBO J.* **17**: 6932–6941.

Miki, H., Yamaguchi, H., Suetsugu, S. and Takenawa, T., 2000. IRSp53 is an essential intermediate between Rac and WAVE in the regulation of membrane ruffling. *Nature* **408**: 732–735.

Mizutani, K., Miki, H., He, H., Maruta, H. and Takenawa, T., 2002. Essential role of neural Wiskott-Aldrich syndrome protein in podosome formation and degradation of extracellular matrix in src-transformed fibroblasts. *Cancer Res.* **62**: 669–674.

Molina, I. J., Kenney, D. M., Rosen, F. S. and Remold-O'Donnell, E., 1992. T cell lines characterize events in the pathogenesis of the Wiskott–Aldrich syndrome. *J. Exp. Med.* **176**: 867–874.

Rohatgi, R., Ma, L., Miki, H., Lopez, M., *et al.*, 1999. The interaction between N-WASP and the Arp2/3 complex links Cdc42-dependent signals to actin assembly. *Cell* **97**: 221–231.

Rohatgi, R., Nollau, P., Ho, H. Y., Kirschner, M. W. and Mayer, B. J., 2001. Nck and phosphatidylinositol 4,5-bisphosphate synergistically activate actin polymerization through the N-WASP-Arp2/3 pathway. *J. Biol. Chem.* **276**: 26448–26452.

Suetsugu, S., Miki, H. and Takenawa, T., 1998. The essential role of profilin in the assembly of actin for microspike formation. *EMBO J.* **17**: 6516–6526.

Suetsugu, S., Miki, H. and Takenawa, T., 1999. Identification of two human WAVE/SCAR homologues as general actin regulatory molecules which associate with Arp2/3 complex. *Biochem. Biophys. Res. Commun.* **260**: 296–302.

Suetsugu, S., Miki, H. and Takenawa, T., 2001. Identification of another Actin-related protein (Arp) 2/3 complex binding site in Neural Wiskott–Aldrich syndrome protein (N-WASP), that complements actin polymerization induced by the Arp2/3 complex activating (VCA) domain of N-WASP. *J. Biol. Chem.* **276**: 33175–33180.

Suetsugu, S., Miki, H. and Takenawa, T., 2002. Spatial and temporal regulation of actin polymerization for cytoskeleton formation through Arp2/3 complex and WASP/WAVE proteins. *Cell Motil. Cytoskeleton* **51**: 113–122.

Symons, M., Derry, J. M., Karlak, B., Jiang, S., *et al.*, 1996. Wiskott–Aldrich syndrome protein, a novel effector for the GTPase CDC42Hs, is implicated in actin polymerization. *Cell* **84**: 723–734.

Wear, M. A., Schafer, D. A. and Cooper, J. A., 2000. Actin dynamics: assembly and disassembly of actin networks. *Curr. Biol.* **10**: R891–895.

Weaver, A., Heuser, J., Karginov, A., Lee, W., *et al.*, 2002. Interaction of cortactin and N-wasp with arp2/3 complex. *Curr. Biol.* **12**: 1270–1278.

Welch, M. D., Iwamatsu, A. and Mitchison, T. J., 1997. Actin polymerization is induced by Arp2/3 protein complex at the surface of *Listeria monocytogenes. Nature* **385**: 265–269.

Yamaguchi, H., Miki, H., Suetsugu, S., Ma, L., *et al.*, 2000. Two tandem verprolin homology domains are necessary for a strong activation of Arp2/3 complex-induced actin polymerization and induction of microspike formation by N-WASP. *Proc. Natl. Acad. Sci. USA* **97**: 12631–12636.

Yamaguchi, H., Miki, H. and Takenawa, T., 2002a. Neural Wiskott–Aldrich Syndrome Protein is involved in hepatocyte growth factor-induced migration, invasion, and tubulogenesis of epithelial cells. *Cancer Res.* **62**: 2503–2509.

Yamaguchi, H., Miki, H. and Takenawa, T., 2002b. Two verprolin homology domains increase the Arp2/3 complex-mediated actin polymerization activities of N-WASP and WAVE1 C-terminal regions. *Biochem. Biophys. Res. Commun.* **297**: 214–219.

Yang, C., Huang, M., DeBiasio, J., Pring, M., *et al.*, 2000. Profilin enhances Cdc42-induced nucleation of actin polymerization. *J. Cell Biol.* **150**: 1001–1012.

Zalevsky, J., Lempert, L., Kranitz, H. and Mullins, R. D., 2001. Different WASP family proteins stimulate different Arp2/3 complex-dependent actin-nucleating activities. *Curr. Biol.* **11**: 1903–1913.

10

Regulation and Function of the Small GTP-binding Protein ARF6 in Membrane Dynamics

Thierry Dubois, Emma Colucci-Guyon, Florence Niedergang, Magali Prigent and **Philippe Chavrier**

ADP-ribosylation factor 6 (ARF6) is the most divergent member of the ARF subgroup of Ras-related small GTP-binding proteins. ARF6 activation, i.e., GTP-loading, occurs at the plasma membrane and is catalysed by Sec7 domain-containing guanine nucleotide exchange factors belonging to the EFA6 and ARNO families. In its active GTP-bound state, ARF6 regulates internalization of membrane components via the endocytic pathway and controls membrane recycling to regions of plasma membrane and cortical actin cytoskeleton remodelling. Recently, several partners of GTP-bound ARF6 have been identified that mediate ARF6 function in plasma membrane organization.

ADP-ribosylation factors (ARFs) form a group of six, small (20-kDa) GTP-binding proteins related to Ras. ARF1 is localized to the Golgi complex where it regulates the recruitment of coat proteins during the formation of transport vesicles (Chavrier and Goud, 1999). Available information regarding ARF3 and ARF5, although limited, points toward a similar role for these proteins in maintaining the secretory pathway (Moss and Vaughan,

Cell Motility. Edited by Anne Ridley, Michelle Peckham and Peter Clark.
 ISBN 0 470 84872 3

1998; Claude *et al.*, 1999). In contrast, ARF6, the least-conserved ARF-family member (67% amino-acid identity with ARF1), is localized to the plasma membrane and early endosomal compartments where it regulates endocytic traffic at the cell periphery (Chavrier and Goud, 1999). Here, we review recent observations that implicate ARF6 as a major regulator of plasma membrane organization, coordinating endocytic traffic with cortical actin cytoskeleton remodelling. In agreement with its proposed function, ARF6 has been found to be essential during cell motility and phagocytosis, two processes requiring membrane movement toward the cell surface and actin cytoskeleton organization.

Intracellular localization of ARF6

In ARFs the classical GDP/GTP nucleotide switch is coupled to a cytosol/membrane transition. The N-terminal myristoylated helix is retracted in ARF:GDP and is released during the GDP-to-GTP conformational switch to insert into the lipid bilayer (Pasqualato *et al.*, 2001). However, despite highly conserved structural features, ARF proteins differ in respect to their steady-state association with membranes: ARF1, ARF3 and ARF5 are found predominantly in the cytosol (GDP-bound), while a major fraction of ARF6 (50% to 100% depending on the cell lines) is membrane-bound (Cavenagh *et al.*, 1996; Prigent M. and Chavrier P., unpublished). These observations could suggest that ARF6 is predominantly in the GTP-bound active conformation. However, using a pull-down assay selectively to precipitate GTP-bound ARF6, we found that ARF6:GTP is low in cultured cells (few percent of total both in Hela cells and mouse RAW 264.7 macrophages). Therefore, we should also consider the possibility that ARF6:GDP is able to associate stably with membranes by interacting with some unknown lipid(s) or protein(s). Moreover, this is not the only example ARF6's special properties. Indeed, a mutant of ARF6 (Q67L) predicted to be defective for GTP hydrolysis (thus mainly GTP-bound) accumulates at the plasma membrane where it induces the formation of actin-based protrusions, while another mutant, ARF6T27N, thought to be in the GDP-bound inactive state, is mostly localized to internal tubulovesicular endosomes (Peters *et al.*, 1995; D'Souza-Schorey *et al.*, 1998). These observations, which suggest differential localization for the active and inactive ARF6 conformations, have been taken as evidence for a role of ARF6 activation for targeted delivery of recycling endosomal vesicles to the plasma membrane (see below and in D'Souza-Schorey *et al.*, 1998; Brown *et al.*, 2001). The mechanism of ARF6 activation has, therefore, become an important question.

Regulation of ARF6 activation

As for all Ras-like proteins, spontaneous GDP release from ARF is a slow reaction that must be catalysed by a guanine nucleotide exchange factor (GEF). A genetic screen for suppressors of ARF2 mutations in *Saccharomyces cerevisiae* led to the identification of the first ARF-specific GEFs called GEA1 and GEA2 (Peyroche *et al.*, 1996). A large number of ARF GEFs have now been characterized (Donaldson and Jackson, 2000). These proteins share a conserved ~200 amino-acid catalytic sec7 domain, and can be separated into two classes based on their sensitivity to inhibition by the drug Brefeldin A (Bfa) (Peyroche *et al.*, 1999). ARNO was initially identified as a Bfa-insensitive ARF1-GEF in mammals (Chardin *et al.*, 1996), and later shown to be weakly active on ARF6 (Frank *et al.*, 1998; Santy and Casanova, 2001). However, our *in vitro* studies have led to the conclusion that ARNO, as well as other ARNO-family members (Donaldson and Jackson, 2000) (see Figure 10.1 for a scheme of ARF GEFs), promotes nucleotide exchange specifically on ARF1 (Franco *et al.*, 1998; Macia *et al.*, 2001). The reason for this disparity remains unclear. In addition, the intracellular localization of ARNO family members is a matter of debate. It has been reported that ARNO-family GEFs accumulate in the cytosol when over-expressed, and are recruited to the plasma membrane in response to receptor tyrosine kinase activation (Venkateswarlu *et al.*, 1998). Membrane translocation requires the ARNO PH domain, which interacts with by phosphoinositide 3-kinase (PI3K) lipid products (Klarlund *et al.*, 1997; Venkateswarlu *et al.*, 1998) (Figure 10.1). These observations, together with the fact that both ARF6 activation and ARNO-GEF activity were insensitive to Bfa, suggested that ARNO-family GEFs promoted ARF6 activation at the plasma membrane in response to

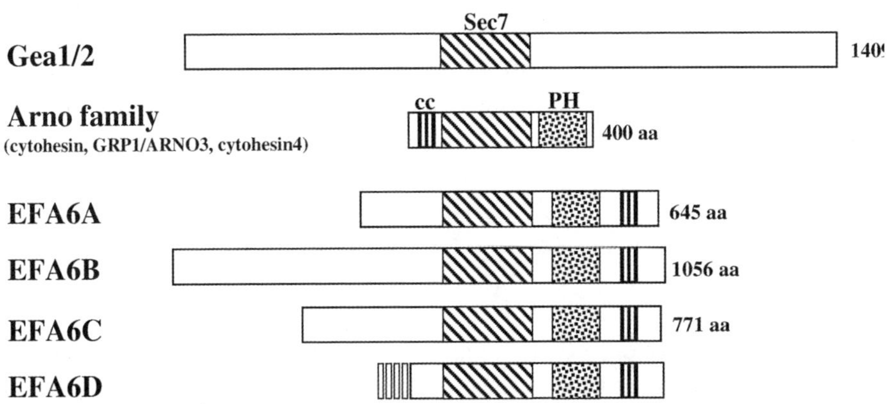

Figure 10.1 Overall structure of sec7 domain-containing ARF GEFs. Sec7 domains, PH domains and putative coiled-coil (cc) regions are depicted (see text). Only a partial sequence of EFA6D/KIAA0942 is available in databases

PI3K activation (Langille *et al.*, 1999). However, it should be noted that ARF1 controls paxillin recruitment to focal adhesions, raising the possibility that ARF1 may be a target for ARNO-family GEFs at peripheral locations (Norman *et al.*, 1998). In addition, other studies found that ARNO associates with the Golgi apparatus, and that this association requires a conserved amino-terminal coiled-coil motif (Figure 10.1) (Lee and Pohajdak, 2000; Mansour *et al.*, 2002). In agreement with this result, over-expression of ARNO GEFs resulted in disruption of the Golgi apparatus and impaired protein transport through the secretory pathway (Monier *et al.*, 1998; Franco *et al.*, 1998). The determination of the precise location(s) of ARNO should help clarify these controversial issues, but awaits the development of specific antibodies able to detect the endogenous protein.

Based on sequence conservation in the Sec7 domain we identified a new family of proteins, called EFA6 (Exchange Factor for ARF6), that promote GDP/GTP exchange specifically on ARF6 in a Bfa-insensitive manner (Franco *et al.*, 1999; Derrien *et al.*, 2002). The EFA6-GEF family consists of four members in humans (called EFA6A to D), all of which share a conserved carboxy-terminal module. In addition to the Sec7 domain, this module contains a PH domain of unknown ligand specificity required for membrane association, as well as a putative coiled-coil motif (Figure 10.1). In contrast to ARNO-family GEFs, over-expressed EFA6A is constitutively bound to the plasma membrane where it induces the formation of actin-based membrane ruffles (Franco *et al.*, 1999). EFA6-mediated actin cytoskeleton reorganization requires the carboxy-terminal region including the coiled-coil motif, and expression of this region alone (PH domain plus coiled-coil) triggers lengthening of actin-rich microvillus-like structures at the surface of fibroblastic cells (Franco *et al.*, 1999; Derrien *et al.*, 2002). Finally, and similarly to the active GTP-bound ARF6Q67L mutant (see below, and D'Souza-Schorey *et al.*, 1995), EFA6A and EFA6B were found to inhibit endocytosis of transferrin (Tfn). All together, these results are consistent with a dual function of EFA6-family GEFs: activation of ARF6 at the plasma membrane through their Sec7 domain and actin cytoskeleton reorganization.

Function of ARF6 in polarized membrane delivery at the plasma membrane

Initial studies by D'Souza-Schorey and colleagues showed that Tfn receptors (Tfn-Rs) accumulated at the surface of ARF6Q67L (GTP-bound)-expressing CHO cells, while they were trapped intracellularly in cells expressing the dominant-negative ARF6T27N mutant, indicating that ARF6 activation is essential for Tfn-R recycling from early endocytic compartments to the cell

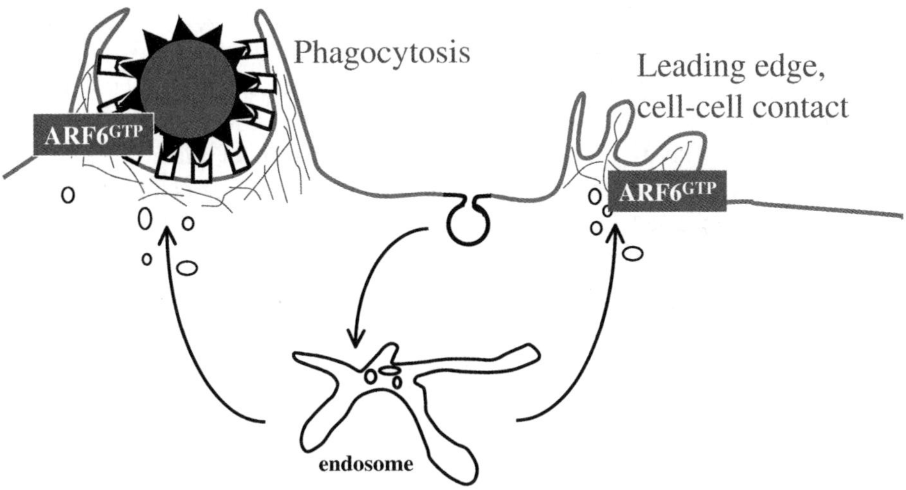

Figure 10.2 A model for ARF6 function in the control of membrane recycling at the plasma membrane. Engagement of surface receptors such as FcRs on macrophages triggers the activation of ARF6 at the plasma membrane. The signalling cascades leading to GTP-loading on ARF6 are not precisely known. They probably involve Sec7 domain-containing GEFs belonging to the ARNO and EFA6 families. Accumulation of GTP-bound ARF6 leads to the recruitment and activation of specific effector proteins, including PLDs and PIP 5K. Production of PA and PIP_2 by these enzymes participates in the remodelling of the plasma membrane and of the cortical actin cytoskeleton. Yet unknown ARF6-specific effectors are probably also involved in controlling the delivery and fusion of recycling vesicles from the endosomes. Membrane recycling is essential for the formation of membrane protrusions such as pseudopods during phagocytosis, or lamellipodia at the leading edge of migrating cells

surface (D'Souza-Schorey *et al.*, 1995). Along similar lines, treatment of HeLa cells over-expressing ARF6 with aluminium fluoride (AlF) was found to trigger the formation of actin-based membrane protrusions that contained ARF6 (Radhakrishna *et al.*, 1996, 1999; Radhakrishna and Donaldson, 1997). Although the mechanism of action of AlF on ARFs is mysterious and may even be indirect through heterotrimeric G proteins, Donaldson and colleagues have proposed that ARF6 activation (by AlF) induces transport of membranes from recycling endosomes to the plasma membrane and influences the actin cytoskeleton along the plasma membrane. Of special note, in contrast to the situation in CHO cells, ARF6 was found to regulate traffic of membrane between the plasma membrane and a non-clathrin-derived early endosomal compartment in HeLa cells (Radhakrishna and Donaldson, 1997). Finally, it has been reported that the active ARF6Q67L mutant stimulates clathrin-mediated endocytosis in epithelial Madin–Darby Canine Kidney cells (Altschuler *et al.*, 1999; Palacios *et al.*, 2001). Interestingly, ARF6-induced activation of the clathrin-dependent pathway was found to affect epithelial cell–cell adhesion by promoting the internalization of E-cadherin, the main adhesive component of adherens junctions (Palacios *et al.*, 2001). All together, these findings are suggestive of cell-type-specific mechanisms associated with ARF6 function in the endocytic pathway.

Evidence is accumulating that uptake and insertion of membrane at the plasma membrane is important for cellular processes such as cell motility and phagocytosis (reviewed in Bretscher and Aguado-Velasco, 1998; Mellman, 2000). The observations that inhibition of ARF6 activity interferes with cell spreading, cell motility, and phagocytosis (Zhang *et al.*, 1998; Song *et al.*, 1998; Santy and Casanova, 2001; Palacios *et al.*, 2001) also support the importance of the recycling pathway for processes requiring membrane protrusion activities. Phagocytosis is triggered by the recognition of ligands exposed on the particle surface by specific receptors on the phagocytic cell (Aderem and Underhill, 1999). Receptor clustering activates a signalling cascade, which in the case of FcγRs involves tyrosine kinases and PI3K (Cox and Greenberg, 2001), leading to the formation of actin-rich pseudopods that progressively engulf the particle in a zipper-like mechanism. It is generally accepted that actin polymerization provides the force necessary for pseudopod extension, and in support of this, Rho-family members are known to be essential for phagocytosis (Castellano *et al.*, 2001). In addition, exocytosis of membrane from internal compartments contributes to pseudopod formation (Booth *et al.*, 2001). In particular, endocytic vesicles bearing VAMP3 (vesicle-associated membrane protein 3) are delivered to the site of phagocytosis suggesting that recycling from endocytic compartments contributes to pseudopod formation (Bajno *et al.*, 2000). ARF6 activity was known to be required during FcγR-mediated phagocytosis (Zhang *et al.*, 1998), however the precise function of ARF6 in phagocytosis remains largely unknown. Recently we began to investigate the role of ARF6 during phagocytosis in macrophages. Our data suggest that local activation of ARF6 at the plasma membrane following FcγR-clustering controls the polarized delivery of membrane components that traffic through the recycling pathway and are essential for pseudopod formation (Niedergang *et al.*, submitted).

Events downstream of ARF6 activation

The identification of the downstream mechanisms by which ARF6:GTP controls membrane trafficking in the endocytic pathway remains a major issue. Two well-characterized ARF effectors are phospholipase D (PLD) and phosphatidylinositol (4)-phosphate 5-kinase (PIP 5K) (Cockcroft, 2001; Honda *et al.*, 1999). PLDs convert phosphatidylcholine into phosphatidic acid (PA), while PIP 5-kinase catalyses the production of phosphatidylinositol (4,5)-bisphosphate (PIP_2). Recently, stimulation of PLD activity by ARF6 has been implicated in the activation of epithelial cell motility (Santy and Casanova, 2001). PLD is also activated and required during FcγR-mediated phagocytosis, and it has been suggested that ARF6 could be involved in PLD activation following FcγR engagement (Kusner *et al.*, 1999; Melendez *et al.*, 2001). There are also clear indications that PIP_2 accumulates in the forming pseudopods at the onset of phagocytosis, however it is not known whether ARF6 regulates this PIP_2 production (Botelho *et al.*, 2000). It is anticipated that activation of PLD and PIP 5K by ARF6 should have a major impact on the organization and property of the cell cortex, as PA and PIP_2 are well-known regulators of membrane trafficking and actin cytoskeleton remodelling. By analogy with other small GTP-binding proteins, it is very likely that ARF6 performs multiple functions at various stages of the recycling pathway including control of vesicle movement, vesicle docking and fusion with the plasma membrane. PLD and PIP 5K are certainly not the only players and it is time to compile a comprehensive list of proteins that interact with GTP-bound ARF6, and to understand the mechanism of their activation by this fascinating small G protein.

Acknowledgements

We are grateful to Dr J. Plastino for critical reading of the manuscript. This work was supported by institutional funding from the CNRS and grants from the Institut Curie, the PRFMMIP (Programme de Recherche Fondamentale en Microbiologie et Maladies Infectieuses et Parasitaires) and the Ligue Nationale contre le Cancer. TD is a recipient of a post-doctoral fellowship from ARC (Association pour la Recherche contre le Cancer).

References

Aderem, A. and Underhill, D. M., 1999. Mechanisms of phagocytosis in macrophages. *Annu. Rev. Immunol.* **17**: 593–623.

Altschuler, Y., Liu, S., Katz, L., Tang, K., Hardy, S., Brodsky, F., Apodaca, G. and Mostov, K., 1999. ADP-ribosylation factor 6 and endocytosis at the apical surface of madin-darby canine kidney cells. *J. Cell Biol.* **147**: 7–12.

Bajno, L., Peng, X. R., Schreiber, A. D., Moore, H. P., Trimble, W. S. and Grinstein, S., 2000. Focal exocytosis of VAMP3-containing vesicles at sites of phagosome formation. *J. Cell Biol.* **149**: 697–706.

Booth, J. W., Trimble, W. S. and Grinstein, S., 2001. Membrane dynamics in phagocytosis. *Semin Immunol.* **13**: 357–364.

Botelho, R. J., Teruel, M., Dierckman, R., Anderson, R., Wells, A., York, J. D., Meyer, T. and Grinstein, S., 2000. Localized biphasic changes in phosphatidylinositol-4,5-bisphosphate at sites of phagocytosis. *J. Cell Biol.* **151**: 1353–1368.

Bretscher, M. S. and Aguado-Velasco, C., 1998. Membrane traffic during cell locomotion. *Curr. Opin. Cell Biol.* **10**: 537–541.

Brown, F. D., Rozelle, A. L., Yin, H. L., Balla, T. and Donaldson, J. G., 2001. Phosphatidylinositol 4,5-bisphosphate and Arf6-regulated membrane traffic. *J. Cell Biol.* **154**: 1007–1017.

Castellano, F., Chavrier, P. and Caron, E., 2001. Actin dynamics during phagocytosis. *Semin. Immunol.* **13**: 347–355.

Cavenagh, M. M., Whitney, J. A., Carroll, K., Zhang C. J., Boman, A. L., Rosenwald, A. G., Mellman, I. and Kahn, R. A., 1996. Intracellular distribution of Arf proteins in mammalian cells. Arf6 is uniquely localized to the plasma membrane. *J. Biol. Chem.* **271**: 21767–21774.

Chardin, P., Paris, S., Antonny, B., Robineau, S., Béraud-Dufour, S., Jackson, C. L. and Chabre, M., 1996. A human exchange factor for ARF contains Sec7- and pleckstrin-homology domains. *Nature* **384**: 481–484.

Chavrier, P. and Goud, B., 1999. The role of ARF and rab GTPases in membrane transport. *Curr. Opin. Cell Biol.* **11**: 466–475.

Claude, A., Zhao, B. P., Kuziemsky, C. E., Dahan, S., Berger, S. J., Yan, J. P., Armold, A. D., Sullivan, E. M. and Melançon, P., 1999. GBF1. A novel golgi-associated bfa-resistant guanine nucleotide exchange factor that displays specificity for adp-ribosylation factor 5. *J. Cell Biol.* **146**: 71–84.

Cockcroft, S., 2001. Signalling roles of mammalian phospholipase D1 and D2. *Cell Mol. Life Sci.* **58**: 1674–1687.

Cox, D. and Greenberg, S., 2001. Phagocytic signaling strategies: Fc(gamma)receptor-mediated phagocytosis as a model system. *Semin. Immunol.* **13**: 339–345.

D'Souza-Schorey, C., Li, G., Colombo, M. I. and Stahl, P. D., 1995. A regulatory role for ARF6 in receptor-mediated endocytosis. *Science* **267**: 1175–1178.

D'Souza-Schorey, C., van Donselaar, E., Hsu, V. W., Yang, C., Stahl, P. D. and Peters, P. J., 1998. ARF6 targets recycling vesicles to the plasma membrane: insights from an ultrastructural investigation. *J. Cell Biol.* **140**: 603–616.

Derrien, V., Couillault, C., Franco, M., Martineau, S., Montcourrier, P., Houlgatte, R. and Chavrier, P., 2002. A conserved C-terminal domain of EFA6-family ARF6-guanine nucleotide exchange factors induces lengthening of microvilli-like membrane protrusions. *J. Cell Sci.* **115**: 2867–2879.

Donaldson, J. G. and Jackson, C. L., 2000. Regulators and effectors of the ARF GTPases. *Curr. Opin. Cell Biol.* **12**: 475–482.

Franco, M., Boretto, J., Robineau, S., Monier, S., Goud, B., Chardin, P. and Chavrier, P., 1998. ARNO3, a Sec7-domain guanine nucleotide exchange factor for ADP ribosylation factor 1, is involved in the control of golgi structure and function. *Proc. Natl. Acad. Sci. USA* **95**: 9926–9931.

Franco, M., Peters, P. J., Boretto, J., van Donselaar, E., Neri, A., D'Souza-Schorey, C. and Chavrier, P., 1999. EFA6, a sec7 domain-containing exchange factor for ARF6, coordinates membrane recycling and actin cytoskeleton organization. *EMBO J.*, **18**: 1480–1491.

Frank, S., Upender, S., Hansen, S. H. and Casanova, J. E., 1998. ARNO is a guanine nucleotide exchange factor for ADP-ribosylation factor 6. *J. Biol. Chem.* **273**: 23–27.

Honda, A., Nogami, M., Yokozeki, T., Yamazaki, M., Nakamura, H., Watanabe, H., Kawamoto, K., Nakayama, K., Morris, A. J., Frohman, M. A. and Kanaho, Y., 1999. Phosphatidylinositol 4-phosphate 5-kinase alpha is a downstream effector of the small G protein ARF6 in membrane ruffle formation. *Cell* **99**: 521–532.

Klarlund, J. K., Guilherme, A., Holik, J. J., Virbasius, J. V., Chawla, A. and Czech, M. P., 1997. Signaling by Phosphoinositide-3,4,5-Trisphosphate Through Proteins Containing Pleckstrin and Sec7 Homology Domains. *Science* **275**: 1927–1930.

Kusner, D. J., Hall, C. F. and Jackson, S., 1999. Fc gamma receptor-mediated activation of phospholipase D regulates macrophage phagocytosis of IgG-opsonized particles. *J. Immunol.* **162**: 2266–2274.

Langille, S. E., Patki, V., Klarlund, J. K., Buxton, J. M., Holik, J. J., Chawla, A., Corvera, S. and Czech, M. P., 1999. ADP-ribosylation factor 6 as a target of guanine nucleotide exchange factor GRP1. *J. Biol. Chem.* **274**: 27099–27104.

Lee, S. Y. and Pohajdak, B., 2000. N-terminal targeting of guanine nucleotide exchange factors (GEF) for ADP ribosylation factors (ARF) to the Golgi. *J Cell Sci.* **113**: 1883–1889.

Macia, E., Chabre, M. and Franco, M., 2001. Specificities for the small G proteins ARF1 and ARF6 of the guanine nucleotide exchange factors ARNO and EFA6. *J. Biol. Chem.* **276**: 24925–24930.

Mansour, M., Lee, S. Y. and Pohajdak, B., 2002. The N-terminal Coiled Coil Domain of the Cytohesin/ARNO Family of Guanine Nucleotide Exchange Factors Interacts with the Scaffolding Protein CASP. *J. Biol. Chem.* **277**: 32302–32309.

Melendez, A. J., Harnett, M. M. and Allen, J. M., 2001. Crosstalk between ARF6 and protein kinase Calpha in Fc(gamma)RI-mediated activation of phospholipase D1. *Curr. Biol.* **11**: 869–874.

Mellman, I., 2000. Quo vadis: polarized membrane recycling in motility and phagocytosis. *J. Cell Biol.* **149**: 529–530.

Monier, S., Chardin, P., Robineau, S. and Goud, B., 1998. Overexpression of the ARF1 exchange factor ARNO inhibits the early secretory pathway and causes the disassembly of the Golgi complex. *J. Cell Sci.* **111**: 3427–3436.

Moss, J. and Vaughan, M., 1998. Molecules in the ARF orbit. *J. Biol. Chem.* **273**: 21431–21434.

Norman, J. C., Jones, D., Barry, S. T., Holt, M. R., Cockcroft, S. and Critchley, D. R., 1998. ARF1 mediates paxillin recruitment to focal adhesions and potentiates rho-stimulated stress fiber formation in intact and permeabilized swiss 3T3 fibroblasts [In Process Citation]. *J. Cell Biol.* **143**: 1981–1995.

Palacios, F., Price, L., Schweitzer, J., Collard, J. and D'Souza-Schorey, C., 2001. An essential role for ARF6-regulated membrane traffic in adherens junction turnover and epithelial cell migration. *EMBO J.* **20**: 4973–4986.

Pasqualato, S., Menetrey, J., Franco, M. and Cherfils, J., 2001. The structural GDP/GTP cycle of human Arf6. *EMBO Reports* **2**: 234–238.

Peters, P. J., Hsu, V. W., Ooi, C. E., Finazzi, D., Teal, S. B., Oorschot, V., Donaldson, J. G. and Klausner, R. D., 1995. Overexpression of wild-type and mutant ARF1 and ARF6: distinct perturbations of nonoverlapping membrane compartments. *J. Cell Biol.* **128**: 1003–1017.

Peyroche, A., Antonny, B., Robineau, S., Acker, J., Cherfils, J. and Jackson, C. L., 1999. Brefeldin A acts to stabilize an abortive ARF-GDP-Sec7 domain protein complex: involvement of specific residues of the Sec7 domain. *Molecular Cell* **3**: 275–285.

Peyroche, A., Paris, S. and Jackson, C. L., 1996. Nucleotide exchange on ARF mediated by yeast Gea1 protein. *Nature* **384**: 479–481.

Radhakrishna, H., Al-Awar, O., Khachikian, Z. and Donaldson, J. G., 1999. ARF6 requirement for Rac ruffling suggests a role for membrane trafficking in cortical actin rearrangements. *J. Cell Sci.* **112**: 855–866.

Radhakrishna, H. and Donaldson, J. G., 1997. ADP-Ribosylation Factor 6 Regulates a Novel Plasma Membrane Recycling Pathway. *J. Cell Biol.* **139**: 49–61.

Radhakrishna, H., Klausner, R. D. and Donaldson, J. G., 1996. Aluminum fluoride stimulates surface protrusions in cells overexpressing the ARF6 GTPase. *J. Cell Biol.* **134**: 935–947.

Santy, L. C. and Casanova, J. E., 2001. Activation of ARF6 by ARNO stimulates epithelial cell migration through downstream activation of both Rac1 and phospholipase D. *J. Cell Biol.* **154**: 599–610.

Song, J., Khachikian, Z., Radhakrishna, H. and Donaldson, J. G., 1998. Localization of endogenous ARF6 to sites of cortical actin rearrangement and involvement of ARF6 in cell spreading. *J. Cell Sci.* **111**: 2257–2267.

Venkateswarlu, K., Oatey, P. B., Tavare, J. M. and Cullen, P. J., 1998. Insulin-dependent translocation of ARNO to the plasma membrane of adipocytes requires phosphatidylinositol 3-kinase. *Curr. Biol.* **8**: 463–466.

Zhang, Q., Cox, D., Tseng, C. C., Donaldson, J. G. and Greenberg, S., 1998. A requirement for ARF6 in fcgamma receptor-mediated phagocytosis in macrophages. *J. Biol. Chem.* **273**: 19977–19981.

11

Chemotaxis of Cancer Cells during Invasion and Metastasis

John Condeelis, Xiaoyan Song, Jonathan M. Backer, Jeffrey Wyckoff and **Jeffrey Segall**

Amoeboid chemotaxis is a basic property of cells engaged in embryogenesis, inflammation, epithelial remodelling and growth, wound healing, angiogenesis and tumour metastasis. The mechanisms used by crawling cells for chemotaxis are just beginning to be explored at the molecular level (Wells *et al.*, 1999; Comer and Parent, 2002; Bailly *et al.*, 2001). Chemotaxis is probably a general property of metastatic cells. A complete understanding of chemotaxis will have far-reaching effects on the treatment of invasive carcinomas by impacting the development of the next generation of cytostatic chemotherapy agents directed against invasive cell motility. In this chapter we consider the earliest events in actin polymerization-based leading edge assembly and how they define the initial asymmetry that is required for chemotaxis of carcinoma cells to EGF.

Chemotaxis to EGF

EGF, acting through the tyrosine kinase receptor family of erbB1-4, is a major growth factor for epithelia in mammals. The level of expression of EGF and

Cell Motility. Edited by Anne Ridley, Michelle Peckham and Peter Clark.
 ISBN 0 470 84872 3

its receptor is well correlated clinically with invasive carcinomas of the breast and prostate (Sherwood and Lee, 1995; Klijn *et al.*, 1994). Growth factors including EGF are present in serum, macrophages, platelets and smooth muscle cells near blood vessels (Calabro *et al.*, 1997; Kume and Gimbrone, 1994). Release of growth factors from these cells or endothelial cells could provide a gradient that would produce a chemotactic response toward blood vessels. This is important since intravasation, the movement of cells into blood vessels, is a major route for egress of carcinoma cells from tumours (Wyckoff *et al.*, 2000). In fact, we have shown a strong correlation between the density of live carcinoma cells in the blood and the number of both single cells and metastases in the lungs establishing intravasation as a key step for metastasis. The behaviour behind this correlation at the single cell level was proposed to be chemotaxis of carcinoma cells within the primary tumour toward growth factors like EGF that are associated with blood vessels, and differences in the ability of carcinoma cells to resist lysis upon intravasation (Wyckoff *et al.*, 2000).

We have investigated the interaction of carcinoma cells with blood vessels in metastatic primary tumours in more detail and determined if carcinoma cells have the intrinsic ability to locomote toward blood vessels and to intravasate without fragmentation. To do this we used GFP expression by carcinoma cells and time-lapse confocal microscopy to image the behaviour of single carcinoma cells near blood vessels. Imaging was carried out in metastatic primary tumours generated by injection of a metastatic mammary carcinoma cell line, MTLn3, into the mammary fat pads of Fisher rats. As shown in Figure 11.1, carcinoma cells are elongated and polarized towards blood vessels suggesting that there is a vessel-mediated attraction for the cells. That intravasation is an active process requiring each cell to cross an intact vessel wall is suggested by the failure of intravenously (IV) injected low molecular weight fluorescent dextran to escape from the vessels during intravasation. Furthermore, time-lapse imaging of carcinoma cells in contact with blood vessels demonstrates the ability of carcinoma cells actively to crawl into the blood vessel as solitary cells using amoeboid locomotion (Farina *et al.*, 1998). These time-lapse microscopy results correlate directly with the large number of GFP-expressing carcinoma cells observed in the circulation of the primary tumours and their absence from the blood vessels of non-metastatic tumours (Wyckoff *et al.*, 2000).

These observations suggest mechanisms by which increased carcinoma cell orientation and locomotion towards blood vessels could increase the efficiency with which they intravasate. The orientation could be induced by chemoattractants diffusing from the blood vessel and its associated cell layers. In culture, metastatic carcinoma cells (MTLn3) show dramatic chemotaxis to growth factors while non-metastatic cells (MTC) do not (Figure 11.2). MTLn3 cells express more EGF receptors than MTC cells

Figure 11.1 Carcinoma cells in the primary tumour are concentrated around blood vessels as a polarized cell layer. Rat mammary carcinoma cells (MTLn3) that stably express GFP to define the cell volume were injected into the mammary fat pad of a rat to form a tumour. Blood vessels were labeled by IV rhodamine-dextran. Imaging of the primary tumour near blood vessels demonstrates that the carcinoma cells (brackets) have accumulated around a blood vessel (dashed lines mark vessel walls) with their pseudopods pointing toward the vessel (arrows). Carcinoma cells more distant from the vessel are not polarized toward the vessel (arrowheads). These results suggest that a chemotactic signal such as EGF is associated with blood vessels in the primary tumour. (This figure has been modified from Wyckoff *et al.*, 2000) (A colour reproduction of this figure can be found in the colour plate section)

(Kaufmann *et al.*, 1994), and experimental expression of the EGF receptor in MTC cells increases chemotactic responses to EGF *in vitro* and metastatic ability *in vivo* (Kaufmann *et al.*, 1994; Wyckoff *et al.*, 1998, 2000). Therefore, chemotaxis to blood vessels, mediated by the EGF receptor, is important in

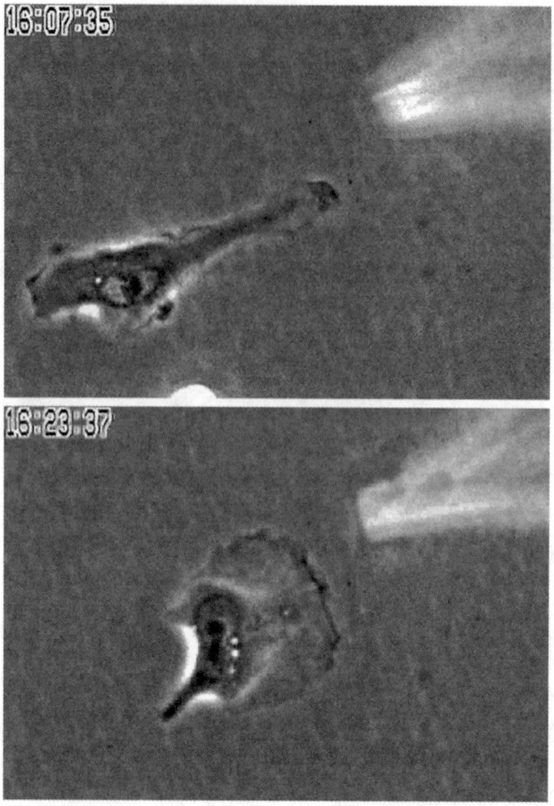

Figure 11.2 Carcinoma cells are chemotactic to EGF *in vitro*. When MTLn3 cells in culture are presented with a pipette filled with EGF they rapidly polarize toward the pipette and move up the concentration gradient. Two frames from a time lapse movie taken 16 min apart illustrate the pipette following activity. In the top frame the cell is moving to the left as the pipette is dropped into the field while in the the bottom frame the cell has reversed direction and is moving up the EGF gradient toward the pipette. (This figure is adapted from Bailly *et al*., 1998)

enhancing metastatic capability in addition to the well-characterized effects of EGF receptor signalling on mitogenesis.

We have tested the hypothesis that chemotaxis to vessels is an important form of egress of carcinoma cells from the primary tumour. We challenged cells within live primary mammary tumours in intact rats using microneedles filled with matrigel and containing chemoattractants to mimic chemotactic signals from blood vessels and associated cell layers (Wyckoff *et al*., 2000). These microneedles, when placed into a mammary tumour, collect large numbers of cells in short time intervals (Figure 11.3). This process is active involving the motility of the cells into the needle. In addition, the migration is dependent on the presence of extracellular matrix within the needle, is greatly

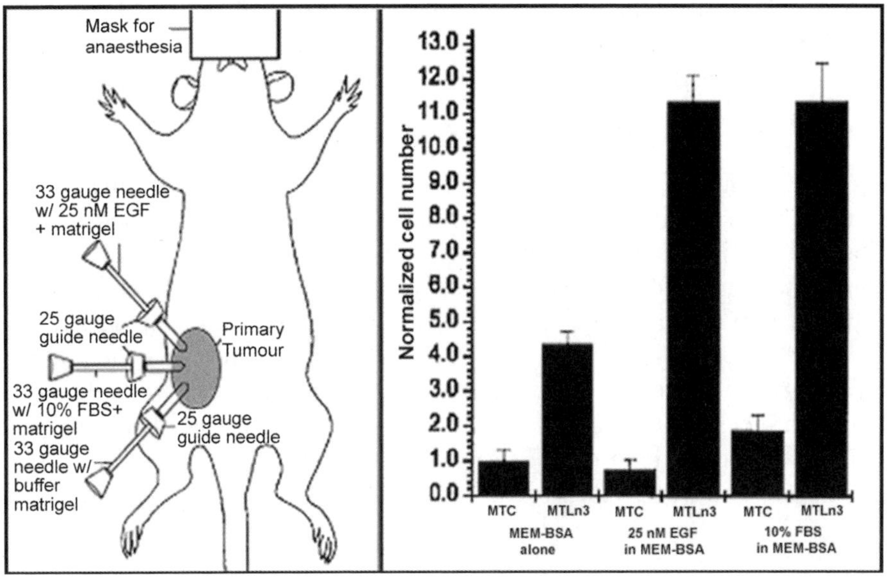

Figure 11.3 Carcinoma cells are chemotactic to EGF in the primary tumour. (Left) Method for using microneedles for *in vivo* cell collection. Microneedles (i.d. 102 μm) filled with matrigel and buffer, 25 nM EGF, or 10% FBS are placed in 25-gauge guide needles that are inserted into the primary tumour of an anesthetized animal. (Right) Carcinoma cells of metastatic (MTLn3) tumours are more efficient than those in nonmetastatic (MTC) tumors at entering the matrigel-containing needles. Maximal response is for cells from MTLn3-generated tumours into EGF- and serum-containing needles. All counts were normalized to MTC cells collected with matrigel in buffer. Error bar = SEM for 4 experiments. (This figure is taken from Wyckoff *et al.*, 2000)

enhanced by the presence of EGF and serum, and is increased in tumours with high metastatic potential (Wyckoff *et al.*, 2000). The conclusion is that the EGF-stimulated chemotaxis of carcinoma cells in the primary tumour is causative for the invasion of the tumour and its metastasis.

Events that define the leading edge during chemotaxis

Hints about how cells read gradients of chemoattractant have been traditionally derived from work with chemotactic amoeboid cells like neutrophils and *Dictyostelium*. Much of the recent work with these cell types implicates phosphoinositide 3-kinase (PI3K) signalling pathways as essential in defining cell polarity during chemotaxis (Servant *et al.*, 2000; Funamoto *et al.*, 2002; Iijima and Devreotes, 2002) (see Chapter by Firtel). Shallow gradients of chemoattractant are amplified and stabilized by the interplay of PI3K and the phosphoinositide 3′-phosphatase (PTEN). This remarkable synergy results from the spatial localization of the two enzymes relative to the leading edge of the cell as it climbs the concentration gradient during chemotaxis. PI3K and its product phosphatidyl inositol (3,4,5) trisphosphate (PIP_3) localize to the leading edge membrane while PTEN and its PIP3 hydrolysis activity localizes to the lateral and basal membranes away from the gradient (Funamoto *et al.*, 2002; Iijima and Devreotes, 2002). This differential distribution suggests a mechanism for how the shallow chemoattractant gradient is amplified inside the cell to permit persistent movement up the gradient. These results also emphasize the importance of a signal upstream of the activation of PI3K and PTEN that causes their correct asymmetric localization. Without this initial asymmetry in signalling, the differential localization of these enzymes would not occur and this would lead to their activation uniformly in the cell which would collapse rather than amplify the gradient of chemoattractant outside.

Is actin polymerization the initial asymmetry generating event?

In *Dictyostelium*, a very early event in signalling in response to the chemoattractant cAMP is an early actin polymerization transient in association with the membrane that peaks by 5 s and is not associated in time with pseudopod extension. Pseudopod extension does not begin until after 30 s of stimulation and is associated with a late actin polymerization transient in time and space (Cox *et al.*, 1992; Eddy *et al.*, 1997). The significance of this early actin polymerization transient may be its association with the activation of PI3K since it shows the same time course and location as PIP_3 accumulation in the leading edge membrane during chemotactic stimulation with cAMP (Funamoto *et al.*, 2002; Iijima and Devreotes, 2002; Eddy *et al.*, 1997). This suggests that the early actin polymerization transient defines a region near the cell membrane as the initial asymmetry compartment that determines the location of PI3K activation.

A similar pattern of early and late actin polymerization transients is observed in carcinoma cells in response to chemotactic stimulation with EGF. Here again the early transient defines the leading edge and precedes lamellipod extension, while the late polymerization transient is spatially and temporally coincident with lamellipodium extension (Figure 11.4) (Bailly *et al.*, 1998; Chan *et al.*, 1998). In vertebrate cells, more is known about the early actin polymerization transient and the case for its involvement in EGF signalling has been explored.

The EGF receptor is an actin-binding protein with a profilin-like actin-binding motif (Hartigh *et al.*, 1992). The actin-binding domain is required for EGF-induced tissue invasion by NIH 3T3 fibroblasts (Heyden *et al.*, 1997).

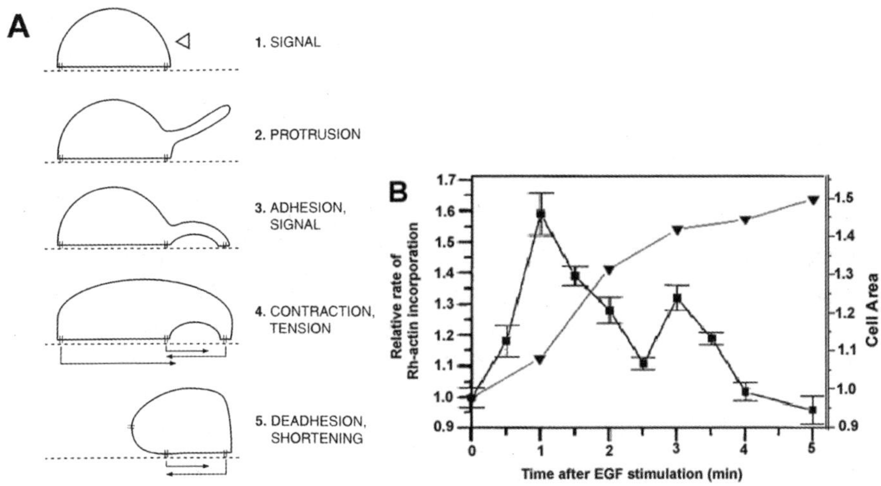

Figure 11.4 The motility cycle of carcinoma cells in response to EGF. (A) The 5-step cycle of cell crawling starts with extension of lamellipodia in response to EGF stimulation. Initial lamellipodium extension occurs off of the substratum and does not require integrin contacts (Bailly *et al.*, 1998). However, once a lamellipodium contacts the substratum, integrin engagement can stabilize the lamellipodium to define it as the dominant lamellipodium. Therefore, the direction of initial extension defines the polarity of the cell and the direction of subsequent cell locomotion. (B) Two actin polymerization transients occur after EGF stimulation as defined by the appearance of free barbed ends *in vivo* (Chan *et al.*, 1998) (■). The early transient peaks by 60 s (Bailly *et al.*, 1999) while the late transient is co-temporal with (▾) lamellipodium extension and peaks at about 180 s at room temperature

The actin-associated EGF receptor is the form of the receptor with high affinity for ligand, and is associated with a number of signalling molecules including phospholipase C (PLC) (Payrastre *et al.*, 1991). In addition, the direct linkage of PLC to the EGF receptor may localize its initial activation to the high-affinity population of receptors which are believed to signal to downstream pathways (Payrastre *et al.*, 1991). The proposal (Wells *et al.*, 1999) that the activation of PLC activity proximal to the EGF receptor causes hydrolysis of PIP_2 and release of cofilin and gelsolin to allow local remodelling of the actin cytoskeleton, is consistent with the identification of a signalling complex containing F-actin and PLC in association with the receptor (Payrastre *et al.*, 1991). Therefore, the high-affinity subpopulation of receptors may be responsible for generating the signal that causes the early actin polymerization transient and formation of the F-actin-dependent signalling complex containing activated PI3K.

EGF receptor-associated actin polymerization is independent of cytosolic Ca^{2+} and therefore unlikely to involve gelsolin activity (Rijken *et al.*, 1995).

However, cofilin in particular is capable of generating new barbed ends quickly and independently of Ca^{2+} leading to actin polymerization proximal to the receptor. Actin polymerization is required for proper signalling by the EGF receptor presumably because only the F-actin-associated receptor assembles a signalling complex that amplifies and moderates the signalling (Rijken *et al.*, 1998). Localized receptor activation would lead to the recruitment and activation of Class I (p85/p110) PI3Ks, through the SH2-domain-mediated binding of p85/p110 to ErbB1/ErbB3 heterodimers or to receptor substrates such as Gab1/2 (Kim *et al.*, 1994; Kong *et al.*, 2000; Rodrigues *et al.*, 2000; Soltoff *et al.*, 1994). Alternatively, p85 binds to adapter proteins such as Grb2 (Wang *et al.*, 1995), which are recruited to autophosphorylated EGF receptors (Lowenstein *et al.*, 1997). EGF-stimulated activation of Ras could also increase PI3K activity at the site of the receptor-associated signalling complex (Rodriguez-Viciana *et al.*, 1994). The initial activation of PI3K by tyrosine-phosphorylated receptors/substrates could lead to the activation of the Rhofamily GTPases Rac and Cdc42, via PIP_3-regulated guanine nucleotide exchange factors (Han *et al.*, 1998; Olson *et al.*, 1996). Activation of Rac or Ccd42 could in turn feed back onto Class IA PI3Ks via binding to the BCR-homology domain of p85, which activates p85/p110 PI3Ks (Zheng *et al.*, 1994; Beeton *et al.*, 1999). Therefore, cofilin-induced actin polymerization could cause an initial asymmetry compartment near the high-affinity receptor that would mark the membrane as a site for subsequent recruitment and activation of PI3K and Rhofamily GTPases. Consistent with this proposal is the identification of an F-actin-dependent positive feedback loop for PIP_3 production involving PI3K and Rhofamily GTPases in neutrophils (Orion *et al.*, 2002).

Do the early and late actin polymerization transients result from different mechanisms?

All actin polymerization *in vivo* is dependent on free barbed ends (Wear *et al.*, 2000). Therefore, to understand the regulation of actin polymerization during cell motility and chemotaxis, one must understand the molecular mechanisms responsible for the appearance of free barbed ends after the stimulation. It is generally agreed that free barbed ends can arise *in vivo* by *de novo* nucleation of filaments from the Arp2/3 complex (see Chapters by Pollard and Machesky), uncapping of the barbed end by the loss of a capping protein from the barbed end, and severing of non-covalent bonds between monomers in F-actin to produce short filaments with free barbed ends. Which combination of these mechanisms is at work during chemotaxis in the different actin polymerization transients has not been studied. However, it is clear that no one mechanism can account for all of the barbed ends and cell behaviour resulting from chemotactic stimulation. In neutrophils, both uncapping of barbed ends and the nucleation activity of the Arp2/3 complex generate free barbed ends in response to FMLP. In this system, interaction between the uncapping and Arp2/3-complex pathways through phosphoinositides appears to be essential for optimum nucleation (Glogauer *et al.*, 2000). In macrophages the stimulation of the Arp2/3 complex by WASp is essential for chemotaxis but a role for uncapping has not yet been studied. In carcinoma cells, both cofilin and the Arp2/3 complex are required for lamellipodium extension: cofilin is needed to form barbed ends and the Arp2/3 complex for the growth of branched filaments (Condeelis, 2001). In addition, there may be a strong cooperation *in vivo* between the cofilin and Arp2/3 complex-dependent pathways, as inhibition of cofilin by microinjection of function-blocking antibodies is sufficient to inhibit the EGF-stimulated generation of all barbed ends (Bailly *et al.*, 2001). Cooperation between cofilin and the Arp2/3 complex has been studied *in vitro* and involves stimulation of the nucleation activity of the Arp2/3 complex by cofilin-severing activity: the Arp2/3 complex prefers to bind to the sides of filaments newly polymerized from the barbed ends generated by cofilin severing (Ichetovkin *et al.*, 2002).

Concerning which of the three mechanisms for barbed end generation is responsible for the early actin polymerization transient, one would predict a mechanism with very rapid kinetics since the early transient peaks by 5 s in *Dictyostelium* and by 35–50 s in carcinoma cells (Bailly *et al.*, 1999). Uncapping by capping protein requires competition between PIP_2 and the barbed end for binding to capping protein. However, PIP_2 is believed to be hydrolysed by the activation of PLC as an early event after stimulation with EGF. This would slow the uncapping of capping protein from barbed ends until after re-synthesis of PIP_2. On the other hand, cofilin-mediated severing would be activated by PIP_2 hydrolysis during the initial PLC activation and cofilin severing is a very fast event (Chan *et al.*, 2000). At least 50% of cofilin in resting carcinoma cells is in the dephosphorylated form (Zebda *et al.*, 2000) and that fraction of cofilin bound to PIP_2 could be activated directly by PIP_2 hydrolysis. Additionally, activation of cofilin at the plasma membrane does not require a recruitment step since the PIP_2-associated cofilin would be pre-localized to PIP_2-rich regions. Finally, cofilin severs crosslinked actin filaments at concentrations 50 to 100-fold lower than previously measured for actin filaments in solution (Figure 11.5) which would make cofilin-mediated severing of tethered filaments at the plasma membrane extremely efficient. In comparison, sequential recruitment of WASp family members to the plasma membrane and then the Arp2/3 complex is a slow process in carcinoma cells taking at least 60 s (Condeelis,

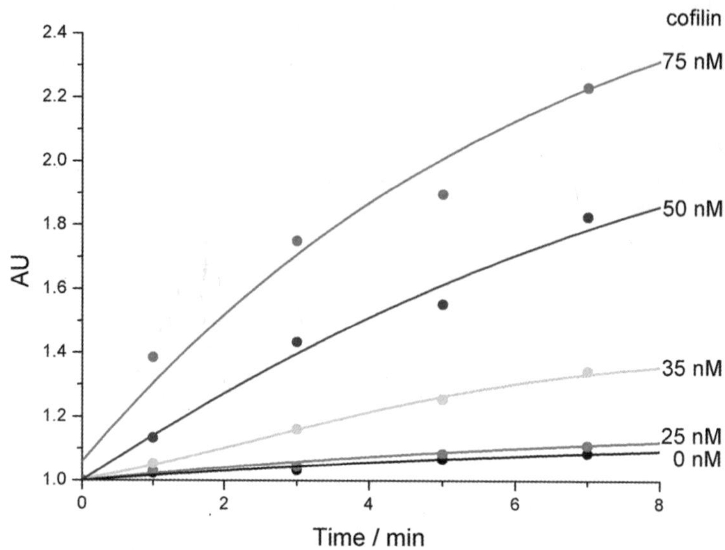

Figure 11.5 Cofilin severs tethered actin filaments at nanomolar concentrations. The severing of actin filaments by the cofilin mutant S3C was observed in the light microscope and quantified in arbitrary units as described previously (Ichetovkin *et al.*, 2002). The severing is detected down to low nM concentrations for the S3C mutant as shown in the figure and down to 9 nM for the wild-type cofilin (not shown) while the K_{50} for severing by wild-type cofilin is measured as μM in solution (Ichetovkin *et al.*, 2002)

unpublished), and thus occurs after the peak of the early actin polymerization transient has passed.

Conclusions

A summary cartoon of the pathways leading to actin polymerization in carcinoma cells in response to the stimulation of EGF receptor is shown in Figure 11.6. The early actin polymerization transient is indicated as pathway 1 and involves cofilin activation by PLC-mediated PIP_2 hydrolysis to release a dephosphorylated population of cofilin from the plasma membrane near activated EGF receptors (Wells *et al.*, 1999). The late actin polymerization transient is indicated as pathway 2 and involves the relatively slow recruitment to the membrane of Rac/Cdc42 and WASp family members leading to the activation of the Arp2/3 complex near to the plasma membrane. Actin filaments polymerized proximal to the receptor by pathway 1 could participate in the assembly of the F-actin dependent EGF receptor signalling complex (Heyden *et al.*, 1997;

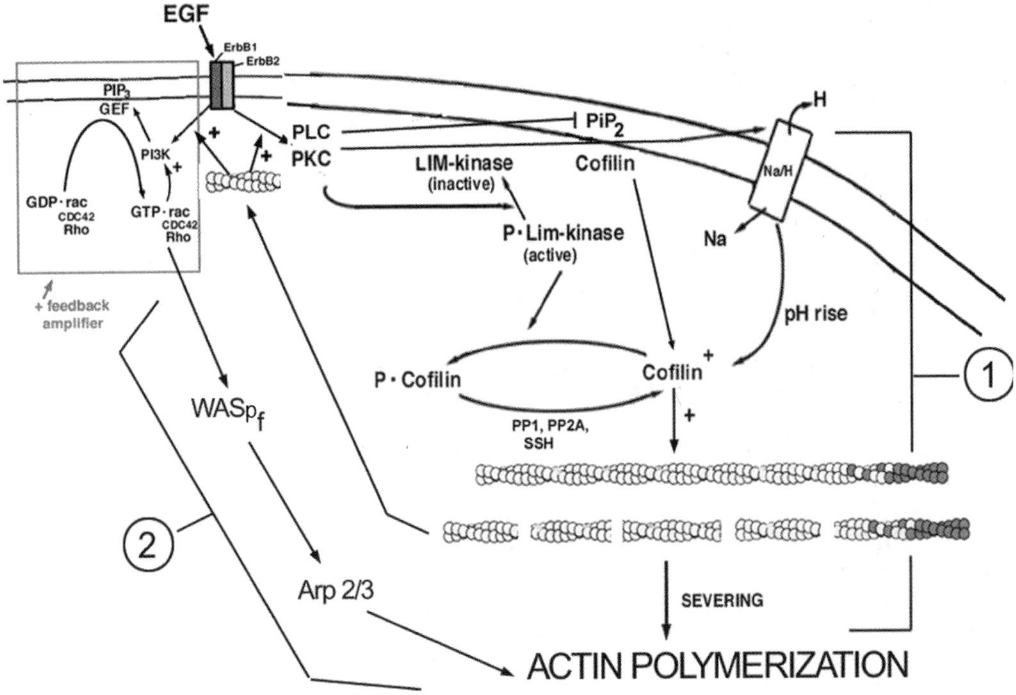

Figure 11.6 Postulated signalling pathways for the early and late actin polymerization transients in response to EGF. Pathways 1 and 2 are described in the text and are proposed to be distinct pathways involving; (1) PLC-stimulated cofilin generation of barbed ends and actin polymerization in a PI3K-independent step. This would be stimulated further by an increase in pH through the action of PKC on Na/H exchangers. (2) A positive feedback loop between PI3K and Rac causing repeated stimulation of WASp family members ($WASp_f$) and the Arp2/3 complex leading to actin polymerization. Actin filaments polymerized proximal to the receptor by pathway × 1 could participate in the assembly of the F-actin-dependent EGF receptor signalling complex (Heyden *et al.*, 1997; Payrastre *et al.*, 1991) and supply filaments to support the F-actin-dependent positive feedback loop for PIP_3 production involving PI3K and Rac (Orion *et al.*, 2002). This positive feedback loop is postulated here to be required for the activation of the late actin polymerization response

Payrastre *et al.*, 1991). This complex is postulated to recruit PI3K and establish the F-actin-dependent positive feedback loop of pathway 2 believed to be required for the maintenance of asymmetry needed for chemotaxis (Orion *et al.*, 2002). In addition, the activation of the Na/H exchanger by the activated protein kinase C (PKC) in this complex would cause activation of cofilin by an increase in pH proximal to the receptor increasing cofilin's severing activity (Bernstein *et al.*, 2000). Further work will be needed to determine if pathways 1 and 2 exist as distinct entities and their relative importance in establishing the cell polarity required for chemotaxis.

References

Bailly, M., Yan, L., Whitesides, G. M., Condeelis, J. S. and Segall, J. S., 1998. Regulation of protrusion shape and adhesion to the substratum during chemotactic responses of mammalian carcinoma cells. *Exp. Cell Res.* **241**: 285–299.

Bailly, M., Macaluso, F., Cammer, M., Segall, J. and Condeelis, J., 1999. The relationship of Arp2/3 complex to actin barbed ends of actin filaments at the leading edge of carcinoma cells after EGF-stimulation. *J. Cell Biol.* **145**: 331–345.

Bailly, M., *et al.*, 2001. The F-actin side binding activity of the Arp2/3 complex is essential for actin nucleation and lamellipod extension. *Curr. Biol.* **11**: 620–625.

Beeton, C. A., Das, P., Waterfield, M. D. and Shepherd, P. R., 1999. The SH3 and BH domains of the p85alpha adapter subunit play a critical role in regulating Class Ia phosphoinositide 3-kinase function *Mol. Cell Biol. Res. Comm.* **1**: 153–157.

Bernstein, B. W., Painter, W. B., Chen, H., Minamide, L. S., *et al.*, 2000. Intracellular pH modulation of ADF/cofilin proteins. *Cell Motil. Cytoskeleton* **47**(4): 319–336.

Calabro, A., Orsini, B., Renzi, D., Papi, L., *et al.*, 1997. Expression of epidermal growth factor, transforming growth factor-alpha and their receptor in the human esophagus. *Histochem. J.* **29**: 745–758.

Chan, A., Raft, S., Bailley, M., Wyckoff, J., *et al.*, 1998. EGF stimulates actin nucleation at the leading edge of the lamellipod in mammary adenocarcinoma cells. *J. Cell Sci.* **111**: 199–211.

Chan, A., Zebda, N., Bailly, M., Segall, J. and Condeelis, J., 2000. Role of cofilin in EGF-stimulated actin polymerization and lamellipod extension. *J. Cell Biol.* **148**: 531–542.

Comer, F. and Parent, C., 2002. PI3K and PTEN: How opposites attract. *Cell* **109**: 541–544.

Condeelis, J., 2001. How is actin polymerization nucleated *in vivo*? *Trends Cell. Biol.* **11**(7): 288–293.

Cox, D., Condeelis, J., Wessels, D., Soll, D., *et al.*, 1992. Targeted disruption of the ABP-120 gene leads to cells with altered motility. *J. Cell Biol.* **116**: 943–955.

Eddy, R., Han, J. and Condeelis, J., 1997. Capping protein terminates but does not initiate chemoattractant-induced actin assembly in Dictyostelium. *J. Cell Biol.* **139**: 1243–1253.

Farina, K., Wyckoff, J., Rivera, J., Lee, H., *et al.*, 1998. Cell motility of tumor cells visualized in living intact tumors using green fluorescent protein. *Cancer Res.* **58**: 2528–2532.

Funamoto, S., Meili, R., Lee, L. and Firtel, R., 2002. Spatial and temporal regulation of 3-phosphoinositides by PI3K and PTEN mediates chemotaxis. *Cell* **109**: 611–623.

Glogauer, M., Hartwig, J. and Stossel, T., 2000. Two pathways through Cdc42 couple the N-formyl receptor to actin nucleation in permeabilized human neutrophils. *J. Cell Biol.* **150**: 785–796.

Han, J., Luby-Phelps, K., Das, B., Shu, X., *et al.*, 1998. Role of substrates and products of PI 3-kinase in regulating activation of Rac-related guanosine triphosphatases by Vav. *Science* **279**: 558–560.

Hartigh, J., Henegouwen, P., Verkleij, A. and Boonstra, J., 1992. EGF receptor is an actin binding protein. *J. Cell Biol.* **119**: 349–355.

Heyden, M., Henegouwen, P., Ruiter, N., *et al.*, 1997. The actin binding domain of the EGF receptor is required for EGF-stimulated tissue invasion. *Exp. Cell Res.* **234**: 521–526.

Ichetovkin, I., Grant, W. and Condeelis, J., 2002. Cofilin produces newly polymerized actin filaments that are preferred for dendritic nucleation by the Arp2/3 complex. *Current Biol.* **12**: 79–84.

Iijima, M. and Devreotes, P., 2002. Tumor suppressor PTEN mediates sensing of chemoattractant gradients. *Cell* **109**: 599–610.

Kaufmann, A. M., Khazaie, K., Wiedemuth, M., Rohde-Schulz, B., *et al.*, 1994. Expression of epidermal growth factor receptor correlates with metastatic potential of 13762NF rat mammary adenocarcinoma cells. *Int. J. Oncol.* **4**: 1149–1155.

Kim, H.-H., Sierke, S. L. and Koland, J. G., 1994. Epidermal growth factor-dependent association of phosphatidylinositol 3-kinase with the *erb*B3 gene product. *J. Biol. Chem.* **269**: 24 747–24 755.

Klijn, J. G., Look, M. P., Portengen, H., *et al.*, 1994. The prognostic value of epidermal growth factor receptor (EGF-R) in primary breast cancer: results of a 10 year follow-up study. *Breast Canc. Res. Treat.* **29**: 73–83.

Kong, M., Mounier, C., Wu, J. and Posner, B. I., 2000. Epidermal growth factor-induced phosphatidylinositol 3-kinase activation and DNA synthesis. Identification of Grb2-associated binder 2 as the major mediator in rat hepatocytes. *J. Biol. Chem.* **275**: 36 035–36 042.

Kume, N. and Gimbrone, M. J., 1994. Lysophosphatidylcholine transcriptionally induces growth factor gene expression in cultured human endothelial cells. *J. Clin. Invest.* **93**: 907–911.

Lowenstein, E. J., Daly, R. J., Batzer, A. G., Li, W., *et al.*, 1992. The SH2 and SH3 domain-containing proteins GRB2 links receptor tyrosine kinases to ras signaling. *Cell* **70**: 431–442.

Olson, M. F., Pasteris, N. G., Gorski, J. L. and Hall, A., 1996. Faciogenital dysplasia protein (FGD1) and Vav, two related proteins required for normal embryonic development, are upstream regulators of Rho GTPases. *Curr. Biol.* **6**: 1628–1633.

Orion, D. Weiner, Neilsen, P. O., Prestwich, G. D., Kirschner, M. W., *et al.*, 2002. A PtdInsP3- and Rho GTPase-mediated positive feedback loop regulates neutrophil polarity. *Nature Cell Biol.* **4**: 509–512.

Payrastre, B., van Bergen en Henegouwen, P. M. P., Breton, M., den Hartigh, J. C., *et al.*, 1991. Phosphoinositide kinase, diacylglycerol kinase, and phospholipase C activities associated to the cytoskeleton: effect of epidermal growth factor. *J. Cell Biol.* **115**: 121–128.

Rijken, P., Post, S., Hage, W., Henegouwen, P., *et al.*, 1995. Actin polymerization localizes to the activated epidermal growth factor receptor in the plasma membrane, independent of the cytosolic free calcium transient. *Exp. Cell Res.* **218**: 223–232.

Rijken, P. J., van Hal, G. J., van der Heyden, M. A. G., Verkleij, A. J. and Boonstra, J., 1998. Actin polymerization is required for negative feedback regulation of epidermal growth factor-induced signal transduction. *Exp. Cell Res.* **243**: 254–262.

Rodrigues, G. A., Falasca, M., Zhang, Z. T., Ong, S. H. and Schlessinger, J., 2000. A novel positive feedback loop mediated by the docking protein Gab1 and phosphatidylinositol 3-kinase in epidermal growth factor receptor signaling. *Mol. Cell. Biol.* **20**: 1448–1459.

Rodriguez-Viciana, P., Warne, P. H., Dhand, R., Vanhaesebroeck, B., *et al.*, 1994. Phosphatidylinositol-3-OH kinase as a direct target of Ras. *Nature* **370**: 527–532.

Servant, G., Weiner, O., Herzmark, P., Balla, T., *et al.*, 2000. Polarization of chemoattractant receptor signaling during neutrophil chemotaxis. *Science* **287**: 1037–1040.

Sherwood, E. R. and Lee, C., 1995. Epidermal growth factor-related peptides and the epidermal growth factor receptor in normal and malignant prostate. *World J. Urol.* **13**: 290–296.

Soltoff, S. P., Carraway, K. L. I., Prigent, S. A., Gullick, W. G. and Cantley, L. C., 1994. ErbB3 is involved in activation of phosphatidylinositol 3-kinase by epidermal growth factor. *Mol. Cell. Biol.* **14**: 3550–3558.

Wang, J., Auger, K. R., Jarvis, L., Shi, Y. and Roberts, T. M., 1995. Direct association of Grb2 with the p85 subunit of phosphatidylinositol 3-kinase. *J. Biol. Chem.* **270**: 12 774–12 780.

Wear, M. A., Schafer, D. and Cooper, J., 2000. Actin dynamics: assembly and disassembly of actin networks. *Curr. Biol.* **10**: 891–895.

Wells, A., Ware, M., Allen, F. and Lauffenburger, D., 1999. Shapping up for shipping out: PLCg signaling of morphological changes in EGF-stimulated fibroblast migration. *Cell Motil. Cytoskel.* **44**: 227–233.

Wyckoff, J., Insel, L., Khazail, K., Lichtner, R., *et al.*, 1998. Suppression of ruffling by the EGF-receptor in chemotactic adenocarcinoma cells. *Exp. Cell Res.* **242**: 100–109.

Wyckoff, J. B., Segall, J. E. and Condeelis, J. S., 2000a. The collection of the motile population of cells from a living tumor. *Cancer Res.* **60**(19): 5401–5404.

Wyckoff, J. B., Jones, J., Condeelis, J. and Segal, J., 2000b. A critical step in metastasis: in vivo analysis of intravasation at the primary tumor. *Cancer Res.* **60**(9): 2504–2511.

Zebda, N., Bernard, O., Lawrence, D. and Condeelis, J., 2000. Phosphorylation of cofilin abolishes EGF-induced actin nucleation at the leading edge and subsequent lamellipod extension. *J. Cell Biol.* **151**: 1119–1127.

Zheng, Y., Bagrodia, S. and Cerione, R. A., 1994. Activation of phosphoinositide 3-kinase by Cdc42Hs binding to p85. *J. Biol. Chem.* **269**: 18 727–18 730.

12

Dynamin and Cytoskeletal-dependent Membrane Processes

James D. Orth, Noah W. Gray, Heather M. Thompson and **Mark A. McNiven**

The mechanochemical GTPase dynamin functions to mediate vesicle formation from several cellular compartments, including the plasma membrane, endosomes and Golgi complex. Dynamin is a multidomain protein that interacts with lipids and numerous additional proteins, including actin-associated proteins such as cortactin. Thus, dynamin is likely to participate in many cell functions that require actin-based membrane dynamics. Indeed, multiple recent observations support the premise that dynamin functions at the interface between cell membranes and the cytoskeleton during membrane ruffling, vesicle trafficking, cytokinesis and synaptogenesis. This chapter summarizes our recent findings supporting these new roles for dynamin and proposes future directions that should provide insight into the molecular basis of membrane dynamics.

Introduction

An overview of dynamin

Dynamin is a large, 100-kDa GTPase that was first discovered as a microtubule-associated protein from bovine brain (Shpetner and Vallee, 1989). Upon cloning, it was realized that dynamin was the mammalian

Cell Motility. Edited by Anne Ridley, Michelle Peckham and Peter Clark.
 ISBN 0 470 84872 3

orthologue of the *Drosophila melanogaster shibire* protein (van der Bliek and Meyerowitz, 1991; Chen *et al.*, 1991). A temperature-sensitive mutant of *shibire* was found to cause paralysis when flies were shifted to the restrictive temperature. Ultrastructural studies of the fly neuromuscular junction revealed that this phenotype was due to a defect in synaptic vesicle recycling (Koenig and Ikeda, 1989). These experiments revealed that numerous vesicular membrane invaginations could form, although final scission was blocked, and the vesicle buds were arrested at the plasma membrane. Interestingly, electron dense collars or rings of protein, later shown to include dynamin, were visible on the necks of the arrested buds. Collectively, these studies in the fly provided insight into the potential function of dynamin and suggested that it acted to sever vesicles from the plasma membrane. Over the past decade, additional studies have demonstrated that there are three alternatively spliced dynamin (Dyn) genes in mammals, Dyn1, Dyn2 and Dyn3, that share at least 70% amino acid identity with each other. Further, the dynamin genes are differentially expressed in adult tissues; Dyn1 is specifically expressed in the brain, Dyn2 is ubiquitously expressed (Cook *et al.*, 1994; Sontag *et al.*, 1994) and Dyn3 expression appears to be limited to the brain, heart, lung and testis (Nakata *et al.*, 1993; Cook *et al.*, 1996).

Domain analysis of the dynamin protein has revealed that each isoform contains an N-terminal GTPase domain, a pleckstrin homology (PH) domain that binds phosphoinositide (PI) lipids (mediating the association of dynamin with membranes) and a C-terminal proline–arginine-rich domain (PRD) that directly binds to Src homology 3 (SH3) domains (Figure 12.1A). The PRD of dynamin binds to several proteins that function to promote vesicle budding,

---

Figure 12.1 (*opposite*) Dynamin participates in many dynamic actin-based membrane processes. (A) This basic schematic of the dynamin proteins highlights its three most important domains. There is an N-terminal GTP hydrolysis domain (GTPase), a centrally located pleckstrin homology domain (PH) that binds to phosphatidylinositol lipids and a proline–arginine rich domain (PRD) at the very C-terminus that binds to numerous proteins, including many that bind and/or regulate the actin cytoskeleton. (B) A fixed cell expressing PIP5KI and Dyn2-GFP co-stained with rhodamine-phalloidin (actin). Several comets can be seen in this cell (arrows). The high-magnification inserts nicely demonstrate that Dyn2-GFP brightly labels the comet ‘head’. Studies revealed that Dyn2 has a role in comet formation and movement. (C) A dividing cell at the late stages of cytokinesis. Dyn2 (red) is highly concentrated at the spindle midzone where it co-localizes with microtubules (green, co-localization = yellow) and plays a role in the late stages of cytokinesis. The high magnification insert shows the striking localization of Dyn2 and microtubules at the midbody matrix. Blue = DAPI. (D) The expression of different Dyn3 spliced-forms induces alternative phenotypes in cultured hippocampal neurons. Although temporally mature, neurons expressing Dyn3baa lack dendritic spines and instead are covered with long filopodia (arrows and D″), which do not assemble functional synapses. As a comparison, a neuron expressing Dyn3aaa does not exhibit any morphological effect on the normal formation of short, mushroom-shaped dendritic spines (D′). (A colour reproduction of this figure can be found in the colour plate section)

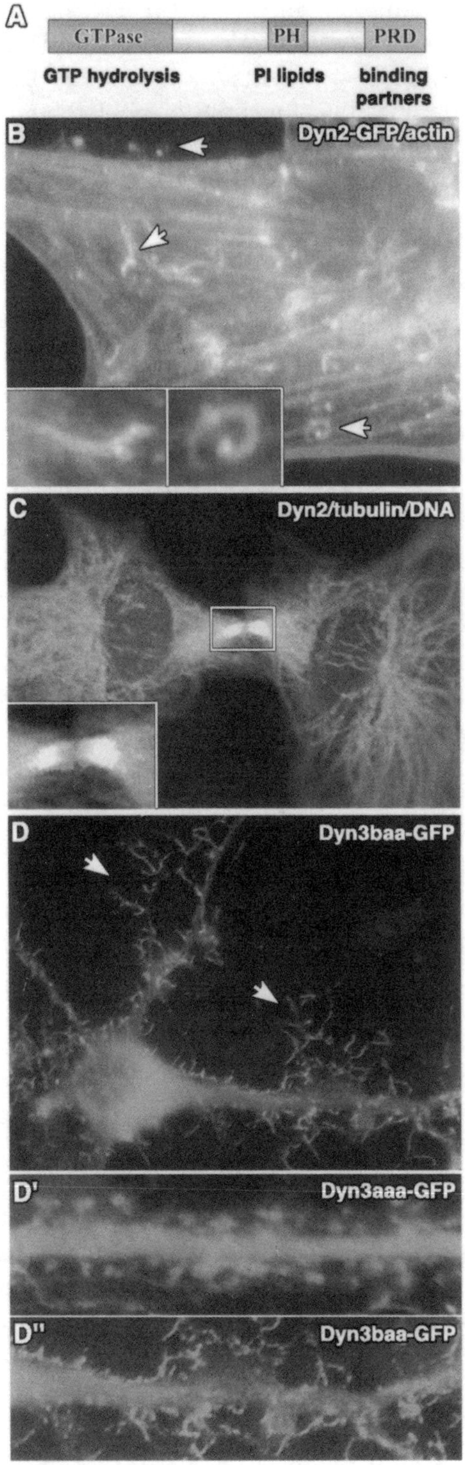
A
GTPase
PH
PRD
GTP hydrolysis
PI lipids
binding partners
B
Dyn2-GFP/actin
C
Dyn2/tubulin/DNA
D
Dyn3baa-GFP
D'
Dyn3aaa-GFP
D''
Dyn3baa-GFP

including endophilin, amphiphysin, intersectin and α-adaptin (McNiven *et al.*, 2000a). Accordingly, much of the work on Dyn1 and Dyn2 has focused on a role in vesicle formation from the plasma membrane and mechanochemical force-generating properties that function to tubulate and sever vesicle necks (Hinshaw, 2000; Zhang and Hinshaw, 2001). In addition to mediating endocytosis, Dyn2 also functions during vesicle formation from the Golgi complex and endosomal sorting centres (Jones *et al.*, 1998; van Dam and Stoorvogel, 2002). Dyn3 is the least studied of the dynamins, but it has been localized to the neuronal postsynaptic density (Gray *et al.*, 2003) and is also hypothesized to function during spermatogenesis in mice (Kamitani *et al.*, 2002).

More recently, dynamin has been proposed to function during actin-dependent processes. This hypothesis is supported by data that demonstrates the dynamin PRD mediates interactions with several actin-binding or actin-associated regulatory proteins, including profilin, cortactin, Abp1, syndapin (PACSIN), Grb2 and Nck (Grb4) (reviewed in Orth and McNiven, 2003). Therefore, it is likely that the membrane modelling activities of dynamin, coupled with its ability to interact with actin, make this protein an important player in actin-membrane processes. In this chapter, we will discuss some data that support a role for dynamin in these processes. The specific cell functions discussed in this chapter include endocytosis, actin-based vesicle transport, cytokinesis and dendritic spine morphogenesis.

Dynamin and the actin-binding protein cortactin interact directly

We began analysing the potential role of dynamin at the interface between actin and membranes as it became clear that dynamin did not localize solely with endocytic structures. Initial studies examining the localization of Dyn2 in fixed and living cells demonstrated a strong enrichment at peripheral membrane ruffles and a co-localization with cortical F-actin. These studies prompted us to investigate whether dynamin could interact directly with F-actin binding proteins. In collaboration with others, we discovered that the SH3 domain of the F-actin binding protein cortactin directly bound to the PRD of Dyn2 (McNiven *et al.*, 2000b). This direct interaction was confirmed using co-immunoprecipitations, GST-fusion pull-downs with peptide competition and blot-overlay experiments. These experiments revealed that the C-terminal half of the Dyn2 PRD was important and specific for binding to the SH3 domain of cortactin but not to the SH3 domain of PLCγ, another dynamin PRD binding partner. Additional studies recently demonstrated that point mutation of tryptophan 525 to a lysine residue in the cortactin SH3 domain disrupted the interaction with Dyn2 (Schafer *et al.*, 2002). This mutation likely introduces a structural defect in the SH3 domain as the same mutation also blocks cortactin binding to CortBP1 (Shank) (Du *et al.*, 1998).

Interestingly, it appears that both cortactin and Dyn2 are recruited to peripheral membrane ruffles and lamellipodia in response to the motogenic growth factor PDGF. Mechanistically, it was demonstrated that the PRD of Dyn2 is required for targeting to the peripheral ruffles and that therefore Dyn2 recruitment to the ruffles is, at least in part, cortactin-dependent. In further support of this, cortactin lacking the SH3 domain is recruited to membrane ruffles and lamellipodia. In contrast to the expression of full-length cortactin, cells expressing the truncated form do not induce enhanced recruitment of Dyn2. It is important to note that Dyn2 has been shown to bind to additional actin regulatory proteins including profilin, Abp1, syndapin, Grb2 and Nck (reviewed in Orth and McNiven, 2003) and thus these proteins may also play a role in Dyn2 targeting to and functioning with the actin cytoskeleton. How Dyn2, together with its actin-associated proteins, regulates membrane ruffling, lamellipodium extension and other actin-membrane processes is currently under intense study.

Participation of dynamin in actin-based membrane dynamics

Dynamin, cortactin and receptor-mediated endocytosis

Perhaps the best characterized function for dynamin is during the endocytosis of clathrin-coated pits (CCPs) from the plasma membrane. The actin cytoskeleton is believed to contribute to the formation of clathrin-coated pits by recruiting proteins that perform a direct role in the process, or by aiding in the deformation of the donor membrane and/or movement of the nascent vesicle away from the membrane (Qualmann *et al.*, 2000; Schafer, 2002). How this cytoskeletal network might interact with the endocytic machinery to regulate vesicle formation and scission remains unclear. A recent study reported that clathrin, dynamin and actin are sequentially recruited to forming CCPs, and that there is a rapid accumulation of actin on the bud or vesicle as it forms and moves away from the plasma membrane (Merrifield *et al.*, 2002). This study demonstrated an accumulation of dynamin and actin at the forming endocytic intermediate but did not directly link the endocytic and cytoskeletal machineries. As Dyn2 participates in the liberation of CCPs from the plasma membrane and directly binds the actin modulating protein cortactin, we hypothesized that cortactin could serve as a bridge between the actin cytoskeleton and endocytic machinery, similar to what has been shown for Abp1, another dynamin binding partner (Kessels *et al.*, 2001).

Cortactin has been shown to localize within the cell cortex and membrane ruffles of cultured cells and is most often found at sites of active actin filament nucleation in cells. Therefore, to test whether cortactin could associate with CCPs and function during endocytosis, we performed both co-localization and endocytic assay experiments with and without perturbation of cortactin function. Immunocytochemistry experiments using either anti-cortactin antibodies or expressed red fluorescent protein-tagged cortactin revealed a dramatic co-localization of cortactin with clathrin and Dyn2 puncta at the plasma membrane (Cao *et al.*, 2003). At the ultrastructural level, immunogold labelling of plasma membrane replicas showed that cortactin localized to both flat (early) and budded (mature) CCPs. Highly curved pits showed an enrichment of cortactin labelling at the vesicle neck, not unlike what has been shown for dynamin.

To test for cortactin's role in endocytosis, cells were microinjected with affinity-purified cortactin antibodies and subsequently challenged to internalize various endocytic ligands. Cortactin antibody-injected cells showed a significant decrease in the uptake of labelled transferrin and LDL, while fluid internalization (examined using dextran) was unchanged. Interestingly, cells expressing the cortactin SH3 domain also exhibited markedly reduced endocytosis of transferrin and LDL. These data demonstrate that cortactin is an important component of the receptor-mediated endocytic (RME) machinery where it regulates the scission of CCPs from the plasma membrane. Thus, cortactin provides a direct link between the dynamic actin cytoskeleton and the membrane 'pinchase' dynamin to support vesicle formation during RME. How the cell coordinates the efforts of cortactin, dynamin, actin and their associated proteins, including syndapin, N-WASp and endocytic coat proteins, to support vesicle formation is currently unknown.

Dynamin and actin-based vesicle trafficking

Actin-based 'comet formation' has emerged as a novel, motor-independent mechanism to move vesicles within cells. These actin-associated vesicles, or comets, are known to form from lipid microdomains in the plasma membrane and from the Golgi complex and consist of a membranous vesicle-head and an actin tail (Figure 12.1B) (Rozelle *et al.*, 2000). While a portion of these actin vesicles represent fluid uptake (macropinosomes), specific protein cargoes remain unknown. Because of their transient nature and rapid movement (0.15 μm/s) they are seldom observed in cells. Stimulation of tyrosine kinase signalling results in a quantifiable difference in comet formation, suggesting that tyrosine phosphorylation is important for this process (Rozelle *et al.*, 2000). A significant step forward was made when it was observed that transient expression of type I phosphatidylinositol phosphate 5-kinases (PIP5KIα) and accumulation of phosphatidylinositol 4,5-bisphosphate (PIP_2) resulted in enhanced comet formation, allowing for more reliable quantitation and mechanistic evaluation of comet formation and movement. Interestingly, transient expression of the activated form of the small GTPase Arf6 also results in comet formation (Schafer *et al.*, 2000).

While imaging cultured cells expressing Dyn2-GFP, we observed small, comet-like structures that had bright Dyn2 puncta associated with their vesicle 'heads' and a more faintly stained 'tail'. In the case of macropinosomes, Dyn2 labelling was confined to the side of the macropinosome opposite the direction of movement, the same as for actin and cortactin (Kaksonen *et al.*, 2000; Orth *et al.*, 2002). These structures appeared identical to the actin comets induced by PIP5KIα and Arf6. Based on these data, we sought to determine the role of Dyn2 in actin comet formation and movement. Cells expressing PIP5KIα and Dyn2-GFP revealed a prominent co-localization of Dyn2, cortactin and actin in comet structures (Figure 12.1B) (Orth *et al.*, 2002). Interestingly, comet formation and motility were normal in wildtype Dyn2-GFP expressing cells but significantly altered in cells expressing Dyn2 mutants. Cells expressing a GTPase-deficient Dyn2 (Dyn2K44A-GFP) contained fewer comets, while those that did form were shorter than normal, and progressed with a considerably decreased velocity. In contrast, comets in cells expressing Dyn2ΔPRD-GFP did not incorporate the mutant Dyn2 protein, suggesting that the PRD has a role in Dyn2 recruitment to the comet. Furthermore, these Dyn2ΔPRD comets were significantly longer and slower than those in control cells. These novel findings were supported by a similar study from Lee and DeCamilli, which showed that Dyn2 is a component of comet structures induced by *Listeria* and PIP5KIα, that the PRD has a role in targeting Dyn2 to the comets and that Dyn2 regulates comet formation and velocity (Lee and De Camilli, 2002). Consistent with these studies, recent *in vitro* data showed that Dyn2 mediates actin polymerization through cortactin, and affects F-actin organization extended from PIP_2-containing liposomes, although how Dyn2 exerts this effect was not defined (Schafer *et al.*, 2002). Together, these studies suggest that Dyn2 functions at the interface of membranes and the actin cytoskeleton to regulate membrane trafficking and F-actin polymerization and organization.

Dynamin and cytokinesis

Cytokinesis involves coordinated interactions between the microtubule and actin cytoskeletons and cellular membranes to form and constrict the cleavage furrow, completing daughter cell separation (for reviews see Glotzer, 2001; Straight and Field, 2000; Robinson and Spudich, 2000). More recently it has also been noted that proteins involved in vesicle trafficking, such as members of the Rab family of small GTPases, vesicle coat proteins and syntaxins, are also present at the cleavage furrow and contribute to the completion of cytokinesis (for reviews see Bowerman and Severson, 1999; O'Halloran, 2000; Finger and White, 2002). Consistent with proteins involved in vesicle trafficking and/or cytoskeletal processes being involved in cytokinesis, we recently demonstrated that Dyn2 localizes to the spindle midzone and plays a role in the late stages of cytokinesis (Thompson *et al.*, 2002). Using fluorescence and immunoelectron microscopy, we noted an intense staining of Dyn2 at the microtubule-rich intercellular bridge proximal to the dense midbody matrix (Figure 12.1C). However, Dyn2 staining was not observed along cytoplasmic microtubule bundles induced by paralitaxel treatment, suggesting a specific localization of Dyn2 to the intercellular bridge during cytokinesis and not to microtubule bundles in general. In support of these morphological data, we were able to detect a 30- to 50-fold enrichment of Dyn2 in isolated spindle midbody extracts as compared with interphase extracts. To test for a functional role for dynamin in cytokinesis, the nematode *Caenorhabditis elegans* was used. Importantly, an antibody recognizing *C. elegans* dynamin localized this protein to the ingressing cleavage furrow as well as to the midbody of dividing embryos. Finally, using either a temperature-sensitive dynamin mutant strain of *C. elegans* or wildtype embryos depleted of dynamin protein through RNAi treatment, we were able to demonstrate that dynamin function is necessary for the late stages of cytokinesis. In both cases, disruption of dynamin function did not block mitosis or the initiation of cleavage furrow ingression. However, just before the final separation of the two daughter cells, cleavage furrow ingression was aborted and the cells became multinucleate. Interestingly, dynamin orthologues in plants (Gu and Verma, 1996; Lauber *et al.*, 1997), fungi (Wienke *et al.*, 1999), flies (Swanson and Poodry, 1981) and zebrafish (Feng *et al.*, 2002) have also been implicated in cytokinesis. Thus, this could be a conserved function for dynamin family members across a variety of species. Though the exact role of dynamin in cytokinesis is not known, it could be involved in multiple aspects of this process, such as membrane remodelling, vesicle trafficking or coordinating interactions between cleavage furrow membranes and the microtubule and actin cytoskeletons. The role of dynamin in cytokinesis and its regulation during this process are areas of future investigation.

Dynamin and dendritic spine morphogenesis

Although most mammalian tissues only express a single isoform of dynamin, Dyn2, neurons produce all three forms. Of these three, previous studies in neurons have focused almost exclusively on the brain-specific isoform Dyn1, and its role in synaptic vesicle recycling (Takei *et al.*, 1996; Shupliakov *et al.*, 1997). Additional experiments have since demonstrated a postsynaptic localization for Dyn2 (Okamoto *et al.*, 2001) where it may play a role during glutamate receptor endocytosis (Carroll *et al.*, 1999). As very little work has focused on the testes- and brain-specific dynamin, Dyn3, it is important to understand whether this third isoform performs distinct functions in the neuron. Recently, we have provided evidence implicating Dyn3 in the actin-dependent process of filopodial outgrowth and dendritic spine maturation (Gray *et al.*, 2003).

In this study we found that Dyn3 spliced-forms induced differential effects on the dendritic spine morphology of transfected cultured neurons. Dendritic spines, small actin-rich protrusions that decorate the dendrites of excitatory pyramidal neurons in the hippocampus and cortex, usually emerge after 2–3 weeks in culture, replacing the filopodia that populate the dendrites prior to complete synaptogenesis. When overexpressed in immature neurons, the Dyn3baa spliced-form caused a proliferation of dendritic filopodia at the expense of mushroom-shaped dendritic spines, even in temporally mature cultures (Figure 12.1D). Contrasting with this, the expression of a second Dyn3 spliced-form, Dyn3aaa, did not have any effect on the normal maturation process of the dendrites and spines (Figure 12.1D, inset). Interestingly, these spliced-forms differ by only 10 amino acids, a cassette that is spliced into Dyn3baa, but is lacking in Dyn3aaa. These Dyn3baa-induced filopodia did not contain any postsynaptic markers, nor did they associate with the presynaptic terminals of non-transfected neurons, indicating that these dendritic protrusions are morphologically and functionally immature. The GTPase activity of Dyn3baa was necessary to induce filopodia, as a dominant-negative mutant (Dyn3baaK44A) did not have the same effect on transfected neurons but actually had the opposite effect of causing an early development of mushroom-shaped spines (N.W.G. and M.A.M., unpublished observations). These results are consistent with studies in non-neuronal cells that demonstrate that GTPase-deficient dynamin affects actin-dependent membrane processes, including podosome dynamics (Ochoa *et al.*, 2000) and actin comet formation and movement (Orth *et al.*, 2002; Lee and De Camilli, 2002).

The concept of a postsynaptic dynamin regulating the growth and development of the dendrites is an intriguing proposition. The study that identified Dyn3 as a postsynaptic dynamin also found this isoform to bind to a biochemical complex containing other postsynaptic proteins, such as the scaffolding molecule Homer and the glutamate receptor mGluR5 (Gray *et al.*, 2003). Overexpression of Homer leads to an accelerated maturation of dendritic spines, with these protrusions becoming more functionally and synaptically active weeks before they normally do (Sala *et al.*, 2001). This was attributed to the possibility that as a scaffolding molecule, an abundance of Homer could provide more opportunities to 'capture' postsynaptic protein and allow for the early development of the postsynaptic density (PSD). Thus, like its binding partner Homer, Dyn3 may function to help regulate the dendritic maturation process, causing the growth of filopodia that will eventually stabilize as spines. As the overexpression of either Homer or Dyn3 have opposite effects on spine morphology, perhaps a balance between the activities of each protein is necessary to allow for proper development to occur.

Future studies will explore the role of Dyn3 during the regulation of the filopodia/spine actin cytoskeleton. The effects of Dyn3 expression on spine morphogenesis outlined above may involve actin, as Dyn3-induced neuronal filopodia are lost following treatment with the actin depolymerizing drug Latrunculin A (N.W.G. and M.A.M., unpublished data). Preliminary data also suggest that, like Dyn2, Dyn3 binds to cortactin, an interaction that could modulate the rate and extent of actin polymerization and branching. As Dyn2 was recently found to stimulate actin polymerization, in conjunction with cortactin and other actin-binding proteins (Schafer *et al.*, 2002), there is a precedent warranting the examination of Dyn3 as a potential actin cytoskeleton modulator in neurons.

Conclusions and perspectives

Here we have provided a short discussion of dynamin and some actin-membrane-dependent processes in which this mechanoenzyme has been shown to function. Additional cellular processes that may apply to the study of the dynamin–actin–membrane interface include growth cone extension, immunological synapse formation and cell adhesion. How does dynamin mediate these processes? The current data develop a hypothesis in which dynamin is recruited to the actin–membrane interface via its PH domain and PRD, where it associates with phospholipids and cytoskeletal proteins to regulate various cell processes. In the case of vesicle formation, dynamin is recruited to the invaginated vesicles where it constricts the membrane neck and, through its interactions with actin nucleation proteins, stimulates a 'burst' of F-actin nucleation. This new actin polymerization results in the severing or 'breaking' of the vesicle neck, liberating it from its donor membrane. However, in non-vesicle forming processes, such as membrane ruffling, cytokinesis and dendritic spine morphogenesis, dynamin's role is less clear and, despite recent advances, needs to be studied further. Several important questions about dynamin's role in cytoskeletal processes remain enigmatic. For example, how are dynamin's interactions with binding partners regulated in time and space? What is the function of dynamin in these complexes? How does a mechanochemical motor, dynamin, mediate the polymerization, stability and organization of microfilaments? Further, it will be important to study the roles of the many spliced-forms of dynamin during essential cellular processes and to test if these distinct forms function differentially. Although many studies on dynamin have focused on its role in vesicle formation, it has been proven to be a multi-faceted protein that participates in numerous cellular processes. Understanding the common themes between these seemingly distinct processes will elucidate the importance of the interface between membranes and the cytoskeleton and how it contributes to cell behaviour.

References

Bowerman, B. and Severson, A. F., 1999. Cell division: plant-like properties of animal cell cytokinesis. *Curr. Biol.* **9**: R658–R660.

Cao, H., Orth, J. D., Chen, J., Weller, S. G., *et al.*, 2003. Cortactin is a component of clathrin-coated pits and participates in receptor-mediated endocytosis. *Mol. Cell Biol.* **23**: 2162–2170.

Carroll, R. C., Beattie, E. C., Xia, H., Luscher, C., *et al.*, 1999. Dynamin-dependent endocytosis of ionotropic glutamate receptors. *Proc. Natl. Acad. Sci. USA* **96**: 14112–14117.

Chen, M. S., Obar, R. A., Schroeder, C. C., Austin, T. W., *et al.*, 1991. Multiple forms of dynamin are encoded by *shibire*, a *Drosophila* gene involved in endocytosis. *Nature* **351**: 583–586.

Cook, T., Mesa, K. and Urrutia, R., 1996. Three dynamin-encoding genes are differently expressed in developing rat brain. *J. Neurochem.* **67**: 927–931.

Cook, T. A., Urrutia, R. and McNiven, M. A., 1994. Identification of dynamin 2, an isoform ubiquitously expressed in rat tissues. *Proc. Natl. Acad. Sci. USA* **91**: 644–648.

Du, Y., Weed, S. A., Xiong, W. C., Marshall, T. D. and Parsons, J. T., 1998. Identification of a novel cortactin SH3 domain-binding protein and its localization to growth cones of cultured neurons. *Mol. Cell Biol.*, **18**: 5838–5851.

Feng, B., Schwarz, H. and Jesuthasan, S., 2002. Furrow-specific endocytosis during cytokinesis of zebrafish blastomeres. *Exp. Cell Res.* **279**: 14–20.

Finger, F. P. and White, J. G., 2002. Fusion and fission: membrane trafficking in animal cytokinesis. *Cell* **108**: 727–730.

Glotzer, M., 2001. Animal cell cytokinesis. *Annu. Rev. Cell Dev. Biol.* **17**: 351–386.

Gray, N. W., Fourgeaud, L., Huang, B., Chen, J., *et al.*, 2003. Dynamin 3 is a component of the postsynapse, where it interacts with mGluR5 and Homer. *Curr. Biol.* **13**: 510–515.

Gu, X. and Verma, D. P., 1996. Phragmoplastin, a dynamin-like protein associated with cell plate formation in plants. *EMBO J.* **15**: 695–704.

Hinshaw, J. E., 2000. Dynamin and its role in membrane fission. *Annu. Rev. Cell Dev. Biol.* **16**: 483–519.

Jones, S. M., Howell, K. E., Henley, J. R., Cao, H. and McNiven, M. A., 1998. Role of dynamin in the formation of transport vesicles from the *trans*-Golgi network. *Science* **279**: 573–577.

Kaksonen, M., Peng, H. B. and Rauvala, H., 2000. Association of cortactin with dynamic actin in lamellipodia and on endosomal vesicles. *J. Cell Sci.* **113** (24): 4421–4426.

Kamitani, A., Yamada, H., Kinuta, M., Watanabe, M., *et al.*, 2002. Distribution of dynamins in testis and their possible relation to spermatogenesis. *Biochem. Biophys. Res. Commun.* **294**: 261–267.

Kessels, M. M., Engqvist-Goldstein, A. E., Drubin, D. G. and Qualmann, B., 2001. Mammalian Abp1, a signal-responsive F-actin-binding protein, links the actin cytoskeleton to endocytosis via the GTPase dynamin. *J. Cell Biol.* **153**: 351–366.

Koenig, J. H. and Ikeda, K., 1989. Disappearance and reformation of synaptic vesicle membrane upon transmitter release observed under reversible blockage of membrane retrieval. *J. Neurosci.* **9**: 3844–3860.

Lauber, M. H., Waizenegger, I., Steinmann, T., Schwarz, H., *et al.*, 1997. The *Arabidopsis* KNOLLE protein is a cytokinesis-specific syntaxin. *J. Cell Biol.* **139**: 1485–1493.

Lee, E. and De Camilli, P., 2002. Dynamin at actin tails. *Proc. Natl. Acad. Sci. USA* **99**: 161–166.

McNiven, M. A., Cao, H., Pitts, K. R. and Yoon, Y., 2000a. Pinching in new places: multiple functions for the dynamin family. *TIBS* **25**: 115–120.

McNiven, M. A., Kim, L., Krueger, E. W., Orth, J. D., *et al.*, 2000b. Regulated interactions between dynamin and the actin-binding protein cortactin modulate cell shape. *J. Cell Biol.* **151**: 187–198.

Merrifield, C. J., Feldman, M. E., Wan, L. and Almers, W., 2002. Imaging actin and dynamin recruitment during invagination of single clathrin-coated pits. *Nat. Cell Biol.* **4**: 691–698.

Nakata, T., Takemura, R. and Hirokawa, N., 1993. A novel member of the dynamin family of GTP-binding proteins is expressed specifically in the tesis. *J. Cell Sci.* **105**: 1–5.

O'Halloran, T. J., 2000. Membrane traffic and cytokinesis. *Traffic* **1**: 921–926.

Ochoa, G. C., Slepnev, V. I., Neff, L., Ringstad, N., *et al.*, 2000. A functional link between dynamin and the actin cytoskeleton at podosomes. *J. Cell Biol.* **150**: 377–390.

Okamoto, P. M., Gamby, C., Wells, D., Fallon, J. and Vallee, R. B., 2001. Dynamin isoform-specific interaction with the shank/ProSAP scaffolding proteins of the postsynaptic density and actin cytoskeleton. *J Biol Chem* **276**: 48458–48465.

Orth, J. D. and McNiven, M. A., 2003. Dynamin at the actin–membrane interface. *Curr. Opin. Cell Biol.* **15**: 31–39.

Orth, J. D., Krueger, E. W., Cao, H. and McNiven, M. A., 2002. The large GTPase dynamin regulates actin comet formation and movement in living cells. *Proc. Natl. Acad. Sci. USA* **99**: 167–172.

Qualmann, B., Kessels, M. M. and Kelly, R. B., 2000. Molecular links between endocytosis and the actin cytoskeleton. *J. Cell Biol.* **150**: F111–F116.

Robinson, D. N. and Spudich, J. A., 2000. Towards a molecular understanding of cytokinesis. *Trends Cell Biol.* **10**: 228–237.

Rozelle, A. L., Machesky, L. M., Yamamoto, M., Driessens, M. H., *et al.*, 2000. Phosphatidylinositol 4,5-bisphosphate induces actin-based movement of raft-enriched vesicles through WASP-Arp2/3. *Curr. Biol.* **10**: 311–320.

Sala, C., Piech, V., Wilson, N. R., Passafaro, M., *et al.*, 2001. Regulation of dendritic spine morphology and synaptic function by Shank and Homer. *Neuron* **31**: 115–130.

Schafer, D. A., 2002. Coupling actin dynamics and membrane dynamics during endocytosis. *Curr. Opin. Cell Biol.* **14**: 76–81.

Schafer, D. A., D'Souza-Schorey, C. and Cooper, J. A., 2000. Actin assembly at membranes controlled by ARF6. *Traffic* **1**: 896–907.

Schafer, D. A., Weed, S. A., Binns, D., Karginov, A. V. and Cooper, J. A., 2002. Dynamin 2 and cortactin regulate actin assembly and filament organization. *Curr. Biol.* **12**: 1852–1857.

Shpetner, H. S. and Vallee, R. B., 1989. Identification of dynamin, a novel mechanochemical enzyme that mediates interactions between microtubules. *Cell* **59**: 421–432.

Shupliakov, O., Low, P., Grabs, D., Gad, H., *et al.*, 1997. Synaptic vesicle endocytosis impaired by disruption of dynamin-SH3 domain interactions. *Science* **276**: 259–263.

Sontag, J.-M., Fykse, E. M., Ushkaryov, Y., Liu, J.-P., *et al.*, 1994. Differential expression and regulation of multiple dynamins. *J. Biol. Chem.* **269**: 4547–4554.

Straight, A. F. and Field, C. M., 2000. Microtubules, membranes and cytokinesis. *Curr. Biol.* **10**: R760–R770.

Swanson, M. M. and Poodry, C. A., 1981. The *shibire*^{ts} mutant of *Drosophila*: a probe for the study of embryonic development. *Dev. Biol.* **84**: 465–470.

Takei, K., Mundigl, O., Daniell, L. and De Camilli, P., 1996. The synaptic vesicle cycle: a single vesicle budding step involving clathrin and dynamin. *J. Cell Biol.* **133**: 1237–1250.

Thompson, H. M., Skop, A. R., Euteneuer, U., Meyer, B. J. and McNiven, M. A., 2002. The large GTPase dynamin associates with the spindle midzone and is required for cytokinesis. *Curr. Biol.* **12**: 2111–2117.

van Dam, E. M. and Stoorvogel, W., 2002. Dynamin-dependent transferrin receptor recycling by endosome-derived clathrin-coated vesicles. *Mol. Biol. Cell* **13**: 169–182.

van der Bliek, A. M. and Meyerowitz, E. M., 1991. Dynamin-like protein encoded by the *Drosophila shibire* gene associated with vesicular traffic. *Nature* **351**: 411–414.

Wienke, D. C., Knetsch, M. L. W., Neuhaus, E. M., Reedy, M. C. and Manstein, D. J., 1999. Disruption of a dynamin homologue affects endocytosis, organelle morphology, and cytokinesis in *Dictyostelium discoideum*. *Mol. Biol. Cell* **10**: 225–243.

Zhang, P. and Hinshaw, J. E., 2001. Three-dimensional reconstruction of dynamin in the constricted state. *Nat. Cell Biol.* **3**: 922–926.

13

Regulation of Microtubule Dynamics in Migrating Cells: a New Role for Rho GTPases

Torsten Wittmann and **Clare M. Waterman-Storer**

Rho GTPases, most notably RhoA, Rac1 and Cdc42Hs, were described as major regulators of actin organization over 10 years ago, and thus have received extensive attention as regulators of cell motility. However, researchers have only recently begun to examine their effects on other components of the cytoskeleton. For example, the microtubule cytoskeleton, which also plays a central role in cell locomotion, displays a characteristic polarized organization in migrating cells. Here, we discuss recent findings on how microtubule organization and dynamics are regulated downstream of Rho GTPases in higher eukaryotic cells, and how this contributes to their function during cell migration. Cdc42 is responsible for the reorientation of the centrosome towards the direction of migration, activation of RhoA appears to selectively stabilize a subset of microtubules, and events downstream of Rac1 regulate microtubule plus end growth.

Introduction

Many metazoan cell types, such as fibroblasts or epithelial cells, can become polarized either as a response to an extracellular stimulus or spontaneously, and migrate in a unidirectional fashion. This ability is essential for cells to

---

Cell Motility. Edited by Anne Ridley, Michelle Peckham and Peter Clark.
 ISBN 0 470 84872 3

function in their natural environment. For example, the development of the nervous system in vertebrates requires many complex patterns of cellular migration. Epithelial cells need to migrate in order to close wounds in the epithelial layer, whereas fibroblast motility is crucial for tissue remodelling. Conversely, improper regulation of cell migration is the basis of many abnormal processes, resulting, for example, in the invasiveness of tumour cells.

Control of actin by Rho GTPases in migrating cells

Migrating vertebrate cells in tissue culture show a unique polarized morphology, a broad, flat lamellum extending in the direction of migration that terminates in a ruffling lamellipodium (the leading edge) (Abercrombie *et al*., 1970; Small *et al*., 2002) and a narrow, retracting tail at the rear of the cell. This morphological polarization depends on the underlying organization of the actin cytoskeleton. Actin filament polymerization is nucleated at the leading edge, which generates a highly crosslinked meshwork of actin filaments in the lamellipodium whose growing ends face the front of the cell (Henson *et al*., 1999; Small *et al*., 1978; Svitkina *et al*., 1997; see chapter by Pollard). The constant growth of these filaments, most likely coupled with the action of a myosin motor, both pushes the leading edge forward and generates a retrograde flow of actin towards the cell centre (Cramer, 1997; Henson *et al*., 1999; Lin *et al*., 1997; Wang, 1985; Waterman-Storer and Salmon, 1999). Contraction of actin bundles in the cell body may be required for retraction of the cell's rear (Cramer *et al*., 1997; Small *et al*., 1998). The coupling of these actin motions inside the cell to adhesions to the extracellular matrix drives cell motion.

More than 10 years ago, it was discovered that activation of members of the Rho family of GTPases can induce changes in the actin cytoskeleton and adhesions that are typically associated with migrating cells. In fibroblasts, RhoA activity results in the generation of contractile actin bundles and large adhesions to the substrate. Rac1 activity induces actin polymerization to drive lamellipodial protrusion and the formation of small adhesions, and Cdc42 generates cell polarity and induces the formation of filopodia (Hall, 1998; Nobes and Hall, 1999; van Aelst and D'Souza-Schorey, 1997).

Rho GTPases act as molecular switches that can be activated by a variety of extracellular stimuli. They cycle between a GTP-bound active form, which can activate downstream effectors, and an inactive GDP-bound form. Rho proteins are tightly regulated by different classes of upstream factors that control the exchange of GDP for GTP and the rate of GTP hydrolysis (Kjoller and Hall, 1999; Symons and Settleman, 2000; van Aelst and D'Souza-Schorey, 1997). Signalling through Rho GTPases is further complicated by their large number of downstream effector molecules that must somehow relay information to the cytoskeleton, which is still poorly understood on a molecular level. For example, it only recently became evident that the induction of actin polymerization downstream of Rac1 and Cdc42 might be due to WASp family proteins, which stimulate actin nucleation through activation of the Arp2/3 complex (Higgs and Pollard, 2001; Takenawa and Miki, 2001; see chapters by Pollard and Insall). Furthermore, it is still an open question how the activity of Rho GTPases is regulated spatially inside the cell to achieve functional polarization of the cytoskeleton to generate directed motility. The discovery that certain membrane lipids that act as activators of Rac1 are localized in a polarized fashion in cells migrating in a chemotactic gradient is a first step towards understanding how cytoskeletal polarity is generated (Firtel and Chung, 2000; Servant *et al.*, 2000; see chapter by Firtel).

The role of microtubules in migrating cells

In contrast to actin, the role of the microtubule cytoskeleton in migrating cells is less well defined. While some small cell types such as neutrophils can migrate in the absence of microtubules, tissue cells such as fibroblasts, endothelial (Figure 13.1) and epithelial cells, for example, require microtubules for directed migration and the regulation of protrusion formation (Bershadsky *et al.*, 1991; Goldman, 1971; Vasiliev *et al.*, 1970; Waterman-

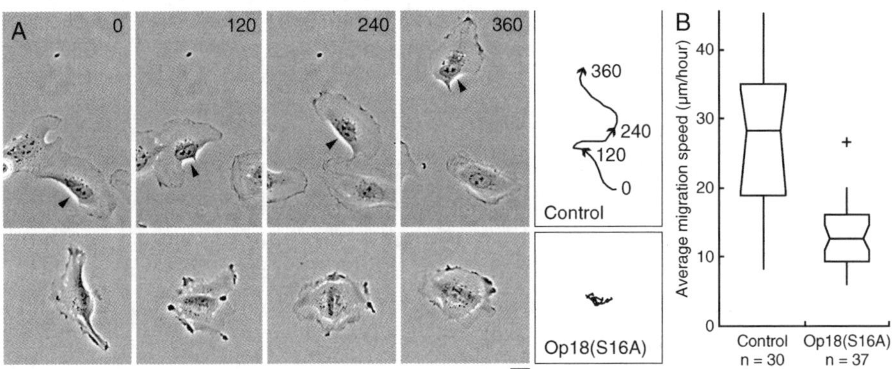

Figure 13.1 Microtubules are required for endothelial cell motility. (A) Human microvascular endothelial cells were injected with control buffer (upper row; arrowhead indicates the injected cell) or phosphorylation-site-deficient, dominant active Op18/stathmin resulting in microtubule depolymerization (lower row) and observed by time-lapse phase contrast microscopy. The box on the right shows the track of the centre of the nucleus at 10-min intervals over a total time of 6 h. Elapsed time is indicated in minutes. Bar, 20 μm. (B) Quantitation of experiments as shown in (A). The average migration speed was calculated from tracks of nuclear movement over a total time of 6 h and is significantly decreased in cells injected with Op18(S16A). n, number of cells analysed

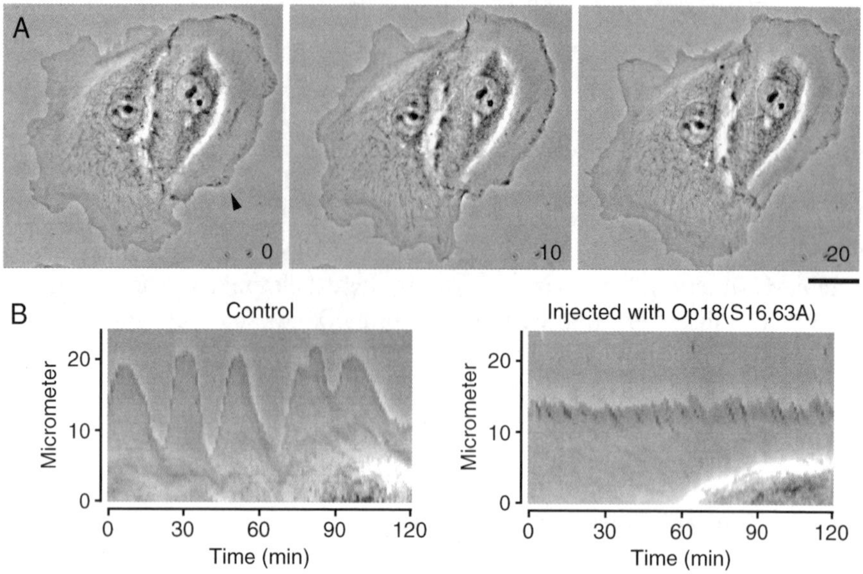

Figure 13.2 Microtubules are required for leading edge protrusion in epithelial cells. (A) The PtK1 cell on the right (arrowhead) was injected with a phosphorylation-site-deficient, dominant active Op18/stathmin protein resulting in depolymerization of most microtubules. The pair of cells was then observed by time-lapse phase contrast microscopy. Shown are three representative frames. Note the extensive protrusive activity of the control cell on the left. Elapsed time is indicated in minutes. Bar, 20 μm. (B) Kymograph analysis shows the position of the leading edge over time of a control PtK1 cell compared with a cell injected with dominant active Op18/stathmin. While the control cell undergoes multiple cycles of protrusion and retraction during the course of 2 h, the edge of the injected cell is almost completely quiescent

Storer *et al.*, 1999; Wittmann and Waterman-Storer, 2001). Accordingly, we also observed that microtubule depolymerization in PtK1 cells, a rat-kangaroo kidney epithelial cell line, by either nocodazole or microinjection of a dominant active mutant of the microtubule-destabilizer Op18/stathmin (Cassimeris, 2002) severely altered leading edge dynamics (Figure 13.2). In control PtK1 cells, large protrusions develop at random sites along the cell's leading edge and extend for several minutes up to 10 μm before retracting again. In the absence of microtubules, however, only minor membrane ruffles form at the extreme periphery of the lamellipodium and are distributed evenly all along the free edge of these cells.

Like actin, microtubule organization and dynamics are polarized in migrating cells. In most cells, microtubules are organized in a radial array by the centrosome in the cell centre, where they are nucleated and most microtubule minus ends are anchored. Microtubule plus ends emanate out to the cell periphery, where they undergo stochastic changes between polymerization and depolymerization, a property known as dynamic instability (Desai and Mitchison, 1997; Mitchison and Kirschner, 1984). A transition from polymerization to depolymerization is called a 'catastrophe', while the opposite is referred to as a 'rescue'. In most migrating cell types, the centrosome is positioned between the nucleus and the leading edge, and the bulk of microtubule plus ends are orientated towards the leading edge. In addition, microtubules in migrating fibroblasts appear to target sites of substrate adhesion and somehow regulate their turnover (Kaverina *et al.*, 1998, 1999; Small and Kaverina, 2003).

The observation that microtubules, like actin, exhibit specific, polarized organization and growth dynamics in migrating cells raises the important question of whether microtubules, like actin, might be regulated downstream of Rho GTPases. Indeed, recent evidence suggests that specific Rho GTPases affect specific aspects of microtubule organization and assembly/disassembly dynamics, and this shall be the focus of the remainder of this chapter.

Centrosome reorientation downstream of Cdc42

The most striking polarization of the microtubule cytoskeleton in many migrating cells is the repositioning of the centrosome, the organizing centre of the radial interphase microtubule network. When fibroblasts are activated to migrate by scratching a wound in a cell monolayer, the centrosome rapidly changes its place from a random orientation relative to the nucleus to a position between the nucleus and the leading edge (Gotlieb *et al.*, 1981; Malech *et al.*, 1977). Centrosome reorientation depends on an intact microtubule cytoskeleton, but is not observed in all cell types and the significance of centrosome reorientation remains unclear (Euteneuer and Schliwa, 1992; Gotlieb *et al.*, 1983; Schliwa and Höner, 1993; Yvon *et al.*, 2002). It was originally speculated that the primary reason for repositioning of the centrosome and the bulk of microtubules towards the direction of migration is the requirement to orientate the secretory apparatus. Indeed, secretion that is preferentially polarized towards the leading edge has been observed in migrating fibroblasts (Bergmann *et al.*, 1983; Hopkins *et al.*, 1994). The requirement of microtubule-based transport for cell locomotion was also demonstrated by microinjection of kinesin-specific antibodies, which inhibited cell motility in a way similar to microtubule depolymerization (Rodionov *et al.*, 1993). However, microtubule-dependent transport in migrating cells might not be limited to membrane organelles, but might include molecules required for the regulation of protrusion formation or focal adhesion turnover (Krylyshkina *et al.*, 2002).

Recently, it has become apparent that Rho GTPases are involved in controlling centrosome orientation. It was first demonstrated in T cells that Cdc42 is required for the reorientation of the centrosome and Golgi apparatus towards antigen-presenting cells possibly to direct lymphokine secretion (Stowers *et al.*, 1995). Similarly, two groups have now shown independently that, in a wound-edge model system, the orientation of the centrosome towards the leading edge of migrating cells also depends on Cdc42 function, as well as the minus-end-directed microtubule-based motor, cytoplasmic dynein. In astrocytes, interactions of integrins with the extracellular matrix at the protruding cell edge lead to polarized recruitment and activation of Cdc42 and an atypical protein kinase C, which in turn might locally activate cytoplasmic dynein (Etienne-Manneville and Hall, 2001). Such localized dynein activity at the cortex in the leading part of the cell could then pull the centrosome forward (Figure 13.3a). In fibroblasts, centrosome reorientation appears to precede protrusion formation. It is solely dependent on Cdc42 function and is

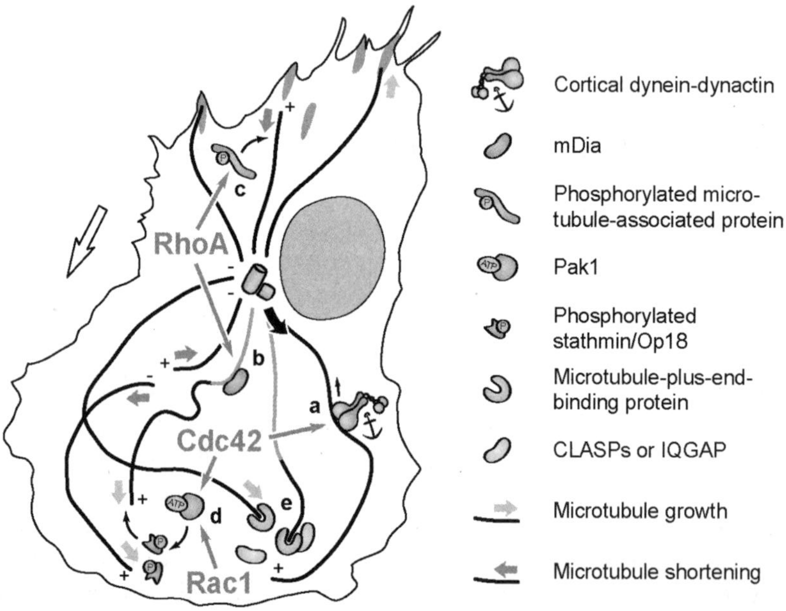

Figure 13.3 Potential mechanisms for how Rho proteins could regulate organization and dynamics of the microtubule cytoskeleton. (a) Cdc42 activity is required for the reorientation of the centrosome towards the direction of migration, which could occur through cortical cytoplasmic dynein activity. (b) Stable, detyrosinated microtubules are induced by RhoA and its downstream effector mDia. (c) RhoA might also cause the phosphorylation of microtubule-associated proteins and thus destabilize microtubules. (d) Rac1 and Cdc42 might decrease the microtubule catastrophe frequency and thus promote microtubule growth through Pak1-dependent phosphorylation of Op18/stathmin. (e) The regulation of microtubule-plus-end-tracking proteins downstream of Rho GTPases might also contribute to increased microtubule growth at the cells leading edge. Microtubules also target focal adhesions and depolymerizing microtubule ends are created in the cell body by retrograde-flow-induced breakage. The open arrow indicates the direction of cell migration. Thick black lines represent microtubules and plus and minus signs indicate microtubule polarity. (A colour reproduction of this figure can be found in the colour plate section)

not blocked by the inhibition of either Rac1 or RhoA (Palazzo *et al.*, 2001a). Thus, it is quite possible that polarized activation of Cdc42 is one of the initial steps in cells that are about to migrate. This is paralleled by its well-established role in yeast, where Cdc42 localization provides the main cue for the polarity of the actin cytoskeleton (Pruyne and Bretscher, 2000).

Microtubule stabilization downstream of RhoA

Partly as a result of centrosome position, the microtubule array itself is polarized and microtubules tend to be aligned along the axis of cell migration with the majority of microtubule plus ends facing the leading edge. It has been observed in fibroblasts that many of these orientated microtubules are composed in large part of tubulin that has been post-translationally detyrosinated at its C-terminus (Gundersen and Bulinski, 1988). Detyrosination is a slow process as compared with microtubule turnover by dynamic instability, and thus serves as an indicator of microtubule stabilization. Detyrosination does not stabilize microtubules by itself, but it has been suggested that dynamic instability in detyrosinated microtubules is blocked by an ATP-sensitive plus end cap (Infante *et al.*, 2000). However, the mechanism by which microtubules become detyrosinated is still mysterious and no tubulin carboxypeptidase has been identified to date. In addition, the intracellular function of detyrosinated microtubules is not well understood, although there is some evidence that certain types of intracellular transport might preferentially occur along detyrosinated microtubule tracks (Lin *et al.*, 2002). Thus, in migrating cells, these stabilized microtubules could serve specifically to deliver cargo to the leading edge.

In fibroblasts, in addition to its effects on actin stress fibre formation, RhoA activation by lysophosphatidic acid increases the number of orientated, detyrosinated microtubules and induces long episodes of pauses in a subset of microtubule plus ends (Cook *et al.*, 1998) (Figure 13.3b). The RhoA-mediated induction of these microtubules has been attributed to its downstream effector mDia. mDia binds to microtubules and might thus directly regulate microtubule dynamic instability (Palazzo *et al.*, 2001b). Overexpression of activated mDia also results in the alignment of microtubules with actin bundles (Ishizaki *et al.*, 2001). However, this might be an indirect effect, since mDia induces stress fibre formation, and microtubules often align along stress fibres (Salmon *et al.*, 2002).

In contrast to this mDia-mediated microtubule stabilization, in neuroblastoma cells, RhoA activation results in increased phosphorylation of the microtubule-associated protein tau, which should promote its dissociation from microtubules and result in their destabilization (Sayas *et al.*, 1999) (Figure 13.3c). Whether RhoA has similar effects on non-neuronal microtubule-associated proteins and whether this affects microtubule dynamics during cell motility is unclear.

Regulation of microtubule dynamic instability downstream of Rac1

In addition to the polarized organization of the microtubule cytoskeleton as a whole, microtubule polymerization dynamics are polarized in migrating cells. Microtubules in the lamella behind the leading edge are moved backwards by actin retrograde flow, indicating a coupling between the actin and microtubule networks (Salmon *et al.*, 2002; Waterman-Storer and Salmon, 1997; Yvon and Wadsworth, 2000). Since microtubule plus ends are often found close to the leading edge, they must undergo net growth as they are continuously swept backwards. Further, as the leading edge protrudes and the rear edge retracts, microtubules grow forward and fill in the advancing cellular space (Waterman-Storer and Salmon, 1997). Thus, although microtubules undergo dynamic instability throughout the cell, microtubule growth is biased towards the leading edge. Indeed, quantitative analysis of the behaviour of individual microtubules indicates that microtubule plus ends in protruding cell edges spend more time growing and undergo far fewer catastrophes than microtubules in central cell regions or at quiescent cell edges (Wadsworth, 1999; Waterman-Storer and Salmon, 1997).

These observations pose the question of how such regional differences in microtubule dynamics are generated. There has so far been no documentation of regional localization or regulation of microtubule-stabilizing factors, such as microtubule-associated proteins, or catastrophe-promoting factors in migrating cells. However, we have found in PtK1 epithelial cells expressing dominant active Rac1(Q61L) that many microtubules reach the very edge of the cell although Rac1(Q61L)-induced actin retrograde flow constantly carries them towards the cell centre (Figure 13.4). In addition, in these cells, exceptionally long microtubules often form extensive bundles that run parallel to the cell edge. This is in contrast to cells expressing dominant negative Rac1(T17N) where few microtubule ends reach the cell edge, and retrograde flow is inhibited (Figure 13.5). Careful measurements of microtubule dynamic instability parameters in cells expressing active Rac1(Q61L) demonstrated a fourfold increase in microtubule net growth, due to a decrease in catastrophe frequency and an increase of the time individual microtubules spend growing as compared with cells expressing dominant negative Rac1(T17N) (Wittmann *et al.*, 2003).

Interestingly, recent data suggest a pathway by which Rac1 and Cdc42 could influence microtubule dynamics. Growth-factor-induced activation of

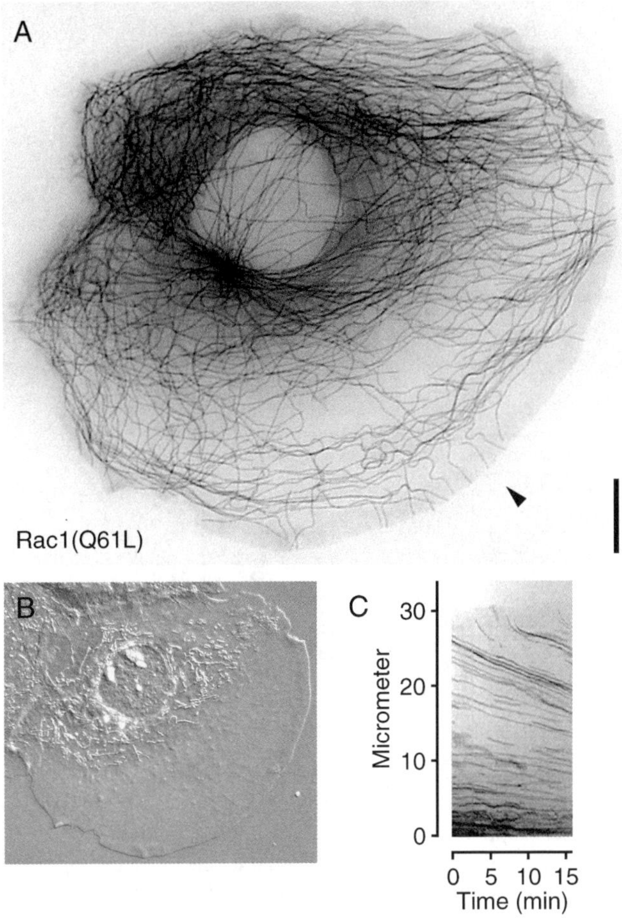

Figure 13.4 Microtubules in a Rac1(Q61L)-expressing PtK1 cell. (A) Microtubules in a live PtK1 cell expressing dominant active Rac1(Q61L) observed by fluorescence microscopy after injection of fluorescently labelled tubulin. The image was contrast inverted and subjected to an unsharp mask filter to better visualize individual microtubules. Dominant active Rac1(Q61L) induces lamellipodia formation all around the free cell edge and many microtubules grow to the very edge of the cell although they are rapidly carried away from the edge by actin-based retrograde flow. Bar, 10 μm. (B) Differential interference contrast image of the same cell. (C) Kymograph analysis (time versus distance plot) of the microtubule time-lapse series along the line indicated by the arrowhead in (A). Oblique dark lines demonstrate microtubule retrograde flow

Rac1 and Cdc42 leads to Pak1-mediated phosphorylation of Op18/stathmin (Daub *et al.*, 2001) (Figure 13.3d). Indeed, phosphorylation of Op18/stathmin upon stimulation with phorbol ester, another stimulator of Rac1 activity, had been observed earlier (Hailat *et al.*, 1990). Op18/stathmin is a tubulin-binding protein that sequesters tubulin dimers and promotes microtubule plus end catastrophes. Phosphorylation inactivates Op18/stathmin, which would

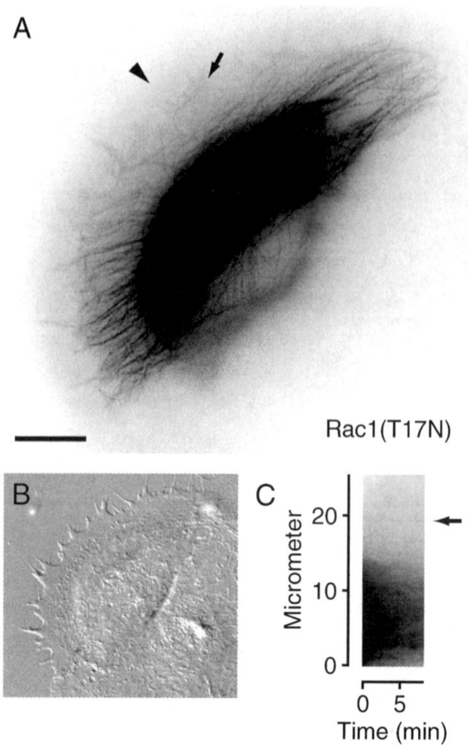

Figure 13.5 Microtubules in a Racl(T17N)-expressing Ptk1 cell. (A) Microtubules in a live PtK1 cell expressing dominant negative Racl(T17N). Since these cells are very contracted it is more difficult to see individual microtubules in the cell periphery. (B) Differential interference contrast image of the same cell. (C) Kymograph analysis (time versus distance plot) of the microtubule time-lapse series along the line indicated by the arrowhead in (A). The arrow in (A) and (C) identifies the same microtubule, which gives rise to the horizontal dark line in the kymograph showing that it basically did not undergo retrograde flow

promote microtubule growth by decreasing the catastrophe frequency (Cassimeris, 2002; Larsson *et al.*, 1997). Accordingly, we have observed specific phosphorylation of Op18/stathmin by recombinant Pak1 *in vitro* at serine 16 and found that Rac1(Q61L)-promoted microtubule growth in Ptk1 cells depends on Pak activity. These observations agree well with the idea that Rac1 is active in the leading edge of a migrating cell, where microtubule plus ends exhibit net growth.

However, constitutively acting Pak1 did not have the same effect on microtubule growth as constitutively active Rac1 (Q61L), indicating that the Pak activity is necessary but not sufficient for Rac1-mediated microtubule growth (Wittman *et al.*, 2003). Therefore other mechanisms locally regulating microtubule dynamics in migrating cells are likely to exist.

Interesting candidates for regional microtubule regulation include a recently identified class of proteins that specifically bind to growing microtubule plus ends (Schroer, 2001; Schuyler and Pellman, 2001). One such protein, adenomatous polyposis coli protein (APC), forms granules that undergo a plus-end-directed movement along microtubules and specifically accumulate in actively protruding areas of cells (Mimori-Kiyosue *et al.*, 2000; Nathke *et al.*, 1996). In addition, APC stabilizes microtubules *in vitro* and *in vivo* (Zumbrunn *et al.*, 2001) and EB1, a protein that can bind APC and tracks growing microtubule plus ends, strongly promotes microtubule growth in *Xenopus* egg extracts (Tirnauer *et al.*, 2002).

CLIP-170, the first protein described to bind to growing microtubule ends, does not seem to have any preference for certain areas of the cell (Perez *et al.*, 1999), but the recently described CLIP-170 associated proteins (CLASPs) preferentially bind to microtubule ends orientated towards the leading edge in serum-stimulated fibroblasts. This polarized localization of CLASPs correlates with the orientation of microtubules in migrating cells, and CLASP2 appears to associate with the ends of acetylated microtubules (Akhmanova *et al.*, 2001). In addition, functional interactions of Rho GTPases and CLIP-170 have been suggested to occur through IQGAP (Fukata *et al.*, 2002) and CLIP-170 appears to promote microtubule plus end rescues (Komarova *et al.*, 2002). Thus, a ternary complex of activated Rac1, CLIP-170 and IQGAP might regulate microtubule dynamics *in vivo*. Consequently, Rho GTPase-mediated regulation of microtubule-plus-end-tracking proteins might also locally regulate microtubule dynamics in migrating cells (Figure 13.3e).

Conclusion

Clearly, microtubules are important for cell migration and during the last years our understanding of how microtubule dynamics might be regulated *in vivo* has grown tremendously. Indeed, recent results suggest that the Rho family of GTPases, well known for their regulatory control of actin organization and dynamics during cell motility, also appear to control the specific microtubule organization and dynamics that occur in motile tissue cells. Thus, Rho proteins may turn out to be master regulators of cytoskeletal dynamics that are required for directed cell locomotion. However, the challenge for the future will be to elucidate the mechanisms by which microtubule dynamics are controlled locally in living cells and more importantly how these local differences might affect cellular behaviour.

References

Abercrombie, M., Heaysman, J. E. and Pegrum, S. M., 1970. The locomotion of fibroblasts in culture. I. Movements of the leading edge. *Exp. Cell Res.* **59**: 393–398.

Akhmanova, A., Hoogenraad, C. C., Drabek, K., Stepanova, T., *et al.*, 2001. Clasps are CLIP-115 and -170 associating proteins involved in the regional regulation of microtubule dynamics in motile fibroblasts. *Cell* **104**: 923–935.

Bergmann, J. E., Kupfer, A. and Singer, S. J., 1983. Membrane insertion at the leading edge of motile fibroblasts. *Proc. Natl. Acad. Sci. USA* **80**: 1367–1371.

Bershadsky, A. D., Vaisberg, E.A., and Vasiliev, J. M., 1991. Pseudopodial activity at the active edge of migrating fibroblast is decreased after drug-induced microtubule depolymerization. *Cell Motil. Cytoskeleton* **19**: 152–158.

Cassimeris, L., 2002. The oncoprotein 18/stathmin family of microtubule destabilizers. *Curr. Opin. Cell Biol.* **14**: 18–24.

Cook, T. A., Nagasaki, T. and Gundersen, G. G., 1998. Rho guanosine triphosphatase mediates the selective stabilization of microtubules induced by lysophosphatidic acid. *J. Cell Biol.* **141**: 175–185.

Cramer, L. P., 1997. Molecular mechanism of actin-dependent retrograde flow in lamellipodia of motile cells. *Front Biosci.* **2**: d260–d270.

Cramer, L. P., Siebert, M. and Mitchison, T. J., 1997. Identification of novel graded polarity actin filament bundles in locomoting heart fibroblasts: implications for the generation of motile force. *J. Cell Biol.* **136**: 1287–1305.

Daub, H., Gevaert, K., Vandekerckhove, J., Sobel, A. and Hall, A., 2001. Rac/Cdc42 and p65PAK regulate the microtubule-destabilizing protein stathmin through phosphorylation at serine 16. *J. Biol. Chem.* **276**: 1677–1680.

Desai, A. and Mitchison, T. J., 1997. Microtubule polymerization dynamics. *Annu. Rev. Cell Dev. Biol.* **13**: 83–117.

Etienne-Manneville, S. and Hall, A., 2001. Integrin-mediated activation of Cdc42 controls cell polarity in migrating astrocytes through PKCzeta. *Cell* **106**: 489–498.

Euteneuer, U. and Schliwa, M., 1992. Mechanism of centrosome positioning during the wound response in BSC-1 cells. *J. Cell Biol.* **116**: 1157–1166.

Firtel, R. A. and Chung, C. Y., 2000. The molecular genetics of chemotaxis: sensing and responding to chemoattractant gradients. *Bioessays* **22**: 603–615.

Fukata, M., Watanabe, T., Noritake, J., Nakagawa, M., *et al.*, 2002. Rac1 and Cdc42 capture microtubules through IQGAP1 and CLIP-170. *Cell* **109**: 873–885.

Goldman, R. D., 1971. The role of three cytoplasmic fibers in BHK-21 cell motility. I. Microtubules and the effects of colchicine. *J. Cell Biol.* **51**: 752–762.

Gotlieb, A. I., May, L. M., Subrahmanyan, L. and Kalnins, V. I., 1981. Distribution of microtubule organizing centers in migrating sheets of endothelial cells. *J. Cell Biol.* **91**: 589–594.

Gotlieb, A. I., Subrahmanyan, L. and Kalnins, V. I., 1983. Microtubule-organizing centers and cell migration: effect of inhibition of migration and microtubule disruption in endothelial cells. *J. Cell Biol.* **96**: 1266–1272.

Gundersen, G. G. and Bulinski, J. C., 1988. Selective stabilization of microtubules oriented toward the direction of cell migration. *Proc. Natl. Acad. Sci. USA* **85**: 5946–5950.

Hailat, N., Strahler, J., Melhem, R., Zhu, X. X., *et al.*, 1990. N-myc gene amplification in neuroblastoma is associated with altered phosphorylation of a proliferation related polypeptide (Op18). *Oncogene* **5**: 1615–1618.

Hall, A.,1998. Rho GTPases and the actin cytoskeleton. *Science* **279**: 509–514.

Henson, J. H., Svitkina, T. M., Burns, A. R., Hughes, H. E., *et al.*, 1999. Two components of actin-based retrograde flow in sea urchin coelomocytes. *Mol. Biol. Cell* **10**: 4075–4090.

Higgs, H. N. and Pollard, T. D., 2001. Regulation of actin filament network formation through ARP2/3 complex: activation by a diverse array of proteins. *Annu. Rev. Biochem.* **70**: 649–676.

Hopkins, C. R., Gibson, A., Shipman, M., Strickland, D. K. and Trowbridge, I. S., 1994. In migrating fibroblasts, recycling receptors are concentrated in narrow tubules in the pericentriolar area, and then routed to the plasma membrane of the leading lamella. *J. Cell Biol.* **125**: 1265–1274.

Infante, A. S., Stein, M. S., Zhai, Y., Borisy, G. G. and Gundersen, G. G., 2000. Detyrosinated (Glu) microtubules are stabilized by an ATP-sensitive plus-end cap. *J. Cell Sci.* **113** (22): 3907–3919.

Ishizaki, T., Morishima, Y., Okamoto, M., Furuyashiki, T., *et al.*, 2001. Coordination of microtubules and the actin cytoskeleton by the Rho effector mDia1. *Nat. Cell Biol.* **3**: 8–14.

Kaverina, I., Rottner, K. and Small, J. V., 1998. Targeting, capture, and stabilization of microtubules at early focal adhesions. *J. Cell Biol.* **142**: 181–190.

Kaverina, I., Krylyshkina, O. and Small, J. V., 1999. Microtubule targeting of substrate contacts promotes their relaxation and dissociation. *J. Cell Biol.* **146**: 1033–1044.

Kjoller, L. and Hall, A., 1999. Signaling to Rho GTPases. *Exp. Cell Res.* **253**: 166–179.

Komarova, Y. A., Akhmanova, A. S., Kojima, S., Galjart, N. and Borisy, G. G., 2002. Cytoplasmic linker proteins promote microtubule rescue in vivo. *J. Cell Biol.* **159**: 589–599.

Krylyshkina, O., Kaverina, I., Kranewitter, W., Steffen, W., *et al.*, 2002. Modulation of substrate adhesion dynamics via microtubule targeting requires kinesin-1. *J. Cell Biol.* **156**: 349–359.

Larsson, N., Marklund, U., Gradin, H. M., Brattsand, G. and Gullberg, M., 1997. Control of microtubule dynamics by oncoprotein 18: dissection of the regulatory role of multisite phosphorylation during mitosis. *Mol. Cell Biol.* **17**: 5530–5539.

Lin, C. H., Espreafico, E. M., Mooseker, M. S. and Forscher, P., 1997. Myosin drives retrograde F-actin flow in neuronal growth cones. *Biol. Bull.* **192**: 183–185.

Lin, S. X., Gundersen, G. G. and Maxfield, F. R., 2002. Export from pericentriolar endocytic recycling compartment to cell surface depends on stable, detyrosinated (glu) microtubules and kinesin. *Mol. Biol. Cell* **13**: 96–109.

Malech, H. L., Root, R. K. and Gallin, J. I., 1977. Structural analysis of human neutrophil migration. Centriole, microtubule, and microfilament orientation and function during chemotaxis. *J. Cell Biol.* **75**: 666–693.

Mimori-Kiyosue, Y., Shiina, N. and Tsukita, S., 2000. Adenomatous polyposis coli (APC) protein moves along microtubules and concentrates at their growing ends in epithelial cells. *J. Cell Biol.* **148**: 505–518.

Mitchison, T. and Kirschner, M., 1984. Dynamic instability of microtubule growth. *Nature* **312**: 237–242.

Nathke, I. S., Adams, C. L., Polakis, P., Sellin, J. H. and Nelson, W. J., 1996. The adenomatous polyposis coli tumor suppressor protein localizes to plasma membrane sites involved in active cell migration. *J. Cell Biol.* **134**: 165–179.

Nobes, C. D. and Hall, A., 1999. Rho GTPases control polarity, protrusion, and adhesion during cell movement. *J. Cell Biol.* **144**: 1235–1244.

Palazzo, A. F., Joseph, H. L., Chen, Y. J., Dujardin, D. L., *et al.*, 2001a. Cdc42, dynein, and dynactin regulate MTOC reorientation independent of Rho-regulated microtubule stabilization. *Curr. Biol.* **11**: 1536–1541.

Palazzo, A. F., Cook, T. A., Alberts, A. S. and Gundersen, G. G., 2001b. mDia mediates Rho-regulated formation and orientation of stable microtubules. *Nat. Cell Biol.* **3**: 723–729.

Perez, F., Diamantopoulos, G. S., Stalder, R. and Kreis, T. E., 1999. CLIP-170 highlights growing microtubule ends in vivo. *Cell* **96**: 517–527.

Pruyne, D. and Bretscher, A., 2000. Polarization of cell growth in yeast. I. Establishment and maintenance of polarity states. *J. Cell Sci.* **113** (3): 365–375.

Rodionov, V. I., Gyoeva, F. K., Tanaka, E., Bershadsky, A. D., *et al.*, 1993. Microtubule-dependent control of cell shape and pseudopodial activity is inhibited by the antibody to kinesin motor domain. *J. Cell Biol.* **123**: 1811–1820.

Salmon, W. C., Adams, M. C. and Waterman-Storer, C. M., 2002. Dual-wavelength fluorescent speckle microscopy reveals coupling of microtubule and actin movements in migrating cells. *J. Cell Biol.* **158**: 31–37.

Sayas, C. L., Moreno-Flores, M. T., Avila, J. and Wandosell, F., 1999. The neurite retraction induced by lysophosphatidic acid increases Alzheimer's disease-like Tau phosphorylation. *J. Biol. Chem.* **274**: 37046–37052.

Schliwa, M. and Höner, B., 1993. Microtubules, centrosomes and intermediate filaments in directed cell movement. *Trends Cell Biol.* **3**: 377–380.

Schroer, T. A., 2001. Microtubules don and doff their caps: dynamic attachments at plus and minus ends. *Curr. Opin. Cell Biol.* **13**: 92–96.

Schuyler, S. C. and Pellman, D., 2001. Microtubule 'plus-end-tracking proteins'. The end is just the beginning. *Cell* **105**: 421–424.

Servant, G., Weiner, O. D., Herzmark, P., Balla, T., *et al.*, 2000. Polarization of chemoattractant receptor signaling during neutrophil chemotaxis. *Science* **287**: 1037–1040.

Small, J. V. and Kaverina, I., 2003. Microtubules meet substrate adhesions to arrange cell polarity. *Curr. Opin. Cell Biol.* **15**: 40–47.

Small, J. V., Isenberg, G. and Celis, J. E., 1978. Polarity of actin at the leading edge of cultured cells. *Nature* **272**: 638–639.

Small, J. V., Rottner, K., Kaverina, I. and Anderson, K. I., 1998. Assembling an actin cytoskeleton for cell attachment and movement. *Biochim. Biophys. Acta* **1404**: 271–281.

Small, J. V., Stradal, T., Vignal, E. and Rottner, K., 2002. The lamellipodium: where motility begins. *Trends Cell Biol.* **12**: 112–120.

Stowers, L., Yelon, D., Berg, L. J. and Chant, J., 1995. Regulation of the polarization of T cells toward antigen-presenting cells by Ras-related GTPase CDC42. *Proc. Natl. Acad. Sci. USA* **92**: 5027–5031.

Svitkina, T. M., Verkhovsky, A. B., McQuade, K. M. and Borisy, G. G., 1997. Analysis of the actin-myosin II system in fish epidermal keratocytes: mechanism of cell body translocation. *J. Cell Biol.* **139**: 397–415.

Symons, M. and Settleman, J., 2000. Rho family GTPases: more than simple switches. *Trends Cell Biol.* **10**: 415–419.

Takenawa, T. and Miki, H., 2001. WASP and WAVE family proteins: key molecules for rapid rearrangement of cortical actin filaments and cell movement. *J. Cell Sci.* **114**: 1801–1809.

Tirnauer, J. S., Grego, S., Salmon, E. D. and Mitchison, T. J., 2002. EB1-microtubule interactions in *Xenopus* egg extracts: role of EB1 in microtubule stabilization and mechanisms of targeting to microtubules. *Mol. Biol. Cell* **13**: 3614–3626.

van Aelst, L. and D'Souza-Schorey, C., 1997. Rho GTPases and signaling networks. *Genes Dev.* **11**: 2295–2322.

Vasiliev, J. M., Gelfand, I. M., Domnina, L. V., Ivanova, O. Y., *et al.*, 1970. Effect of colcemid on the locomotory behaviour of fibroblasts. *J. Embryol. Exp. Morphol.* **24**: 625–640.

Wadsworth, P., 1999. Regional regulation of microtubule dynamics in polarized, motile cells. *Cell Motil. Cytoskeleton* **42**: 48–59.

Wang, Y. L., 1985. Exchange of actin subunits at the leading edge of living fibroblasts: possible role of treadmilling. *J. Cell Biol.* **101**: 597–602.

Waterman-Storer, C. M. and Salmon, E. D., 1997. Actomyosin-based retrograde flow of microtubules in the lamella of migrating epithelial cells influences microtubule dynamic instability and turnover and is associated with microtubule breakage and treadmilling. *J. Cell Biol.* **139**: 417–434.

Waterman-Storer, C. M. and Salmon, E., 1999. Positive feedback interactions between microtubule and actin dynamics during cell motility. *Curr. Opin. Cell Biol.* **11**: 61–67.

Waterman-Storer, C. M., Worthylake, R. A., Liu, B. P., Burridge, K. and Salmon, E. D., 1999. Microtubule growth activates Rac1 to promote lamellipodial protrusion in fibroblasts. *Nat. Cell Biol.* **1**: 45–50.

Wittmann, T. and Waterman-Storer, C. M., 2001. Cell motility: can Rho GTPases and microtubules point the way? *J. Cell Sci.* **114**: 3795–3803.

Wittmann, T., Bokoch, G. and Waterman-Storer, C. M., 2003. Regulation of leading edge microtubule and actin dynamics downstream of Rac1. *J. Cell Biol.* **161**: 845–851.

Yvon, A. M. and Wadsworth, P., 2000. Region-specific microtubule transport in motile cells. *J. Cell Biol.* **151**: 1003–1012.

Yvon, A. M., Walker, J. W., Danowski, B., Fagerstrom, C., *et al.*, 2002. Centrosome reorientation in wound-edge cells is cell type specific. *Mol. Biol. Cell* **13**: 1871–1880.

Zumbrunn, J., Kinoshita, K., Hyman, A. A. and Nathke, I. S. (2001) Binding of the adenomatous polyposis coli protein to microtubules increases microtubule stability and is regulated by GSK3 beta phosphorylation. *Curr. Biol.* **11**: 44–49.

14

Calpain Regulation of Cell Migration

Anna Huttenlocher

Cell migration requires complex adhesive interactions between a cell and its substrate. The integrin-mediated adhesions established during cell movement are both highly regulated and dynamic. The calcium-dependent protease calpain regulates both adhesive complex disassembly and rear retraction, and protrusion events at the leading edge. Temporal and spatial regulation of calpain activity is likely to be critical for effective directional migration. The complex regulation of motility pathways by calpain is supported by the differential effects of calpain inhibition on fibroblast and neutrophil motility. Challenge for future investigation will be to identify critical effectors of calpain during cell motility and to understand the mechanisms that regulate calpain activity both temporally and spatially during cell migration.

Cell migration is a fundamental process involved in both normal and pathological conditions such as wound healing, inflammation and tumour invasion and metastasis. Integrin-mediated adhesion to extracellular matrix proteins is a critical regulator of cell migration. Integrin receptors are a family of heterodimeric cell surface adhesion receptors that link the extracellular matrix to the actin cytoskeleton. Upon engagement, integrins cluster in the membrane to form organized adhesive contact sites, such as focal adhesions (Hynes, 1992; Schoenwaelder and Burridge, 1999). Adhesive contact sites are highly dynamic structures that undergo regulated assembly, translocation and

---

Cell Motility. Edited by Anne Ridley, Michelle Peckham and Peter Clark.
 ISBN 0 470 84872 3

disassembly in both stationary and migratory cells (Zamir *et al.*, 2000). To migrate, cells must coordinately assemble and disassemble integrin-containing adhesive contacts (Huttenlocher *et al.*, 1995; Lauffenburger and Horwitz, 1996). While progress has been made in understanding mechanisms involved in cell migration, we still have a limited understanding of how regulated adhesive complex assembly and disassembly occurs and contributes to directional cell migration. This chapter focuses on the mechanisms of integrin-mediated cell migration and the role that the calcium-dependent protease calpain plays in this process.

Basic steps of cell movement

Cell migration requires a regulated and dynamic interaction between the cell and its surrounding substrate. To migrate cells respond to directional cues and extend a leading edge, lamellipodia or filopodia, which stabilizes in the direction of cell movement. For cell translocation to occur, the leading edge of the cell stabilizes an adhesive complex, which generates the force and traction required for cell movement. Subsequently, the cell must release adhesions at the rear to allow for directional progress. Cell migration may therefore be separated into distinct stages: (1) membrane protrusion with stabilization of cell–substratum adhesion; (2) generation of contractile force and (3) detachment at the cell's rear (Stossel, 1993). The classic three-step migration pattern, i.e., with distinct cell protrusion, cell body contraction and rear detachment, describes the migration patterns of fibroblasts. Although representative of many cell types, the mechanisms that govern the movement of the more rapidly moving cells of the immune system, such as neutrophils, appear to be distinct. In contrast to fibroblasts, neutrophils efficiently coordinate adhesion formation at the cell front and rear release, thereby demonstrating a gliding movement (Cox and Huttenlocher, 1998).

External factors that regulate cell migration

Cell migration involves the integration of external cues, including both migration-promoting and migration-inhibiting signals. The external cues that modulate cell motility are diverse and include growth factors, extracellular matrix components and cell–cell interactions. The extracellular and growth factor environment in the adult organisms is generally migration-inhibiting, with the majority of cells non-migratory. However, in response to perturbations, such as tissue wounding, the extracellular environment may become permissive and contain migration-promoting signals that stimulate the migration of fibroblasts and other cell types. It is important to consider that signals that may be permissive for the migration of one cell type may not be permissive, or in fact may be inhibitory, for the migration of other cell types. For example, leukocytes can navigate through diverse extracellular environments as a normal part of immune surveillance. This basic principle raises interesting questions about what makes a leukocyte migratory in what may otherwise be non-permissive tissues and the changes that occur in metastatic cancer that trigger the invasive properties of cells.

Central to this premise is the existence of migration-inhibiting cues. An important migration-inhibiting external signal is mediated by the extracellular matrix mileu. For example, certain extracellular matrix components at high density, such as fibronectin, may be inhibitory to cell migration. In fact, many cell types exhibit a biphasic relationship between adhesion and migration rate, with optimum speed occurring at an intermediate cell–substratum adhesiveness (Huttenlocher *et al.*, 1996). Migration is inhibited at lower than optimal cell–substratum adhesiveness presumably because cells cannot form stable adhesions; conversely, at higher than optimal conditions, previous studies have suggested that migration may be inhibited because cells cannot effectively release adhesions at the cell's rear (Huttenlocher *et al.*, 1995). However, recent studies suggest that ligand-density dependent regulation of intracellular signalling and cell polarization/protrusion may also be an important mechanism by which high density of ligand inhibits cell migration (Cox *et al.*, 2001). More specifically, these studies demonstrated that high fibronectin density down-regulates signalling pathways via Rac and Cdc42, critical for cell protrusion and polarization, thereby inhibiting cell migration.

A related, but less understood mechanism, may also contribute a migration-inhibiting cue. This mechanism involves contact-mediated inhibition of cell movement, with suppression of motility upon contact between specific cell types. Contact-mediated inhibition of cell migration was initially detailed by Abercrombie *et al.* to describe the inhibition of migration and motile activity that occurs after cell–cell contact in migrating fibroblasts (Trinkaus, 1984). In fibroblasts, after contact between ruffling membrane surfaces of adjacent cells the regions become quiescent and the cells appear to attach to each other, while other regions of the cells continue to ruffle. Subsequently, the cells will detach and migrate in the opposite direction. Contact inhibition is more extreme in epithelial cells, which tend to remain attached via cadherin-mediated cell–cell adhesion following contact. The contact inhibition and more stable contacts that form between epithelial cells provide an important mechanism to limit the invasive characteristics of carcinomas (Frixen *et al.*, 1991). Although cadherin-mediated cell–cell adhesion inhibits the invasiveness of tumours, its role in contact-mediated inhibition of motility has not been clearly defined, although it is possible that cross-talk between integrins and cadherins modulates contact-mediated inhibition of motile activity (Huttenlocher *et al.*, 1998).

Integrin receptors and focal adhesions

Integrin receptors are transmembrane $\alpha\beta$ heterodimers that bind to and recognize the extracellular matrix (ECM) and/or receptors on the surface of other cells (Hynes, 1992). Multiple members of the integrin family have been identified and activities of these receptors have been correlated with diverse functions that include cell migration, differentiation and proliferation. Integrins play a central role in cell migration by performing both an adhesive function, linking the ECM to intracellular cytoskeletal proteins, and a signal transduction function. The focal adhesion, which forms upon integrin clustering, represents the adhesive complex that forms between the cell and its surrounding ECM (Burridge *et al.*, 1988; Critchley, 2000). The focal adhesion serves a structural function by linking actin filaments to the ECM via integrin receptors and associated proteins. The β integrin cytoplasmic domain has been found to associate directly with intracellular cytoskeletal proteins, including talin and α-actinin, and is required for the localization of integrins to focal adhesions (Zamir and Geiger, 2001). In addition to serving a structural function, focal adhesions also serve a signalling function by associating with many signalling molecules including focal adhesion kinase (FAK), protein kinase C (PKC) and Src (Clark and Brugge, 1995; Giancotti and Ruoslahti, 1999; Juliano, 2002). Many studies have demonstrated that integrin-mediated adhesion to the ECM regulates the activity of the Rho family of GTPases (Ren et al., 1999; Schwartz and Ginsberg, 2002). We have also demonstrated that density of ligand can regulate signalling of the Rho family GTPases, with a biphasic dependent regulation of the activity of Rac and Cdc42, but not Rho, in CHOK1 cells adherent to a fibronectin-coated surface (Cox *et al.*, 2001).

Calpain

The calpains are intracellular calcium-dependent cysteine proteases that are widely expressed in all mammalian cells as well as in invertebrates and fungi. The calpain–calpastatin system plays an important role in extra-lysosomal intracellular proteolysis. The calpain system has been implicated in basic cellular processes including apoptosis, cell proliferation and more recently cell migration. Studies have suggested that calpain may modulate basic cellular processes by affecting both cytoskeletal associations and the functions of intracellular signalling pathways (Sato and Kawashima, 2001). A role for the calpain–calpastatin system has also been shown for a number of pathological conditions including cardiac ischaemia, arthritis, muscular dystrophy and Alzheimer's disease (Saido *et al.*, 1994).

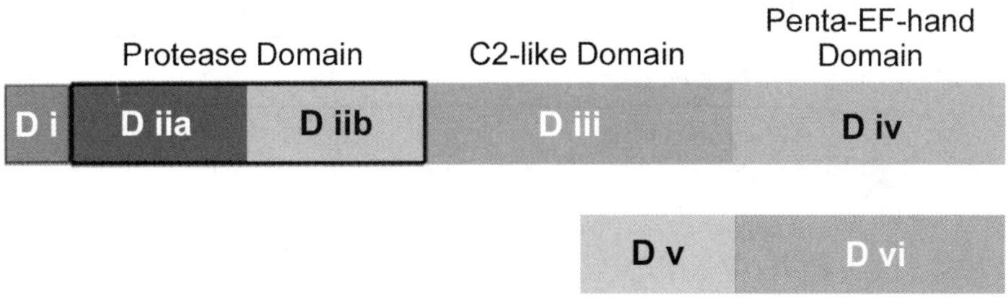

Figure 14.1 Schematic of the domain structure of m- and μ-calpain 80 kDa large subunit (top) and 30 kDa shared regulatory subunit (bottom). The N-terminal domain i is cleaved during activation, domains iia and iib contain the protease domain, domain iii contains lipid binding and phosphorylation sites and domain iv contains EF hands. The regulatory subunit domain vi is homologous to domain iv of the large subunit and contains EF hands with a potential role in calcium binding. (A colour reproduction of this figure can be found in the colour plate section)

There are two ubiquitous calpain isoforms that are defined by their *in vitro* requirement of calcium for activation. μ-calpain (calpain I) requires calcium concentrations in the micromolar range for activation and m-calpain (calpain II) requires calcium concentrations in the millimolar range (Glading *et al.*, 2002; Perrin and Huttenlocher, 2002). Both isoforms contain an 80 kDa catalytic subunit and a 30 kDa regulatory domain (Figure 14.1). The regulatory domain is identical for both μ- and m-calpain. A critical role for calpain in normal embryonic development is supported by the observation that targeted disruption of the small calpain regulatory subunit in mice results in embryonic lethality at day 10 (Arthur *et al.*, 2000).

Calpains act on many substrates *in vitro* including enzymes, cytoskeleton-associated proteins (talin, filamin, α-actinin) (Beckerle *et al.*, 1987), integrin cytoplasmic domains (Du *et al.*, 1995) and the signalling molecules protein kinase C (PKC), focal adhesion kinase (FAK) and Src (Carragher *et al.*, 2002; Cooray *et al.*, 1996; Glading *et al.*, 2002) (Table 14.1). However, few physiological substrates for calpain activity *in vivo* have been identified. There is recent evidence that talin is an important substrate for calpain activity *in vivo* (Dourdin *et al.*, 2001).

Other recent studies suggest that the integrin cytoplasmic domain may be an important substrate for calpain. For example, a calpain-cleaved form of the β3 integrin cytoplasmic domain is found in peripheral adhesive complexes in spreading cells (Bialkowska *et al.*, 2000). Calpain cleavage modifies its substrates rather than performing a degradative function. There is evidence that calpain, through such modifications, plays an important role in regulating intracellular signalling pathways and modulates reversible, dynamic processes such as cell migration (Glading *et al.*, 2002; Inomata *et al.*, 1996). However

Table 14.1 Putative focal adhesion and signalling calpain substrates (reviewed in Glading *et al.*, 2002 and Sato *et al.*, 2001)

Adhesion complex	Signalling proteins
Talin	PKC
Filamin	RhoA
Paxillin	pp60Src
α-actinin	ZAP 70
β1 integrin	PTP1-B
β3 integrin	EGFR
β4 integrin	PLC-β
Vinculin	
Ezrin	
Spectrin	
FAK	

unlike other mechanisms in signal transduction such as phosphorylation events, proteolysis involves an irreversible change to its substrate, a change that to be effective during migration would have to be tightly regulated and limited in its scope.

Many external factors can regulate calpain activity. There is evidence that integrin ligation and signalling through the epidermal growth factor (EGF) receptor can activate calpain activity in certain cell types (Glading *et al.*, 2001), while signalling through G-protein coupled receptors (GPCR) can inhibit its activity (Shiraha *et al.*, 1999). Calpain activity is regulated by calpastatin, an endogenous calpain inhibitor. Calpastatin is highly specific for μ- and m-calpain and is ubiquitously expressed in cells. Calpain activity may also be regulated by membrane phospholipids, with earlier studies indicating that PIP_2 activates calpain activity (Arthur and Crawford, 1996; Saido *et al.*, 1992). There is some evidence that the ubiquitous calpain isoforms m- and μ-calpain may have distinct intracellular distributions. For instance, while m-calpain has been found to localize to focal adhesions in some cell types (Beckerle *et al.*, 1987), μ-calpain distribution tends to be diffuse throughout the cytoplasm, although some studies have demonstrated localization to the cell periphery in integrin-containing complexes (Bialkowska *et al.*, 2000). Recent studies have demonstrated that m-calpain, but not μ-calpain, may be segregated into lipid rafts, providing an additional mechanism to localize calpain activity. A further mechanism that regulates calpain activity is phosphorylation. For example, phosphorylation by protein kinase A (PKA) can inhibit calpain activity, while recent evidence suggests that ERK-mediated phosphorylation can activate calpain (Glading *et al.*, 2002; Shiraha *et al.*, 2002).

Mechanisms of cell detachment and focal adhesion disassembly: a role for calpain

There is evidence that the rate of detachment at the cell's rear is the rate-limiting step of cell migration speed under many conditions. Release of integrin-mediated adhesions involves the formation of a contractile force and the contribution of signalling mechanisms that allows for adhesive release. Contraction supplies the force required to break integrin-mediated adhesions, while signalling mechanisms may determine whether the integrin–ligand or integrin–cytoskeletal bond is preferentially broken at the cell's rear. Mechanisms of cell detachment are distinct in different cell types and under different extracellular matrix environments. Under conditions of low adhesiveness between the cell and the extracellular matrix, preferential release of the integrin–ligand bond is most likely. In contrast, under conditions of higher adhesiveness, there is evidence that cleavage of the integrin–cytoskeletal linkage occurs with higher frequency in fibroblasts. This results in integrin being left behind on the substrate after cell detachment. Slower moving cells, such as fibroblasts, favour breaking the integrin–cytoskeletal bond, whereas rapidly moving cells may have mechanisms that release the integrin–ligand bond instead. This latter mechanism retains integrins on the cell surface and/or allows the integrins to be taken up into endocytic vesicles with trafficking to the cell centre.

Several lines of evidence support a role for actin-myosin generated contraction in adhesion disassembly at the cell's rear during migration. Amoebae deficient in myosin II exhibit decreased migration which is exaggerated on more adhesive substrata (Eddy *et al.*, 2000; Wessels *et al.*, 1988). Also, mitogen-activated protein kinase (MAPK) was found to enhance the migration of fibroblast-like cells through stimulating myosin light chain phosphorylation with an associated increase in contraction (Klemke *et al.*, 1997). Another signalling mechanism contributing to rear release involves the small GTP-binding protein, Rho. Inhibition of Rho reduces rear release in migrating leukocytes (Worthylake *et al.*, 2001). It is possible that Rho mediates rear release by stimulating contractility (Chrzanowska-Wodnicka and Burridge, 1996), or alternatively, the effects of Rho on rear retraction may be through modulation of the adhesive complex stability.

A role for calcium transients in adhesive release at the rear of migrating neutrophils has been clearly demonstrated (Cox and Huttenlocher, 1998). Calcium depletion or calcium buffering inhibits neutrophil migration on fibronectin and vitronectin by specifically inhibiting rear release (Hendey and Maxfield, 1993). Further evidence for calcium transients regulating adhesive release is provided by the observation that in highly adhesive keratocytes, stretch-activated calcium channels result in influx of calcium at the cell rear immediately preceding adhesive release (Lee *et al.*, 1997). Calcium-dependent mechanisms that have previously been implicated in cell detachment include a role for the phosphatase calcineurin in neutrophils on vitronectin (Hendey *et al.*, 1992).

More recent studies from our laboratory and other laboratories implicate a central role for the protease calpain as a calcium-dependent molecule involved in cell detachment. We have previously reported that inhibition of calpain activity reduces the migration rate of CHO cells *in vitro* (Huttenlocher *et al.*, 1997). This decreased migration rate after calpain inhibition is the result of an inhibition of rear retraction. Calpain inhibition also reduces the amount of integrin that remains behind on the substratum after rear detachment (Palecek *et al.*, 1998) suggesting that calpain may weaken the integrin-cytoskeletal bonds in this region, facilitating rear release. Interestingly, calpain inhibition does not seem to affect cell detachment of neutrophils under most conditions (Lokuta *et al.*, 2003). This observation suggests that rapid (less adhesive) and slow moving (more adhesive) cells may use different mechanisms to regulate adhesive release at the cell's rear. In fact, at low cell–substratum adhesiveness, calpain inhibitors do not significantly affect migration or detachment rates of fibroblasts, suggesting that at lower adhesiveness, contractile forces are sufficient to sever the integrin–ligand bond. However, at higher cell–substratum adhesiveness, contractile forces may not be sufficient to allow adhesive release and calpain-mediated proteolysis may also be required.

The effects of calpain on rear detachment are likely to be mediated through calpain-dependent remodelling of focal adhesions. Inhibiting calpain in fibroblasts alters the morphology of focal adhesions in comparison to control cells (Huttenlocher *et al.*, 1997). Focal adhesions are larger, more stable and located at the periphery of the cell after calpain inhibition instead of being distributed throughout the cell interior. Actin stress fibres that normally cross the cell and terminate in focal adhesions are also altered when calpain is inhibited. After calpain inhibition the actin stress fibres are primarily cortical. Similar changes are observed in calpain-deficient embryonic fibroblasts (Capn4$^{-/-}$ fibroblasts), supporting a central role for calpain in the regulation of focal adhesions and the actin cytosekeleton (Dourdin *et al.*, 2001). Previous studies have supported a role for calpain in focal adhesion disassembly. For example, fragments of collagen promote focal adhesion disassembly by the calpain-dependent cleavage of FAK, paxillin and talin (Carragher *et al.*, 1999). In live studies, we have recently shown that calpain inhibition modulates focal complex composition and disassembly. Calpain inhibition blocks localization of α-actinin to the focal adhesion and prevents subsequent translocation of the focal adhesion to the cell interior as well as focal complex disassembly (Bhatt *et al.*, 2002). It has been proposed that cleavage of some component of the focal adhesion, talin for example, may alter the composition of the focal adhesion thus allowing the recruitment of different proteins, modifying the cytoskeletal linkage, or destabilizing the focal complex. As a result, calpain-mediated proteolysis promotes complex translocation, disassembly and rear retraction.

Calpain during adhesion formation and directional migration

There is also evidence supporting a role for calpain in adhesion formation and cell spreading (Potter *et al.*, 1998; Stewart *et al.*, 1998). Inhibition of calpain activity reduces lymphocyte (Stewart *et al.*, 1998) and endothelial cell spreading (Kulkarni *et al.*, 1999) on extracellular matrix. In contrast, fibroblast spreading is not significantly affected by calpain inhibition and calpain inhibition can, in fact, enhance the adhesiveness and spreading of primary neutrophils (Lokuta *et al.*, 2003). These differences highlight the importance of cellular context in calpain-mediated modulation of the actin cytoskeleton and cell migration (Figure 14.2).

In addition to physically altering protein complexes such as focal adhesions, calpain is also likely to participate in cell motility by affecting intracellular signalling pathways. Studies by Kulkarni *et al.* (1999) suggest that calpain may mediate integrin-mediated signalling and cell spreading upstream of Rho family members by activating Rac and Rho activity (Kulkarni *et al.*, 1999). Specifically, they hypothesize that calpain inhibition reduces Rac and Rho activity and the effects of calpain inhibition may be rescued by activating Rac and Rho. In addition, in endothelial cells calpain is required for the formation of focal contacts, small clusters of integrins that form early in the cell spreading process. These early integrin clusters contain cleaved $\beta 3$ integrin, a known substrate of m-calpain and μ-calpain (Kulkarni and Fox, 2000). Focal complexes and focal adhesions, which form later in spreading, were not found to contain either μ-calpain or cleaved β integrin.

Recent studies show that the focal complexes formed in cells that are deficient in calpain activity are different from those formed in control cells (Bhatt *et al.*, 2002). It is possible that these abnormal complexes display distinct signalling properties that contribute to the changes in cell protrusion observed after calpain inhibition. In support of this possibility, calpain-deficient fibroblasts (Capn4$^{-/-}$ fibroblasts) not only exhibit reduced migration rates, but also have increased cell protrusions (Dourdin *et al.*, 2001). Specifically, Capn4$^{-/-}$ embryonic fibroblasts demonstrate enhanced formation of membrane projections, suggesting that calpain may normally function to suppress protrusion and the formation of membrane projections. In accordance with this possibility we find that the activity of Cdc42 and Rac is increased after calpain inhibition in neutrophils (Lokuta *et al.*, 2003).

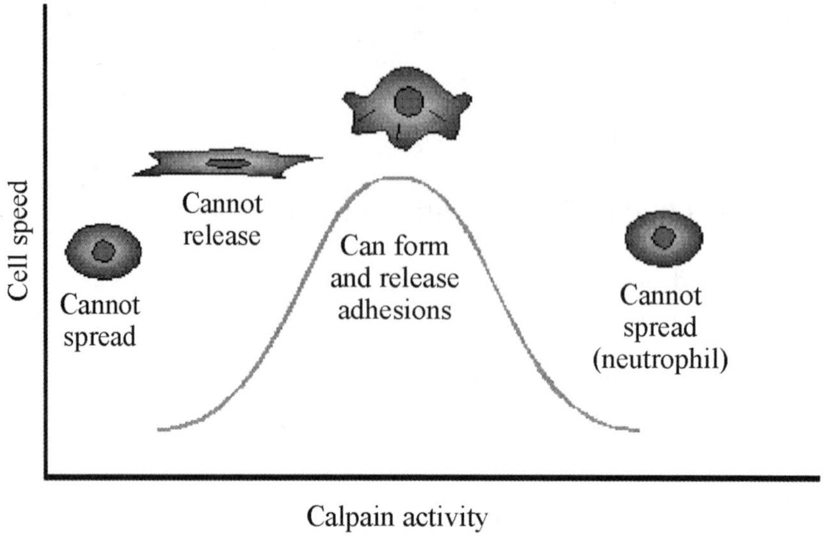

Figure 14.2 Biphasic dependence of cell migration speed on calpain activity. Neutrophils have high endogenous calpain activation and treatment with cell permeable calpain inhibitors induces cell adhesion and random migration. In contrast, fibroblasts have a medium amount of calpain activity in adherent cells that is supportive of cell migration. Calpain inhibition results in an inhibition of rear retraction and migration speed. At its extreme, calpain inhibition can reduce cell spreading as depicted. This is most notable in lymphocytes and endothelial cells

Our recent studies also support an important role for calpain in mechanosensing and re-orientation in response to cell–cell contact. Fibroblasts normally re-orientate after contact with a neighbouring cell with protrusion away from the contacting cell. In Capn4$^{-/-}$ fibroblasts or after calpain inhibition in CHOK1 cells, we find that cells continue to migrate over contacting cells without contact inhibition or re-orientation after contact (Perrin, unpublished observation).

Recent observations also support a role for calpain in regulating neutrophil chemotaxis (Lokuta *et al.*, 2003). Chemotaxis is directional migration in response to a gradient of chemoattractant (Allen *et al.*, 1998; Devreotes and Zigmond, 1988; Zigmond *et al.*, 1981). The gradients that stimulate directional cell migration may be very shallow, with as little as a 2% gradient of chemoattractant between the cell front and rear (Parent and Devreotes, 1999). Recent evidence suggests that shallow gradients of chemoattractant are translated into larger gradients of intracellular signalling proteins that are required for cell polarization and directional cell migration. It is not clear if positive signals alone are sufficient for the optimum response and chemotaxis of cells to external signals. The participation of negative signals, that inhibit

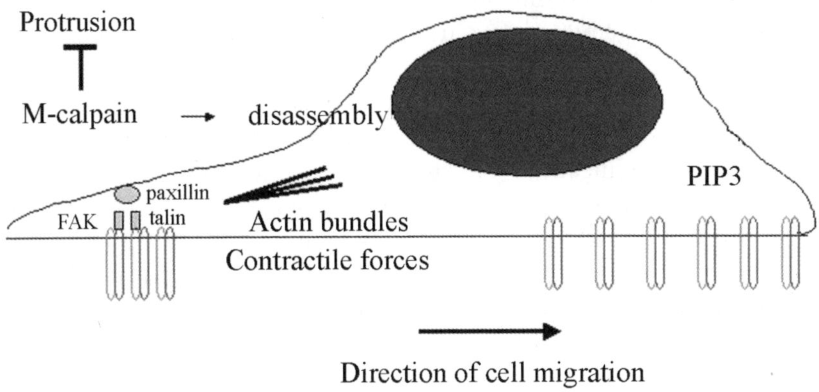

Figure 14.3 Model of calpain activation in a migrating cell. Indirect evidence suggests that calpain may be active at the cell's rear where it facilitates focal complex disassembly and rear retraction

protrusion in a direction away from the chemoattractant, may also potentially participate during chemotaxis. There has been some recent progress in understanding how cells migrate directionally in a gradient of chemoattractant. Although an even distribution of chemoattractant receptor have been found on the surface of cells during chemotaxis (Jin *et al.*, 2000; Servant *et al.*, 1999), recent studies has demonstrated that a gradient in chemoattractant induces an asymmetry in the localization of PIP_3. In response to a gradient of chemoattractant PIP_3 localizes to the leading edge of migrating neutrophils (Servant *et al.*, 2000) (see Chapter 17). The findings suggest that a gradient of chemoattractant induces localized changes in membrane organization and recruitment of signalling proteins. We observe a dose-dependent enhancement of neutrophil chemokinesis or random migration after calpain inhibition (Lokuta *et al.*, 2003). We also find that at higher concentrations of cell-permeable inhibitors neutrophil chemotaxis and directional persistence toward chemoattractant is diminished, supporting a role for calpain in directional sensing in response to a gradient of chemoattractant. Together, the findings raise the intriguing possibility that enhanced calpain activity at the rear of chemotaxing cells may act as a negative signal that serves to re-direct migration in the direction of chemoattractant (Figure 14.3).

Conclusions

Despite recent progress, our understanding of the mechanisms that regulate directional migration and the integration of diverse external cues remains limited. The findings from our laboratory and other laboratories support a critical role for the calcium-dependent protease calpain during migration and in the regulation of integrin-containing adhesive contact sites. It is likely that the effects of calpain on cell motility are to a large extent mediated by its effects on the composition and organization of focal adhesions. However, how the irreversible modification of substrates by calpain participates in intracellular signalling and the dynamic regulation of cell motility remains a challenge. The tight regulation of its activity with only localized activation is likely to be critical for its participation in cell migration. Its activation is likely to be required for cell motility and cell retraction, while at the same time critical for mediating stop signals that inhibit the protrusion machinery and cell movement. The different roles of calpain in the migration of fibroblasts and neutrophils underlie the complexity of its regulation and participation during cell migration. The coming years promise to be informative and exciting as we gain understanding of how migration is regulated, both temporally and spatially, and how mediators such as calpain participate in this regulation.

References

Allen, W. E., Zicha, D., Ridley, A. J. and Jones, G. E., 1998. A role for Cdc42 in macrophage chemotaxis. *J. Cell Biol.* **141**: 1147–1157.

Arthur, J. S. and Crawford, C., 1996. Investigation of the interaction of m-calpain with phospholipids: calpain–phospholipid interactions. *Biochim. Biophys. Acta* **1293**: 201–206.

Arthur, J. S., Elce, J. S., Hegadorn, C., Williams, K. and Greer, P. A., 2000. Disruption of the murine calpain small subunit gene, Capn4: calpain is essential for embryonic development but not for cell growth and division. *Mol. Cell Biol.* **20**: 4474–4481.

Beckerle, M. C., Burridge, K., DeMartino, G. N. and Croall, D. E., 1987. Colocalization of calcium-dependent protease II and one of its substrates at sites of cell adhesion. *Cell* **51**: 569–577.

Bhatt, A., Kaverina, I., Otey, C. and Huttenlocher, A., 2002. Calpain regulates adhesive complex composition and disassembly. *J. Cell Science* **115**: 3415–3425.

Bialkowska, K., Kulkarni, S., Du, X., Goll, D. E., Saido, T. C. and Fox, J. E., 2000. Evidence that beta3 integrin-induced Rac activation involves the calpain-dependent formation of integrin clusters that are distinct from the focal complexes and focal adhesions that form as Rac and RhoA become active. *J. Cell Biol.* **151**: 685–696.

Burridge, K., Fath, K., Kelly, T., Nuckolls, G. and Turner, C., 1988. Focal adhesions: transmembrane junctions between the extracellular matrix and the cytoskeleton. *Annu. Rev. Cell Biol.* **4**: 487–525.

Carragher, N. O., Levkau, B., Ross, R. and Raines, E. W., 1999. Degraded collagen fragments promote rapid disassembly of smooth muscle focal adhesions that correlates with cleavage of pp125(FAK), paxillin, and talin. *J. Cell Biol.* **147**: 619–630.

Carragher, N. O., Westhoff, M. A., Riley, D., Potter, D. A., *et al.*, 2002. v-Src-induced modulation of the calpain–calpastatin proteolytic system regulates transformation. *Mol. Cell Biol.* **22**: 257–269.

Chrzanowska-Wodnicka, M. and Burridge, K., 1996. Rho-stimulated contractility drives the formation of stress fibers and focal adhesions. *J. Cell Biol.* **133**: 1403–1415.

Clark, E. A. and Brugge, J. S., 1995. Integrins and signal transduction pathways: the road taken. *Science* **268**: 233–239.

Cooray, P., Yuan, Y., Schoenwaelder, S. M., Mitchell, C. A., *et al.*, 1996. Focal adhesion kinase (pp125FAK) cleavage and regulation by calpain. *Biochem. J.* **318**: 41–47.

Cox, E. A. and Huttenlocher, A., 1998. Regulation of integrin-mediated adhesion during cell migration. *Microsc. Res. Tech.* **43**: 412–419.

Cox, E. A., Sastry, S. K. and Huttenlocher, A., 2001. Integrin-mediated adhesion regulates cell polarity and membrane protrusion through the Rho family of GTPases. *Mol. Biol. Cell* **12**: 265–277.

Critchley, D. R., 2000. Focal adhesions – the cytoskeletal connection. *Curr. Opin. Cell. Biol.* **12**: 133–139.

Devreotes, P. N. and Zigmond, S., 1988. Chemotaxis in eukaryotic cells: a focus on leukocytes and *Dictyostelium*. *Annu. Rev. Cell. Biol.* **4**: 649–686.

Dourdin, N., Bhatt, A. K., Dutt, P., Greer, P. A., *et al.*, 2001. Reduced cell migration and disruption of the actin cytoskeleton in calpain-deficient embryonic fibroblasts. *J. Biol. Chem.* **276**: 48 382–48 388.

Du, X., Saido, T. C., Tsubuki, S., Indig, F. E., *et al.*, 1995. Calpain cleavage of the cytoplasmic domain of the integrin beta 3 subunit. *J. Biol. Chem.* **270**: 26 146–26 151.

Eddy, R. J., Pierini, L. M., Matsumura, F. and Maxfield, F. R., 2000. Ca2+-dependent myosin II activation is required for uropod retraction during neutrophil migration. *J. Cell Sci.* **113**: 1287–1298.

Frixen, U., Behrens, J., Sachs, M., Eberle, G., *et al.*, 1991. E-cadherin mediated cell–cell adhesion prevents invasiveness of human carcinoma cell lines. *J. Cell Biol.* **111**: 173–185.

Giancotti, F. G. and Ruoslahti, E., 1999. Integrin signaling. *Science* **285**: 1028–1032.

Glading, A., Uberall, F., Keyse, S., Lauffenburger, D. and Wells, A., 2001. Membrane proximal ERK signaling is required for M-calpain activation downstream of epidermal growth factor receptor signaling. *J. Biol. Chem.* **276**: 23341.

Glading, A., Lauffenburger, D. and Wells, A., 2002. Cutting to the chase: calpain protease in cell motility. *Trends Cell Biol.* **12**: 46–54.

Hendey, B. and Maxfield, F. R., 1993. Regulation of neutrophil motility and adhesion by intracellular calcium transients. *Blood Cells* **19**: 143–161; discussion 161–164.

Hendey, B., Klee, C. B. and Maxfield, F. R., 1992. Inhibition of neutrophil chemokinesis on vitronectin by inhibitors of calcineurin. *Science* **258**: 296–299.

Huttenlocher, A., Sandborg, R. R. and Horwitz, A. F., 1995. Adhesion in cell migration. *Curr. Opin. Cell Biol.* **7**: 697–706.

Huttenlocher, A., Ginsberg, M. H. and Horwitz, A. F., 1996. Modulation of cell migration by integrin-mediated cytoskeletal linkages and ligand-binding affinity. *J. Cell Biol.* **134**: 1551–1562.

Huttenlocher, A., Palecek, S. P., Lu, Q., Zhang, W., *et al.*, 1997. Regulation of cell migration by the calcium-dependent protease calpain. *J. Biol. Chem.* **272**: 32 719–32 722.

Huttenlocher, A., Lakonishok, M., Kinder, M., Wu, S., *et al.*, 1998. Integrin and cadherin synergy regulates contact inhibition of migration and motile activity. *J. Cell Biol.* **141**: 515–526.

Hynes, R. O., 1992. Integrins: versatility, modulation and signaling in cell adhesion. *Cell* **69**: 11–25.

Inomata, M., Hayashi, M., Ohno-Iwashita, Y., Tsubuki, S., *et al.*, 1996. Involvement of calpain in integrin-mediated signal transduction. *Arch. Biochem. Biophys.* **328**: 129–134.

Jin, T., Zhang, N., Long, Y., Parent, C. A. and Devreotes, P. N., 2000. Localization of the G protein betagamma complex in living cells during chemotaxis. *Science* **287**: 1034–1036.

Juliano, R. L., 2002. Signal transduction by cell adhesion receptors and the cytoskeleton: functions of integrins, cadherins, selectins, and immunoglobulin-superfamily members. *Annu. Rev. Pharmacol. Toxicol.* **42**: 283–323.

Klemke, R. L., Cai, S., Giannini, A. L., Gallagher, P. J., *et al.*, 1997. Regulation of cell motility by mitogen-activated protein kinase. *J. Cell Biol.* **137**: 481–492.

Kulkarni, S. and Fox, J. E., 2000. Localization of calpain by immunofluorescence in adherent cells. *Methods Mol. Biol.* **144**: 151–159.

Kulkarni, S., Saido, T. C., Suzuki, K. and Fox, J. E., 1999. Calpain mediates integrin-induced signaling at a point upstream of Rho family members. *J. Biol. Chem.* **274**: 21 265–21 275.

Lauffenburger, D. A. and Horwitz, A. F., 1996. Cell migration: a physically integrated molecular process. *Cell* **84**: 359–369.

Lokuta, M., Nuzzi, P. and Huttenlocher, A., 2003. Calpain regulates neutrophil chemotaxis. *Proc. Natl. Acad. Sci. USA* **100**: 4006–4011.

Palecek, S. P., Huttenlocher, A., Horwitz, A. F. and Lauffenburger, D. A., 1998. Physical and biochemical regulation of integrin release during rear detachment of migrating cells. *J. Cell Sci.* **111**: 929–940.

Parent, C. A. and Devreotes, P. N., 1999. A cell's sense of direction. *Science* **284**: 765–770.

Perrin, B. and Huttenlocher, A., 2002. Calpain. *Int. J. Bioch. Cell Biol.* **34**: 722–725.

Potter, D. A., Tirnauer, J. S., Janssen, R., Croall, D. E., *et al.*, 1998. Calpain regulates actin remodeling during cell spreading. *J. Cell Biol.* **141**: 647–662.

Ren, X. D., Kiosses, W. B. and Schwartz, M. A., 1999. Regulation of the small GTP-binding protein Rho by cell adhesion and the cytoskeleton. *Embo J.* **18**: 578–585.

Saido, T. C., Shibata, M., Takenawa, T., Murofushi, H. and Suzuki, K., 1992. Positive regulation of mu-calpain action by polyphosphoinositides. *J. Biol. Chem.* **267**: 24 585–24 590.

Saido, T. C., Sorimachi, H. and Suzuki, K., 1994. Calpain: new perspectives in molecular diversity and physiological–pathological involvement. *Faseb J.* **8**: 814–822.

Sato, K. and Kawashima, S., 2001. Calpain function in the modulation of signal transduction molecules. *Biol. Chem.* **382**: 743–751.

Schoenwaelder, S. M. and Burridge, K., 1999. Bidirectional signaling between the cytoskeleton and integrins. *Curr. Opin. Cell Biol.* **11**: 274–286.

Schwartz, M. A. and Ginsberg, M. H., 2002. Networks and crosstalk: integrin signalling spreads. *Nat. Cell Biol.* **4**: E65–E68.

Servant, G., Weiner, O., Neptune, E., Sedat, J. and Bourne, H., 1999. Dynamics of a chemoattractant receptor in living neutrophils during chemotaxis. *Mol. Biol. Cell.* **10**: 1163–1178.

Servant, G., Weiner, O., Herzmark, P., Balla, T., Sedat, J. and Bourne, H., 2000. Polarization of chemoattractant receptor signaling during neutrophil chemotaxis. *Science* **287**: 1037–1040.

Shiraha, H., Glading, A., Gupta, K. and Wells, A., 1999. IP-10 inhibits epidermal growth factor-induced motility by decreasing epidermal growth factor receptor-mediated calpain activity. *J. Cell Biol.* **146**: 243–254.

Shiraha, H., Glading, A., Chou, J., Jia, Z. and Wells, A., 2002. Activation of m-calpain (calpain II) by epidermal growth factor is limited by protein kinase A phosphorylation of m-calpain. *Mol. Cell Biol.* **22**: 2716–2727.

Stewart, M. P., McDowall, A. and Hogg, N., 1998. LFA-1-mediated adhesion is regulated by cytoskeletal restraint and by a Ca2+-dependent protease, calpain. *J. Cell Biol.* **140**: 699–707.

Stossel, T. P., 1993. On the crawling of animal cells. *Science* **260**: 1086–1094.

Trinkaus, J., 1984. *Cells into Organs: the forces that shape the embryo*. Reading, MA: Prentice Hall.

Wessels, D., Soll, D. R., Knecht, D., Loomis, W. F., *et al.*, 1988. Cell motility and chemotaxis in *Dictyostelium* amebae lacking myosin heavy chain. *Dev. Biol.* **128**: 164–177.

Worthylake, R. A., Lemoine, S., Watson, J. M. and Burridge, K., 2001. RhoA is required for monocyte tail retraction during transendothelial migration. *J. Cell Biol.* **154**: 147–160.

Zamir, E. and Geiger, B., 2001. Molecular complexity and dynamics of cell–matrix adhesions. *J. Cell Sci.* **114**: 3583–3590.

Zamir, E., Katz, M., Posen, Y., Erez, N., *et al.*, 2000. Dynamics and segregation of cell-matrix adhesions in cultured fibroblasts. *Nat. Cell Biol.* **2**: 191–196.

Zigmond, S. H., Levitsky, H. I. and Kreel, B. J., 1981. Cell polarity: an examination of its behavioral expression and its consequences for polymorphonuclear leukocyte chemotaxis. *J. Cell Biol.* **89**: 585–592.

15

Role of Villin in the Dynamics of Actin Microfilaments

Rafika Athman, Sylvie Robine and **Daniel Louvard**

Villin is a tissue specific actin-binding protein, present in intestinal and proximal kidney brush borders. *In vitro*, villin has been shown to bundle and sever F-actin in a calcium-dependent manner. Our previous *in vivo* data on villin knock-out mice showed that, while this protein is not necessary for the bundling of F-actin, it is important for the reorganization of the actin cytoskeleton elicited by stresses during both physiological and pathological conditions (Ferrary *et al.*, 1999). These data suggest that villin may be involved in the actin cytoskeleton remodelling necessary for many cellular processes requiring cytoskeleton plasticity. Our further studies enabled us to show that villin is involved in the actin dynamics occurring during different cellular events. Indeed, we showed that villin expression results in increased HGF induced cell motility and morphogenesis evaluated by tubulogenesis assays (Athman *et al.*, in press). We also showed that villin is involved in the actin remodelling events necessary for the infectious process of *Shigella flexneri in vivo* as well as in cell culture models (Athman *et al.*, submitted). Altogether, we show that villin acts as an enhancer of the actin cytoskeleton dynamics elicited by HGF stimulation and infection. Our studies thus show that villin is not only a structural actin-binding protein as shown by its role in the architecture of the intestinal brush border but also an actin remodelling protein playing a key role in the regulation of actin dynamics through its multiple properties.

Cell Motility. Edited by Anne Ridley, Michelle Peckham and Peter Clark.
 ISBN 0 470 84872 3

Introduction

A unique feature of intestinal epithelial cells is the presence of a brush border composed of numerous membrane extensions called microvilli, fingerlike extensions which play a key role in the absorptive function of these specialized cells. They are abundant on those epithelial cells that require a very large surface area to function efficiently. Each microvillus is about 0.08 μm wide and 1 μm long, making the cell's absorptive area 20 times greater than it would be without them. The plasma membrane that covers these microvilli is highly specialized, bearing a thick extracellular coat of polysaccharide and digestive enzymes. The core of each intestinal microvillus is a rigid bundle of 20 to 30 parallel actin filaments that extend from the tip of the microvillus down into the cell cortex. The actin filaments in the bundle are all orientated with their plus ends pointing away from the cell body. They are held together at regular intervals by actin-bundling proteins. Fimbrin, a ubiquitous actin binding protein, is involved in actin microfilament organization in microspikes and filipodia and it also contribute to the bundling of actin filaments into microvilli. By contrast, villin is only found in microvilli of the digestive and urinary tracts where it is known to bundle actin filaments, thus playing a key role in the maintenance of brush border architecture.

Villin, a structural actin-binding protein

Structure and function

Villin, first isolated from chicken intestinal epithelial cells and later from mammalian species, is an acidic polypeptide with a molecular mass of 92.5 kDa, occurring in monomeric form. Villin is only detected in a few epithelial cells from the gastrointestinal, the urogenital and the respiratory tracts.

The gelsolin family to which villin belongs contains several members: gelsolin, adseverin, CapG, advillin and supervillin (Figure 15.1). These proteins all contain three to six homologous evolutionarily conserved domains. The typical villins have six of these and in addition they have a so-called 'head piece' at the C-terminus that is the largest single difference between gelsolin and villins (for review see Friederich *et al.*, 1990). Villin (Northrop *et al.*, 1986), gelsolin (Kwiatkowski and Yin, 1987), adseverin (Lueck *et al.*, 1998) and CapG (Mishra *et al.*, 1994) have the common property of binding to barbed ends of actin filaments with high affinity. Villin, gelsolin and adseverin can sever actin filaments whereas CapG lacks this activity (Southwick and DiNubile, 1986). Among all actin-binding proteins, villin is unique in presenting capping, bundling and severing properties in a single protein.

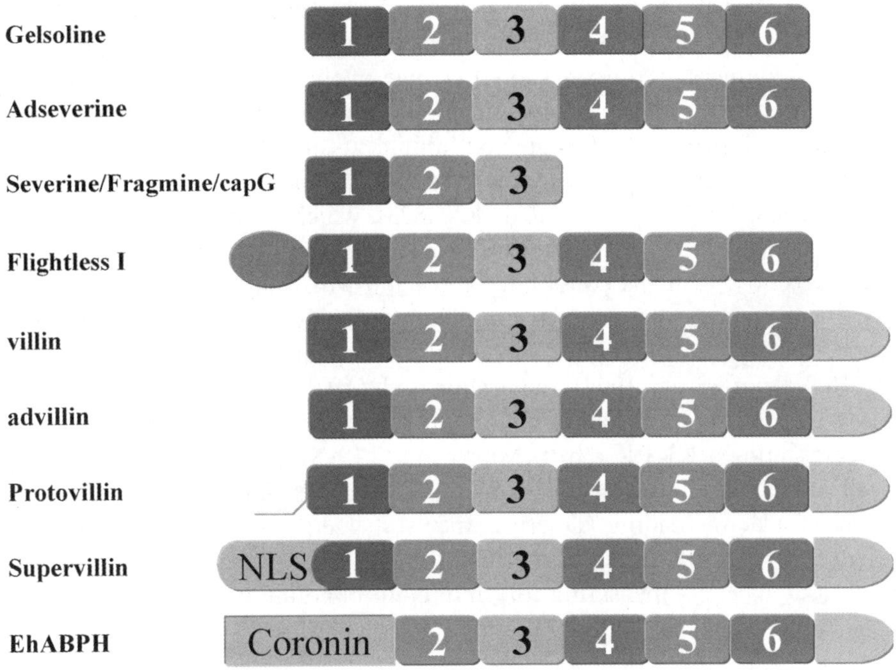

Figure 15.1 Domain structure of the Gelsolin/villin superfamily. (A colour reproduction of this figure can be found in the colour plate section)

The solution structure of the amino-terminal domain of villin (14T, amino acids 1–126) has been determined by heteronuclear resonance spectroscopy. A central β-sheet with four antiparallel strands and a fifth parallel strand on one edge forms the core structure which is surrounded by amphipathic helices, two on one side and a two-stranded parallel β-sheet with another helix on the other side. Mutational analyses of villin suggest that the actin-binding region is localized near the parallel strand at the edge of the central β-sheet. Recently, the NMR structure of the 35 C-terminal residues of chicken villin was established (Simenel *et al.*, 1995). This sequence forms an autonomously folding subdomain which comprises three helices. The hydrophobic side chains of the three helices contribute to a compact hydrophobic core structure.

Villin contains at least three actin-binding sites, two of which are Ca^{2+} dependent and located on the core domain (Kd value: 7 μM). The third one, Ca^{2+} independent, is situated on the headpiece domain (Kd values: 0.3 μM). Villin sequences involved in F-actin binding have been mapped using synthetic peptides. A cluster of charged amino acids located at the interdomain region of domains 1–2 (amino acids 133–147) is part of the F-actin binding site of the core domain. Two rapidly exchanging Ca^{2+}-binding sites have been determined, one located in the core domain ($Kd = 3.5 \times 10^{-6}$ M) and one in the headpiece ($Kd = 7.4 \times 10^{-6}$ M). A third, non-exchangeable high affinity site, is located in the core domain. Binding of Ca^{2+} to that located in the headpiece induces a conformational change of the core part of the molecule and thereby may induce a change in the molecule's biochemical behaviour.

Binding of PIP_2 to this sequence inhibits its interaction with actin. The actin-binding site of the headpiece domain comprises a charged motif of amino acids, KKEK (amino acids 820–823), which is conserved throughout species (Friederich *et al.*, 1992, 1999) and essential for its biological activities. More recently, proteins containing a domain with similarities with the head piece and which is likely to provide actin-binding activity, have been identified. This includes the two subunits of dematin (Rana *et al.*, 1993), abLIM (Roof *et al.*, 1997), supervillin (p205) (Pestonjamasp *et al.*, 1997). Moreover, among the gelsolin/villin family members, advillin (p92) is the only one presenting a KKEK motif (Marks *et al.*, 1998) and a conserved lysine at position 815 in the headpiece domain. These sites have been shown to be critical for actin binding (Doering and Matsudaira, 1996). The presence of advillin in intestinal brush border could explain the absence of phenotype in villin KO mice, compensating for villin actin bundling property.

Villin expression and brush border morphogenesis

Villin is a marker for a few cell types in adults and in embryos, and a differentiation marker for epithelial cells displaying a brush border (for reviews see Louvard, 1989; Heintzelman and Mooseker, 1992). During early embryonic development in chicken and mouse, villin is detectable in cells of the primitive gut which are precursors of the adult intestine. In adults, villin synthesis increases during the differentiation process of the enterocyte, which takes place when the enterocytes migrate along from crypt to the tip of the intestinal villus. Moreover, villin expression is maintained in primary tumours or metastases deriving from tissues which normally express villin. To recapitulate the tissue-specific expression pattern of the endogenous villin gene, we have shown that a 9 kb regulatory region of the mouse villin gene contains the necessary cis-acting elements and can be used to drive the expression of heterologous genes in immature and differentiated epithelial cells of the small and/or large intestinal mucosa. *C-met* was reported directly to interact with signalling proteins PLCγ (Ponzetto *et al.*, 1994). Furthermore, it has been reported that villin interacts with several signalling molecules including PLCγ (Panebra *et al.*, 2001), phosphatidylinositol 4,5-biphosphate (PIP_2) (Janmey *et al.*, 1987) and calcium (Janmey and Matsudaira, 1988). *In vitro*, villin tyrosine phosphorylation has been shown to enhance actin severing and to inhibit villin actin bundling property (Zhai *et al.*, 2001). These findings suggest that villin could play an essential role in the actin cytoskeleton

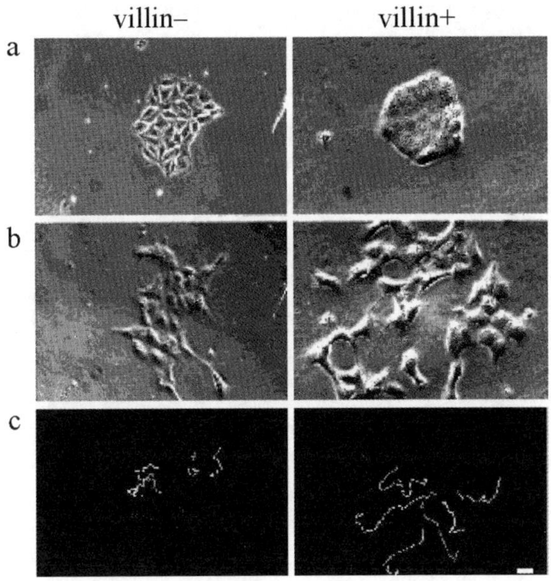

Figure 15.2 HGF-induced cell motility in villin expressing MDCK cells. Phase contrast images of villin – or villin + MDCK cells before (panel a) and after 9 h of HGF (10 IU/ml) stimulation (panel b). Panel c illustrates cell tracking after 9 h of treatment

dynamics in response to specific physiological stimuli. We have developed several approaches to evaluate the role of villin in cellular events that require actin cytoskeleton plasticity and remodelling, such as cell motility, cell morphogenesis and bacterial infections. These studies were performed using primary cultures of enterocytes derived from $vil^{+/+}$ and $vil^{-/-}$ mice and Tet Off MDCK cells expressing the villin transgene in a doxycycline-controlled manner. We show that villin plays a role as a potentiator of the actin cytoskeleton dynamics elicited by extracellular signals. Indeed, we have found that villin is an enhancer of HGF-induced cell motilty and morphogenesis (Figure 15.2 and Table 15.1). Moreover, its role as a potentiator of actin dynamics upon growth factor stimulation takes place through the PLCγ signalling pathway (Athman *et al*., in press).

The actin cytoskeleton remodelling is also required in other cellular processes such as bacterial infection. *Shigella flexneri* is the causative agent of bacillary dysentery. This gram-negative pathogen induces the reorganization of the host actin cytoskeleton during the infection process. The main target of this bacterium being the enterocyte, we took advantage of our villin KO model to study the infectious process in intestinal primary cultures from $vil^{+/+}$ and $vil^{-/-}$. This work in progress has been performed in collaboration with Dana Philpott and Philippe Sansonetti (Institut Pasteur, Paris). Briefly, we have shown, using primary cultures of enterocytes, that villin is involved in bacterial entry and cell-to-cell dissemination. Villin is present in the actin

Table 15.1 HGF-induced cell motility in primary cultures of enterocytes or MDCK cells expressing villin

	Mean velocity (μm/min)	
	$Vil^{-/-}$	$Vil^{+/+}$
Primary cultures of enterocytes	0.25	0.38
MDCK cells	0.29	0.52

Velocity was calculated using tracks of individual cells after videorecording for 9 h (see Figure 15.2 for a typical experiment).

comet formed by intracellular bacteria. Together, these results indicate that villin plays an important role in the different steps of *Shigella* infection through its ability to remodel the actin cytoskeleton reorganization elicited by the invasion process (Athman *et al.*, submitted).

Villin as a regulator of actin dynamics

Background on actin dynamics

Actors of actin dynamics

Proteins of the gelsolin family are regulated by calcium (which activates severing and/or capping) and certain inositol phospholipids (which cause uncapping, the release of actin filament barbed ends, and thus allows for actin polymerization from the barbed ends). In the case of gelsolin itself, there is both a widely-expressed intracellular form and a higher molecular weight extracellular form found in plasma; this latter form appears to play a role in scavenging extracellular actin released from lysed cells. CapZ is a calcium-independent heterodimeric protein that caps barbed ends of filaments in platelets, which appears to be coordinated with severing and capping by gelsolin family proteins. ADF/cofilin family proteins are calcium-independent, and also promote actin filament disassembly; however, the activity of these proteins differs from that of gelsolin family proteins. ADF/cofilin proteins appear to accelerate depolymerization from the pointed ends of actin filaments and weakly to sever filaments without capping. Arp2/3 complex and associated factors appear to promote actin filament assembly by *de novo* nucleation (or stabilization of nuclei), complementing assembly from barbed ends of existing filaments following barbed-end uncapping. This complex forms a cap on filament pointed ends and may also help to establish a branched actin filament geometry by binding to the sides of existing filaments.

Proteins that promote disassembly of actin filaments or cap filament barbed ends play roles in actin dynamics and cell motility. Actin assembly can be initiated by: (1) *de novo* nucleation (or stabilization of nuclei) with the involvement of the Arp2/3 complex and (2) by Rac and phosphoinositide-regulated release of capping and/or severing proteins of the gelsolin/villin family (CapG, gelsolin, villin and adseverin), and the CapZ family of capping proteins, from the barbed ends of actin filaments. Actin filament severing by gelsolin family or ADF/cofilin family proteins could also in itself be a mechanism for generating more free barbed ends from existing filaments.

How can villin enhance actin dynamics during cell motility?

There are a number of different types of actin-based motility. Different cells seem to move using different strategies. However, there are some general properties of cell movement. In general, the leading edge of a moving cell is the main site of actin assembly and crosslinking (or gelation, a type of 'solidification' of the cytoplasm). The leading edge of an animal cell can display a variety of types of protrusion: lamellipodia, filopodia (also known as microspikes) and pseudopodia. In general, behind the protrusive structure, there is a region of active actin disassembly, where filaments are shortened and crosslinks disrupted. Actin monomers resulting from this disassembly appear to then flow forward, where they can be added to the barbed ends of polymerizing filaments at the leading edge. Treadmilling allows continuous polymerization at barbed (plus) ends while disassembling continuously at pointed (minus) ends, without net increase in polymer mass. This type of polymerization can exert force and push out the plasma membrane during protrusion or power the movement of intracellular bacteria (e.g. *Listeria*, *S. flexneri*).

ADF/cofilin proteins appear weakly to sever filaments without capping and could perhaps generate new free barbed ends to support polymerization. These proteins are most active not directly adjacent to the leading edge of motile cells but somewhat behind it, and these ends would be rapidly capped since there are high concentrations of capping proteins in the typical cell. The ends would have to be uncapped or prevented from being capped in the first place (by phosphoinositides) to support further elongation, which again would make uncapping a major control point for generation of free barbed ends from existing filaments. Certainly for villin that severs filaments, capping is indeed an essential activity which also means that subsequent uncapping is necessary for existing filaments to elongate. Actin disassembly generally occurs behind the leading edge and can be mediated by proteins of both the calcium-activated gelsolin family and the calcium-independent ADF/cofilin family (also known as actophorin, destrin or depactin). The mechanism of action of these proteins is distinct but overlapping with that of severing proteins of the gelsolin/villin family. Phosphoinisotides appear to inhibit the function of gelsolin family, CapZ family and ADF/cofilin family members while calcium positively regulates the function of proteins of the gelsolin family. Phosphorylation but not calcium appears to regulate the activity of ADF/cofilin proteins (for general review see Cooper and Schafer, 2000).

Although the major proteins involved in the actin dynamics are well identified, it is not clear how these proteins are involved in the integration of signals that initiate motility or maintain chemotaxis toward or away from a stimulus source. Further studies remain to be conducted in order better to understand the role of the actin dynamics in specialized cells such as enterocytes. This process may be important for intestinal cell plasticity in response to extracellular signals or cell damage. We propose that the epithelial–mesenchymal transition process requires both the recruitment of apical actin and actin-binding proteins bearing multifunctional properties, such as villin which, with enrichment at the cell leading edge, provide the dynamic process necessary for cell propulsion. As a general principle we suggest that the signalling events leading to actin dynamics are mediated by proteins which are specific to the cell lineage considered and are therefore recruited to assure dynamic functions, thus increasing the efficiency of the cellular response. This adaptative process is necessary for epithelial cells to make a rapid and efficient response to extracellular signals during physiological or pathological situations without any *de novo* protein synthesis. Conversely, we wonder how a protein whose expression is restricted to a defined tissue can be integrated into signalling pathways but may play an active role when expressed in another cellular context? MDCK cells, routinely used for cell motility and morphogenesis experiments, are able easily to perform the epithelium–mesenchyme transition by using some components of the minimal molecular machinery described in the previous section. Addition of villin to a cell that does not express this multifunctional protein can lead to an increased turnover of actin monomers in the polymerization/depolymerization cycle of actin that controls the dynamics of cell motility, thus acting as a potentiator (Figure 15.3).

Perspectives

The role of villin in actin dynamics can be due to its severing and/or capping property. The severing property of villin has been shown to be regulated by its phosphorylation state (Zhai *et al.*, 2002). Indeed phosphorylated villin is known to increase actin severing, *in vitro*. It would thus be interesting to analyse the consequence of its phosphorylation status in the regulation of brush border actin dynamics. Moreover, actin dynamics have been shown to

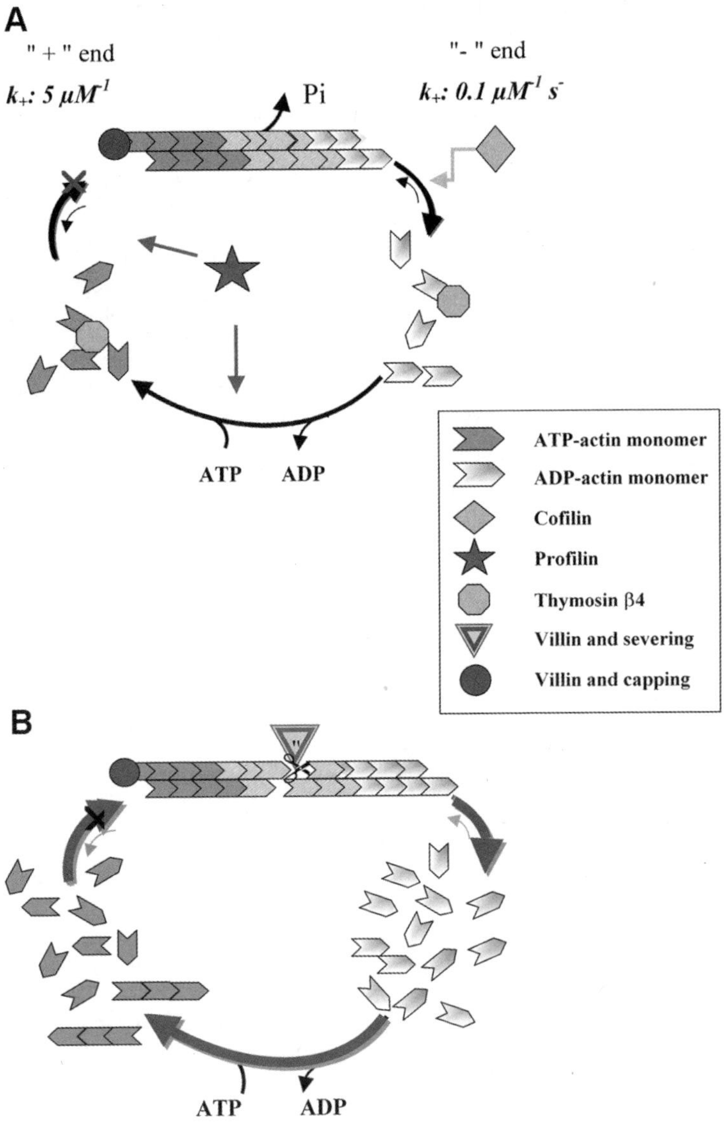

Figure 15.3 Actin dynamics is controlled by actin-binding proteins (A) and is enhanced in presence of villin (B). (A colour reproduction of this figure can be found in the colour plate section)

regulate important cellular functions such as ionic transport. Many studies have shown that some transporters or ionic channels are regulated by the actin cytoskeleton organization. The involvement of an actin remodelling protein such as villin is to be further investigated in this process (Cantiello, 1995; Cantiello *et al.*, 1991). Future studies need to be performed to unravel the signalling pathways and protein complexes that couple the structural activities of villin to signal transduction processes.

Our *in vivo* studies performed on vil$^{+/+}$ and vil$^{-/-}$ mice by administration of Dextran Sodium Sulfate (DSS), an abrasive agent of the intestinal epithelium, have shown a higher probability of death of villin null mice, indicating their higher susceptibility to cell damage. This allowed us to suggest a role for villin in epithelial cell plasticity in response to cell injury. This model of intestinal injury, close to other animal models of rectolitis, could lead, in a long-term project, to a useful investigation of villin status in human colonic disease such as Inflammatory Bowel Disease (IBD). A genetic predisposition has been suggested, and many environmental factors, including bacterial, viral and, perhaps, dietary antigens, can trigger an ongoing enteric inflammatory cascade during IBDs. This raises the question of how epithelial cells interact with components of the immune system and manage to promote the epithelial mesenchyme transition in response to injury. The investigation of villin expression levels or mutations in the IBD context can contribute to an understanding of IBD pathogenesis.

References

Cantiello, H. F., 1995. Role of the actin cytoskeleton on epithelial Na+ channel regulation. *Kidney Int.* **48**: 970–984.

Cantiello H. F., Stow, J. L., *et al.*, 1991. Actin filaments regulate epithelial Na+ channel activity. *Am. J. Physiol.* **261**: C882–C888.

Cooper, J. A. and Schafer, D. A., 2000. Control of actin assembly and disassembly at filament ends. *Curr. Opin. Cell Biol.* **12**: 97–103.

Costa de Beauregard, M. A., Pringault, E., *et al.*, 1995. Suppression of villin expression by antisense RNA impairs brush border assembly in polarized epithelial intestinal cells. *Embo J.* **14**: 409–421.

Doering, D. S. and Matsudaira, P., 1996. Cysteine scanning mutagenesis at 40 of 76 positions in villin headpiece maps the F-actin binding site and structural features of the domain. *Biochem.* **35**: 12677–12685.

Ferrary, E., Cohen-Tannoudji, M., *et al.*, 1999. *In vivo*, villin is required for Ca^{2+} dependent F-actin disruption in intestinal brush-borders. *J. Cell Biol.* **146**: 819–829.

Friederich, E., Huet, C., *et al.*, 1989. Villin induces microvilli growth and actin redistribution in transfected fibroblasts. *Cell* **59**: 461–475.

Friederich, E., Pringault, E., *et al.*, 1990. From the structure to the function of villin, an actin-binding protein of the brush border. *Bioessays* **12**: 403–408.

Friederich, E., Vancompernolle, K., *et al.*, 1992. An actin-binding site containing a conserved motif of charged amino acid residues is essential for the morphogenic effect of villin. *Cell* **70**: 81–92.

Friederich, E., Vancompernolle, K., *et al.*, 1999. Villin function in the organization of the actin cytoskeleton. Correlation of *in vivo* effects to its biochemical activities *in vitro*. *J. Biol. Chem.* **274**: 26751–26760.

Heintzelman, M. B. and Mooseker, M. S., 1992. Assembly of the intestinal brush border cytoskeleton. *Curr. Top. Dev. Biol.* **26**: 93–122.

Janmey, P. A. and Matsudaira, P. T., 1988. Functional comparison of villin and gelsolin. Effects of Ca^{2+}, KCl, and polyphosphoinositides. *J. Biol. Chem.* **263**: 16738–16743.

Janmey, P. A., Iida, K., *et al.*, 1987. Polyphosphoinositide micelles and polyphosphoinositide-containing vesicles dissociate endogenous gelsolin-actin complexes and promote actin assembly from the fast-growing end of actin filaments blocked by gelsolin. *J. Biol. Chem.* **262**: 12228–12236.

Kwiatkowski, D. J. and Yin, H. L., 1987. Molecular biology of gelsolin, a calcium-regulated actin filament severing protein. *Biorheology* **24**: 643–647.

Louvard, D., 1989. The function of the major cytoskeletal components of the brush border. *Curr. Opin. Cell Biol.* **1**: 51–57.

Lueck, A., Brown, D., *et al.*, 1998. The actin-binding proteins adseverin and gelsolin are both highly expressed but differentially localized in kidney and intestine. *J. Cell Sci.* **111**: 3633–3643.

Marks, P. W., Arai, M., *et al.*, 1998. Advillin (p92): a new member of the gelsolin/villin family of actin regulatory proteins. *J. Cell Sci.* **111**: 2129–2136.

Mishra, V. S., Henske, E. P., *et al.*, 1994. The human actin-regulatory protein cap G: gene structure and chromosome location. *Genomics* **23**: 560–565.

Northrop, J., Weber, A., *et al.*, 1986. Different calcium dependence of the capping and cutting activities of villin. *J. Biol. Chem.* **261**: 9274–9281.

Panebra, A., Ma, S. X., *et al.*, 2001. Regulation of phospholipase C-gamma(1) by the actin-regulatory protein villin. *Am. J. Physiol. Cell Physiol.* **281**: C1046–C1058.

Pestonjamasp, K. N., Pope, R. K., *et al.*, 1997. Supervillin (p205): A novel membrane-associated, F-actin-binding protein in the villin/gelsolin superfamily. *J. Cell Biol.* **139**: 1255–1269.

Pinson, K. I., Dunbar, L., *et al.*, 1998. Targeted disruption of the mouse villin gene does not impair the morphogenesis of microvilli. *Dev. Dyn.* **211**: 109–121.

Pinto, D., Robine, S., *et al.*, 1999. Regulatory sequences of the mouse villin gene that efficiently drive transgenic expression in immature and differentiated epithelial cells of small and large intestines. *J. Biol. Chem.* **274**: 6476–6482.

Ponzetto, C., Bardelli, A., *et al.*, 1994. A multifunctional docking site mediates signaling and transformation by the hepatocyte growth factor/scatter factor receptor family. *Cell* **77**: 261–271.

Rana, A. P., Ruff, P., *et al.*, 1993. Cloning of human erythroid dematin reveals another member of the villin family. *Proc. Natl. Acad. Sci. USA* **90**: 6651–6655.

Roof, D. J., Hayes, A., *et al.*, 1997. Molecular characterization of abLIM, a novel actin-binding and double zinc finger protein. *J. Cell Biol.* **138**: 575–588.

Simenel, C., Rose, T., *et al.*, 1995. Conformational behaviour of a synthetic peptide of the C-terminus of villin that interacts with actin: an NMR, CD and stimulated annealing study. *Int. J. Pept. Protein Res.* **45**: 574–586.

Southwick, F. S. and DiNubile, M. J., 1986. Rabbit alveolar macrophages contain a Ca^{2+}-sensitive, 41,000-dalton protein which reversibly blocks the 'barbed' ends of actin filaments but does not sever them. *J. Biol. Chem.* **261**: 14191–14195.

Zhai, L., Zhao, P., *et al.*, 2001. Tyrosine phosphorylation of villin regulates the organization of the actin cytoskeleton. *J. Biol. Chem.* **276**: 36163–36167.

Zhai, L., Kumar, N., *et al.*, 2002. Regulation of actin dynamics by tyrosine phosphorylation: identification of tyrosine phosphorylation sites within the actin-severing domain of villin. *Biochemistry* **41**: 11750–11760.

16

Scar, WASp and the Arp2/3 Complex in *Dictyostelium* Migration

Simone Blagg and **Robert Insall**

The Arp2/3 complex plays a key role in the polymerization of actin filaments in moving cells. Activation of the Arp2/3 complex is in turn controlled by the members of the WASp/Scar family of proteins. The two defining members of this family, N-WASp and Scar, are regulated differently, and appear to play different roles in cell movement. Inactive N-WASp is inhibited by an intramolecular interaction, whereas recent data suggests that Scar is held inactive within a complex of four other proteins. All four proteins are present in *Dictyostelium*, suggesting this method of Scar regulation is important and conserved through evolution. We have been analysing this system using *Dictyostelium* genetics, a system which has provided valuable insights into many aspects of the cytoskeleton.

Introduction

It has been known for two decades that actin polymerization plays a major role in driving forward the leading edge of migrating cells (Tilney *et al.*, 1983). However, the molecular mechanism of this process is only now being elucidated.

Cell Motility. Edited by Anne Ridley, Michelle Peckham and Peter Clark.
 ISBN 0 470 84872 3

Cell movement requires cycles of polymerization and depolymerization of actin. These are regulated by Rho family GTPases, which relay signals (both endogenous or exogenous) to the actin polymerizing machinery of the cell, in particular the Arp2/3 complex (Machesky and Gould, 1999). Activation of Arp2/3 induces polymerization of new actin filaments adjacent to the plasma membrane, usually at the leading edge. These new actin filaments push the membrane forwards and determine the direction of cell locomotion.

Our understanding of cell motility has developed rapidly within the last decade. The discovery and characterization of the actin nucleating machinery of the cell, including the Arp2/3 complex and its regulators, has been a particularly important advance.

The Arp2/3 complex and WASp family proteins

In living cells, the Arp2/3 complex is the major factor responsible for *de novo* actin nucleation. From its initial purification and characterization (Machesky *et al.*, 1994; Mullins *et al.*, 1997), to its connection first with *Listeria* motility (Welch *et al.*, 1997) and more recently Rho-family GTPases (Machesky and Insall, 1998), the Arp2/3 complex has been found to play a pivotal role in actin assembly. The complex consists of seven proteins: two actin related proteins, Arp2 and Arp3, a WD repeat protein, p40, and four novel proteins termed p34, p20, p21 and p16 (Kelleher *et al.*, 1995). The crystal structure of this complex has now been resolved (Robinson *et al.*, 2001), providing further insights into the molecular relationships between these subunits.

The rate-determining step for actin polymerization is nucleation, the initial formation of a small filament. Once a short filament is formed, it extends at a rapid and largely unregulated rate (Pollard *et al.*, 2000). The Arp2/3 complex acts as a catalyst for nucleation, and thereby controls the overall rate of actin polymerization. The two actin related proteins in the complex are thought to form a dimer similar to an actin dimer, which can act as a stable nucleus for actin polymerization (Kelleher *et al.*, 1995; Machesky *et al.*, 1994; Mullins *et al.*, 1997). Arp2/3 alone does not stimulate actin filament production. Rather, it is activated by a range of accessory proteins when cells are stimulated. Members of the WASp (Wiskott–Aldrich Syndrome protein) family of proteins are the best understood Arp2/3 activators. These interact with the p21 subunit of the complex (Machesky and Insall, 1998), inducing a conformational change which enables the complex to nucleate actin (Zalevsky *et al.*, 2001a). WASp family proteins in general are thought to help coordinate actin reorganization by coupling Rho family GTPases and other upstream signalling molecules to the activation and mobilization of the Arp2/3 complex (Machesky and Gould, 1999).

The mammalian WASp/Scar family currently consists of five members. WASp is apparently only expressed in haematopoietic cell lines, while N-WASp is a more widely expressed relative of WASp, and three Scar isoforms (which have also been named WAVEs). All family members contain a proline-rich domain, a WH2 (WASp homology 2) actin monomer binding domain and a conserved acidic domain, which lies carboxy-terminal to the WH2 domain. It is this acidic domain which binds to the Arp2/3 complex (Machesky and Insall, 1998). The C-terminal WH2 and acidic domains are referred to together as the WA domain. At the amino terminus, WASp contains a WH1 (WASp homology 1) domain, while Scar contains a Scar homology 1 domain, which is not homologous to WASp, but is conserved within the Scar family (Bear *et al.*, 1998). The binding interaction of this Scar homology domain remains unknown.

WASp was first identified as the product of the gene mutated in Wiskott–Aldrich syndrome (WAS), an X-linked disease which affects platelets and lymphocytes (Derry *et al.*, 1994). It was subsequently found that WASp interacts with Cdc42, providing a link between this GTPase and the cytoskeleton (Kirchhausen and Rosen, 1996; Symons *et al.*, 1996). Phosphatidylinositol 4,5-bisphosphate (PIP_2) has also been found to be important in normal WASp function; PIP_2 and Cdc42 are capable of synergistically activating N-WASp (Rohatgi *et al.*, 2000). It is thought that PIP_2 can reduce the affinity between the N- and C-termini of N-WASp, thereby relieving an autoinhibitory interaction, and that it can also indirectly activate Cdc42 (Rohatgi *et al.*, 2000). Additionally, a recent study has demonstrated a Cdc42-independent pathway of N-WASp activation, which involves the synergistic action of PIP2 and the adaptor protein Nck (binding through its SH3 domains to the N-WASp polyproline domain), leading to Arp2/3 dependent actin polymerization (Rohatgi *et al.*, 2001). In agreement with this work, it had been previously shown that Nck coordinates the assembly of an actin nucleation complex at the surface of vaccinia virus by binding to a tyrosine-phosphorylated viral protein through its SH2 domain, and recruiting N-WASp through its SH3 domain (Frischknecht *et al.*, 1999).

Scar, also known as WAVE (WASp/verprolin homologous protein) in mammalian cells (Miki *et al.*, 1998), has three isoforms in humans (Suetsugu *et al.*, 1999). Recombinant Scar1 has been found to interact with cAMP-dependent protein kinase (PKA) and Abl kinases when expressed in HEK-293 cells, and endogenous Scar has been found to co-purify with these proteins in brain extracts (Westphal *et al.*, 2000). This study therefore proposes that Scar1 can be classified as an A-kinase anchoring protein (AKAP), that can tether PKA to defined subcellular sites, and puts forward the idea of Scar1 as an actin-associated scaffolding protein that recruits PKA and Abl.

The WASp family and Arp2/3 complex members are conserved in all eukaryotes, and are clearly essential to normal cell function. Furthermore, recent *Drosophila* studies demonstrate that static structures (such as bristles), as well as the actin protrusions of motile cells, are constructed using the WASp family and Arp2/3 complex (Hudson and Cooley, 2002; Miller, 2002; Zallen *et al.*, 2002).

However, the WASp family proteins are not the only regulators of Arp2/3 complex activity. Members of a related group, the Ena/VASP family, have also been shown to promote actin assembly in mammalian cells (Laurent *et al.*, 1999). Similarly, cortactin (in mammalian cells) and Abp1p (in *Saccharomyces*) have been shown to be capable of activating the Arp2/3 complex and stabilize newly synthesized actin filaments (Olazabal and Machesky, 2001). Studies on yeast have also shown that the cortactin-like filament binding protein Abp1p, which contains acidic domains similar to those of the WASp family, may function by recruiting Arp2/3 to the sides of actin filaments (Goode *et al.*, 2001).

Regulation of WASp and Scar

WASp contains a GBD (GTPase Binding Domain), also known as a CRIB (Cdc42 and Rac Interactive Binding) domain, through which it binds Cdc42. This interaction is thought to couple extracellular signals to activation of the Arp2/3 complex and stimulation of actin polymerization (Ma *et al.*, 1998). WASp appears to be regulated in part by auto-inhibition; isolated, native WASp is inactive until stimulated (Higgs and Pollard, 2000). It has been found that the N-terminal domain (including the GBD domain) binds the C-terminal domain (the WA domain which binds Arp2/3 and actin) thereby inhibiting its function. Cdc42 binding to the GBD domain relieves this inhibition, presumably by causing a conformational change, thereby activating WASp and, subsequently, the Arp2/3 complex (Higgs and Pollard, 2000; Rohatgi *et al.*, 2000). PIP_2 appears to fulfil a similar activating role, leading to suggestions that WASp plays a part in integrating lipid and GTPases signals, though it is not clear how far changes in PIP_2 levels can explain alterations in the actin cytoskeleton.

The control of Scar has been harder to dissect, in particular because of the lack of obvious signal-related domains. Scar1 has been found to be recruited to the tips of protruding lamellipodia, but not of filopodia, in living mammalian cells (Hahne *et al.*, 2001), suggesting that Scar is somehow coupled to Rac. However, direct binding to Rac has not been found (Miki *et al.*, 1998). It therefore seems likely that the activity of Scar proteins is controlled through binding to other proteins.

Unlike N-WASp, Scar is not autoinhibited, and pure protein is thought to be constitutively active. It therefore appears that other binding proteins act to inhibit Scar activity *in vivo*. Recent work has identified a complex of proteins which are bound to unstimulated Scar, forming an inactive complex (Eden *et al.*, 2002). These proteins have been identified as NCKAP1, PIR121, Abi2 and HSPC300, which encodes a 9 kDa protein of unknown function.

Nck Associated protein 1 (NCKAP1) was first identified as Nap125 in a screen for Nck binding proteins (Kitamura *et al.*, 1996) and has an association with Alzheimer's disease, where its expression was found to be down-regulated (Suzuki *et al.*, 2000; Yamamoto *et al.*, 2001). NCKAP1 was found to bind Rac indirectly through a 140 kDa protein (Kitamura *et al.*, 1997), which is now presumed to be PIR121. Abl interactor 2 (Abi2) is an SH3 domain containing protein which interacts with the c-Abl tyrosine kinase and is implicated in cytoskeletal function, cell migration and transformation (Dai *et al.*, 2001; Dai and Pendergast, 1995). This protein has also been shown to localize to sites of actin polymerization (Stradal *et al.*, 2001). Abi2 may be an important link between Scar and Abl, which were found to co-precipitate together (Westphal *et al.*, 2000). It is also homologous to hNap1bp, a protein previously identified as a binding partner of NCKAP1 (Yamamoto *et al.*, 2001).

PIR121 (p53-inducible messenger RNA) (Saller *et al.*, 1999) is thought to bind directly to both Scar and Rac. It may therefore be the missing link in the signalling process involved in membrane ruffling and lamellipodia formation. Eden *et al.* (2002) and Kitamura *et al.* (1996, 1997) suggest that Rac1 and Nck can cause dissociation of the Scar complex by binding to PIR121, and by doing so causing the release of active Scar and HSPC300. These proteins, which remain bound to one another, could in turn nucleate via the Arp2/3 complex. Although there is as yet no biological evidence for this model, it does fit well with the various confusing strands of data, so it remains our preferred working hypothesis. Various pieces of evidence are in agreement with the idea of an inhibitory Scar complex. For example, NCKAP1 and PIR121 have been identified as proteins that bind to Nck and activated Rac (Kitamura *et al.*, 1996, 1997; Kobayashi *et al.*, 1998), whereas Scar and HSPC300 do not bind these proteins.

The proteins thought to be present in a complex with Scar in mammalian cells are also found in *Dictyostelium discoideum*. Recent work by our group has demonstrated that disruption of the PIR121 gene leads to severe defects in shape, motility and chemotaxis of this amoeba.

Recently, an unrelated study in *C. elegans* has also suggested the association of PIR121 and NCKAP1 (in this case named GEX-2 and GEX-3 respectively) and their role in development (Soto *et al.*, 2002). This study finds that GEX-2 and GEX-3 are essential for proper tissue morphogenesis, as these two proteins together regulate cell migration and shape change. This provides more evidence from an entirely different perspective that NCKAP1 and PIR121 act together to regulate actin polymerization, and consequently cell movement, *in vivo*.

Control of actin dynamics in *Dictyostelium*

Over the years, the study of the social amoeba *D. discoideum* has given great insights into mammalian biochemical systems and mammalian disease mechanisms. Its similarities to higher eukaryotic systems make it an ideal organism to study for processes such as actin dynamics, and analysis of the near-complete genomic sequence has made such similarities even more apparent (Glockner *et al.*, 2002).

The ease with which *Dictyostelium* genes can be manipulated has led to its becoming a powerful tool for the study of cell motility. Numerous mutants with gene disruptions have been generated, providing a mass of information about the functions of the pathways leading to actin polymerization and cell locomotion. For instance, *Dictyostelium* mutants lacking both isoforms of profilin have greatly decreased motility, increased F-actin concentration, impaired cytokinesis and developmental defects, highlighting the role of profilin as an actin sequestering protein *in vivo* (Haugwitz *et al.*, 1994). Capping protein is another important factor in actin polymerization and functions by terminating growth of actin filament ends. The phenotype of mutants with decreased levels of capping protein, including increased F-actin content and slow movement, helped to prove the function of this protein, previously suggested only *in vitro*, in living cells (Hug *et al.*, 1995).

Interfering with upstream signalling events has also proved useful in generating information about the molecular organization of signalling pathways. The *Dictyostelium* genome encodes a surprisingly large number of small GTPases, including at least 13 Rac subfamily members, all homologous to mammalian Rac1 (Wilkins and Insall, 2001). One study has outlined the consequences of overactive Rac1B and the absence of Rac1 GAP (Chung *et al.*, 2000). Both mutants have similar phenotypes, including inefficient chemotaxis, high number of lateral pseudopods and low cell speed. This shows that the Rac subfamily of proteins is involved in the regulation of cell movement through the actin cytoskeleton in *Dictyostelium* as in mammals. Another study disrupted the DGAP1 gene, the protein product of which connects Rac signalling to the actin cytoskeleton, and discovered such mutants also had increased F-actin contents and large leading edges (Faix *et al.*, 1998).

The majority of known proteins involved in mammalian actin dynamics have also been identified in *Dictyostelium*, allowing cell movement to be meaningfully researched at a lower rung on the evolutionary ladder. The major players of the actin nucleating machinery of the cell, including all seven members of the Arp2/3 complex (Insall *et al.*, 2001), WASp and Scar, are all present in *Dictyostelium* and are highly homologous to their mammalian counterparts. However, it is clear that there are differences; for example, Cdc42 activity has yet to be demonstrated in *Dictyostelium* (Wilkins and

Insall, 2001), despite obvious filopodia which closely resemble those in mammalian cells, raising questions about the precise signalling pathways required for formation of such structures. Nevertheless, the underlying biochemical processes of actin polymerization and its regulation appear to be largely conserved in humans and the highly motile social amoebas.

The Arp2/3 complex and its regulation in *Dictyostelium*

The Arp2/3 complex has been identified and purified in *Dictyostelium*, and its members have been shown to be closely related to their mammalian counterparts (Insall *et al.*, 2001). It redistributes in response to chemotactic stimuli, and is implicated in phagocytosis, macropinocytosis and chemotaxis.

Discovery of Scar1

Scar was first discovered in *Dictyostelium*, as a suppressor of the developmental phenotypes caused by deletion of a cAMP receptor. The gene encodes a WASp-related protein capable of activating Arp2/3 complex, and thus actin nucleation (Bear *et al.*, 1998; Machesky and Insall, 1998; Machesky *et al.*, 1999). G-protein coupled cAMP receptors (cARs) transduce extracellular cAMP chemoattractant signals. In cAR^- strains of *Dictyostelium*, the multicellular development of the organism arrests before tip formation, but this phenotype can be rescued if Scar is also removed genetically from these cells (hence Scar, for Suppressor of cAR2). Scar null cells are considerably smaller than wild-type cells, have reduced levels of filamentous actin, and have abnormal cell morphology and actin distribution during chemotaxis (Bear *et al.*, 1998).

Since this initial work, there has been great interest in the function of Scar over the last few years. It has been found that Scar regulates many processes, including macropinocytosis and phagocytosis in *Dictyostelium* (Seastone *et al.*, 2001).

Other Arp2/3 regulators

The WASp family proteins are present in *Dictyostelium* in the form of a single Scar protein, and a single WASp protein. However, these proteins are not the only factors responsible for promotion of Arp2/3-dependent actin polymerization. The *Dictyostelium* CARMIL (Capping protein, Arp2/3 and Myosin I Linker) protein contains an acidic domain, capable of activating Arp2/3 (Jung *et al.*, 2001). In addition to binding and activating Arp2/3, this protein also appears to interact with small myosins (myoB and myoC) and capping protein, unearthing a potentially vital pathway which links many components of the actin polymerization machinery of the cell.

Evolutionary implications

The evolutionary lineage of the social amoebas is now thought to have diverged from that of mammals shortly after plants separated from fungi and animals. Despite this ancient separation, *Dictyostelium* cells move in a highly similar fashion to crawling cells from mammals and other metazoa, using the same basic framework (actin filaments, nucleated by the Arp2/3 complex under the control of WASp and Scar1; Devreotes and Zigmond, 1988) and regulated by many of the same controls (Racs and heterotrimeric G-proteins coupled to serpentine receptors). It is therefore very informative to consider which systems are conserved in nearly all eukaryotes which crawl using actin, and which have appeared more recently. Well-conserved proteins could be considered to be core components, while proteins which are found in mammals but not elsewhere would be more likely to represent refinements and enlargements of the actin repertoire.

All of the components which were required to reconstitute *Listeria* motility *in vitro* (Loisel *et al.*, 1999) are present in *Dictyostelium*. Actin, all seven members of the Arp2/3 complex, cofilin and capping protein are all present and highly related to their mammalian homologues. Likewise, the three additional proteins which enhanced the efficiency of the reaction (profilin, α-actinin and VASP; Loisel *et al.*, 1999) are highly conserved. The core components defined *in vitro* by Loisel *et al.* are thus consistent with our evolutionary definition.

More interesting are the regulators of the Arp2/3 complex. Scar1 and all four proteins defined by Eden *et al.* (PIR121, NAP1, HSPC300 and Abi2) are strikingly conserved. The other protein which has been described as a Scar/WAVE regulator, IRSp53 (Miki *et al.*, 2000), has not been seen, despite genome sequencing which has been estimated to have hit 95% of all genes. If, as seems likely, it is not found once the genome is completed, it is unlikely to be essential for the coupling between the well-conserved Scar and Arp2/3 complex. WASp is also present in *Dictyostelium*, although there is as yet no sign of the putative WASp regulator WIP. However, WIP-related proteins are extremely divergent: the analagous regulator for N-WASp, CR16, is only barely recognizable as related (Ho *et al.*, 2001). This is odd, considering the strong conservation of the rest of the system and the presumed essential role for WIP.

Of the less well-understood regulators of the Arp2/3 complex, CARMIL is only known in amoebae, although it has clear homologues in metazoans. It will be very interesting to see the roles of these homologues. Likewise, the cortactin relative ABP1 has been identified in the *Dictyostelium* genome. It therefore seems that most of the main proteins of current interest are evolutionarily ancient, and likely to be important for actin regulation in nearly all cells.

One major exception is yeast. *Saccharomyces cerevisiae* has no Scar1 or other members of its regulatory complex, no CARMIL, and the WASp homologue (Las17p/Bee1p) does not contain a CRIB domain. This is extraordinary considering that yeasts diverged from the mammalian lineage far more recently than *Dictyostelium*. The most plausible hypothesis is that yeasts underwent a wholesale alteration in their actin cytoskeleton at some point in evolution. Under extreme pressure to reduce the quantity of DNA, it must have been worthwhile to lose the more than 10 kbp of coding sequence needed for the Scar complex. Likewise, in *Dictyostelium* CARMIL couples small myosins to the Arp2/3 complex (Jung *et al.*, 2001). In *Saccharomyces*, unlike any other organisms, the Myosin I proteins bind directly to Arp2/3 (Lechler *et al.*, 2000), which may have allowed another set of genes to be lost.

Coupling signalling pathways to Arp2/3 dependent actin polymerization

The exact mechanisms by which signals lead to the creation of actin structures in specific areas of the cell, and eventual cell movement, are incompletely understood at present, although knowledge in this area is rapidly increasing. The elucidation of the events that lead to different actin structures is already providing a far clearer picture of how cells move.

How do WASp family members activate the Arp2/3 complex?

Various recent studies have contributed significantly to our understanding of the molecular basis of the control of Arp2/3 by WASp family members.

Activation of WASp proteins exposes Arp2/3 binding sites in the WA domain, allowing binding to Arp2/3 (Rohatgi *et al.*, 1999). In the case of unstimulated WASp, intramolecular reactions with the GTPase-binding domain occlude C-terminal residues, rendering the complex unable to bind Arp2/3. This inhibition of the protein is relieved when activated by Cdc42 and PIP_2 binding, which cause a dramatic conformational change in the structure of WASp. This leads to disruption of the hydrophobic core and exposure of the C-terminal WA domain, allowing WASp to bind to the Arp2/3 complex (Higgs and Pollard, 2000; Kim *et al.*, 2000; Rohatgi *et al.*, 2000). The WA domain (WH2 and acidic domain, also called the VCA domain by some authors) of WASp has been used to study the binding of WASp to actin; this domain binds to a single actin monomer with a K_d of 0.6 μM, and to the Arp2/3 complex with a K_d of 0.6 μM (Marchand *et al.*, 2001). Although WASp and Scar proteins bind to the Arp2/3 complex and to actin with nearly identical affinities, they do not activate it to the same extent. Scar induces a substantially slower rate of nucleation, while WASp and N-WASp activities are 16 and 70-fold higher, respectively, relative to Scar 1. This difference appears to be due to differing numbers of acidic amino acids at the C-terminus (Zalevsky *et al.*, 2001b).

A number of extra factors have been implicated in the correct functioning of WASp family proteins. It is clear that the control of the actin cytoskeleton is more complex than is often described, for example by assigning single GTPases to specific structures. Cdc42 is frequently described as leading to filopodia production via N-WASp, with Rac simply leading to lamellipodia via Scar. In *Dictyostelium*, multiple filopodia are formed in the absence of Cdc42, while different activated Rac proteins induce actin structures of different shape and stability, in different parts of the cell.

How does the actin-polymerization machinery of the cell produce distinct actin structures?

All motile cells, from single-celled organisms such as *Dictyostelium* to motile cells from the human body, use their actin polymerization to produce a variety of different actin structures with different uses. Although the identities of most of the major players in actin filament production are now known, the ways that the relatively small number of proteins interact to produce different structures is a much more complex problem. At present it is unknown if organization of actin is determined by the mechanism of actin polymerization itself, or if the actin filaments are organized by actin binding proteins after polymerization.

The Arp2/3 complex is found at the sides of pre-existing actin filaments, and activators like the WASp family cause it to generate a branched network (Blanchoin *et al.*, 2000; Mullins *et al.*, 1998). However, the precise pathways that lead to the generation of various different actin structures such as filopodia, lamellipodia and membrane ruffles are unknown. Perhaps the scaffolding properties of the WASp family adaptor proteins are important, allowing interaction with a number of proteins resulting in greater sensitivity to different signals and leading to a larger number of outcomes. Also, other components of the large complexes which include the WASp family, such as profilins, may be involved in modulating or bringing together cytoskeletal elements and signalling molecules such as Rho GTPases, in specific areas of the cell (Witke *et al.*, 1998).

Creation of specific actin structures in specific areas may also involve clustering of proteins such as WASp, which may help control the spatial distribution of actin. In addition to the possibility of controlling distribution of actin, such clustering has been shown to enhance actin polymerization *in vivo* (Castellano *et al.*, 2001).

References

Bear, J. E., Rawls, J. F. and Saxe, C. L., 3rd, 1998. SCAR, a WASP-related protein, isolated as a suppressor of receptor defects in late *Dictyostelium* development. *J. Cell Biol.* **142**: 1325–1335.

Blanchoin, L., Amann, K. J., Higgs, H. N., Marchand, J. B., *et al.*, 2000. Direct observation of dendritic actin filament networks nucleated by Arp2/3 complex and WASP/Scar proteins. *Nature* **404**: 1007–1011.

Castellano, F., Le Clainche, C., Patin, D., Carlier, M. F. and Chavrier, P., 2001. A WASp–VASP complex regulates actin polymerization at the plasma membrane. *EMBO J.* **20**: 5603–5614.

Chung, C. Y., Lee, S., Briscoe, C., Ellsworth, C. and Firtel, R. A., 2000. Role of Rac in controlling the actin cytoskeleton and chemotaxis in motile cells. *Proc. Natl. Acad. Sci. USA* **97**: 5225–5230.

Dai, Z. and Pendergast, A. M., 1995. Abi-2, a novel SH3-containing protein interacts with the c-Abl tyrosine kinase and modulates c-Abl transforming activity. *Genes Dev.* **9**: 2569–2582.

Dai, Z., Kerzic, P., Schroeder, W. G. and McNiece, I. K., 2001. Deletion of the Src homology 3 domain and C-terminal proline-rich sequences in Bcr-Abl prevents Abl interactor 2 degradation and spontaneous cell migration and impairs leukemogenesis. *J. Biol. Chem.* **276**: 28954–28960.

Derry, J. M., Ochs, H. D. and Francke, U., 1994. Isolation of a novel gene mutated in Wiskott–Aldrich syndrome. *Cell* **78**: 635–644.

Devreotes, P. N. and Zigmond, S. H., 1998. Chemotaxis in eukaryotic cells: a focus on leukocytes and *Dictyostelium*. *Ann. Rev. Cell. Biol.* **4**: 649–686.

Eden, S., Rohatgi, R., Podtelejnikov, A. V., Mann, M. and Kirschner, M. W., 2002. Mechanism of regulation of WAVE1-induced actin nucleation by Rac1 and Nck. *Nature* **418**: 790–793.

Faix, J., Clougherty, C., Konzok, A., Mintert, U., *et al.*, 1998. The IQGAP-related protein DGAP1 interacts with Rac and is involved in the modulation of the F-actin cytoskeleton and control of cell motility. *J. Cell Sci.* **111**: 3059–3071.

Frischknecht, F., Moreau, V., Rottger, S., Gonfloni, S., *et al.*, 1999. Actin-based motility of vaccinia virus mimics receptor tyrosine kinase signalling. *Nature* **401**: 926–929.

Glockner, G., Eichinger, L., Szafranski, K., Pachebat, J. A., *et al.*, 2002. Sequence and analysis of chromosome 2 of *Dictyostelium discoideum*. *Nature* **418**: 79–85.

Goode, B. L., Rodal, A. A., Barnes, G. and Drubin, D. G., 2001. Activation of the Arp2/3 complex by the actin filament binding protein Abp1p. *J. Cell Biol.* **153**: 627–634.

Hahne, P., Sechi, A., Benesch, S. and Small, J. V., 2001. Scar/WAVE is localised at the tips of protruding lamellipodia in living cells. *FEBS Lett.* **492**: 215–220.

Haugwitz, M., Noegel, A. A., Karakesisoglou, J. and Schleicher, M., 1994. *Dictyostelium* amoebae that lack G-actin-sequestering profilins show defects in F-actin content, cytokinesis, and development. *Cell* **79**: 303–314.

Higgs, H. N. and Pollard, T. D., 2000. Activation by Cdc42 and PIP(2) of Wiskott–Aldrich syndrome protein (WASp) stimulates actin nucleation by Arp2/3 complex. *J. Cell Biol.* **150**: 1311–1320.

Ho, H. Y., Rohatgi, R., Ma, L. and Kirschner, M. W., 2001. CR16 forms a complex with N-WASP in brain and is a novel member of a conserved proline-rich actin-binding protein family. *Proc. Natl. Acad. Sci. USA* **98**: 11306–11311.

Hudson, A. M. and Cooley, L., 2002. A subset of dynamic actin rearrangements in *Drosophila* requires the Arp2/3 complex. *J. Cell Biol.* **156**, 677–687.

Hug, C., Jay, P. Y., Reddy, I., McNally, J. G., *et al.*, 1995. Capping protein levels influence actin assembly and cell motility in *Dictyostelium. Cell* **81**: 591–600.

Insall, R., Muller-Taubenberger, A., Machesky, L., Kohler, J., *et al.*, 2001. Dynamics of the *Dictyostelium* Arp2/3 complex in endocytosis, cytokinesis, and chemotaxis. *Cell Motil. Cytoskeleton* **50**: 115–128.

Jung, G., Remmert, K., Wu, X. F., Volosky, J. M. and Hammer III, J. A., 2001. The *Dictyostelium* CARMIL protein links capping protein and the Arp2/3 complex to type I myosins through their SH3 domains. *J. Cell Biol.* **153**: 1479–1497.

Kelleher, J. F., Atkinson, S. J. and Pollard, T. D., 1995. Sequences, structural models, and cellular localization of the actin-related proteins Arp2 and Arp3 from *Acanthamoeba. J. Cell Biol.* **131**: 385–397.

Kim, A. S., Kakalis, L. T., Abdul-Manan, N., Liu, G. A. and Rosen, M. K., 2000. Autoinhibition and activation mechanisms of the Wiskott–Aldrich syndrome protein. *Nature* **404**: 151–158.

Kirchhausen, T. and Rosen, F. S., 1996. Disease mechanism: unravelling Wiskott–Aldrich syndrome. *Curr. Biol.* **6**: 676–678.

Kitamura, T., Kitamura, Y., Yonezawa, K., Totty, N. F., *et al.*, 1996. Molecular cloning of p125Nap1, a protein that associates with an SH3 domain of Nck. *Biochem. Biophys. Res. Commun.* **219**: 509–514.

Kitamura, Y., Kitamura, T., Sakaue, H., Maeda, T., *et al.*, 1997. Interaction of Nck-associated protein 1 with activated GTP-binding protein Rac. *Biochem. J.* **322**(3): 873–878.

Kobayashi, K., Kuroda, S., Fukata, M., Nakamura, T., *et al.*, 1998. p140Sra-1 (specifically Rac1-associated protein) is a novel specific target for Rac1 small GTPase. *J. Biol. Chem.* **273**: 291–295.

Laurent, V., Loisel, T. P., Harbeck, B., Wehman, A., *et al.*, 1999. Role of proteins of the Ena/VASP family in actin-based motility of *Listeria monocytogenes. J. Cell Biol.* **144**: 1245–1258.

Lechler, T., Shevchenko, A. and Li, R., 2000. Direct involvement of yeast type I myosins in Cdc42-dependent actin polymerization. *J. Cell Biol.* **148**: 363–373.

Loisel, T. P., Boujemaa, R., Pantaloni, D. and Carlier, M. F., 1999. Reconstitution of actin-based motility of *Listeria* and *Shigella* using pure proteins. *Nature* **401**: 613–616.

Ma, L., Rohatgi, R. and Kirschner, M. W., 1998. The Arp2/3 complex mediates actin polymerization induced by the small GTP-binding protein Cdc42. *Proc. Natl. Acad. Sci. USA* **95**: 15362–15367.

Machesky, L. M. and Gould, K. L., 1999. The Arp2/3 complex: a multifunctional actin organizer. *Curr. Opin. Cell Biol.* **11**: 117–121.

Machesky, L. M. and Insall, R. H., 1998. Scar1 and the related Wiskott–Aldrich syndrome protein, WASP, regulate the actin cytoskeleton through the Arp2/3 complex. *Curr. Biol.* **8**: 1347–1356.

Machesky, L. M., Atkinson, S. J., Ampe, C., Vandekerckhove, J. and Pollard, T. D., 1994. Purification of a cortical complex containing two unconventional actins from

Acanthamoeba by affinity chromatography on profilin-agarose. *J. Cell Biol.* **127**: 107–115.

Machesky, L. M., Mullins, R. D., Higgs, H. N., Kaiser, D. A., *et al.*, 1999. Scar, a WASp-related protein, activates nucleation of actin filaments by the Arp2/3 complex. *Proc. Natl. Acad. Sci. USA* **96**: 3739–3744.

Marchand, J. B., Kaiser, D. A., Pollard, T. D. and Higgs, H. N., 2001. Interaction of WASP/Scar proteins with actin and vertebrate Arp2/3 complex. *Nat. Cell Biol.* **3**: 76–82.

Miki, H., Suetsugu, S. and Takenawa, T., 1998. WAVE, a novel WASP-family protein involved in actin reorganization induced by Rac. *EMBO J.* **17**: 6932–6941.

Miki, H., Fukuda, M., Nishida, E. and Takenawa, T., 1999. Phosphorylation of WAVE downstream of mitogen-activated protein kinase signaling. *J. Biol. Chem.* **274**: 27605–27609.

Miki, H., Yamaguchi, H., Suetsugu, S. and Takenawa, T., 2000. IRSp53 is an essential intermediate between Rac and WAVE in the regulation of membrane ruffling. *Nature* **408**: 732–735.

Miller, K. G., 2002. Extending the Arp2/3 complex and its regulation beyond the leading edge. *J. Cell Biol.* **156**: 591–593.

Mullins, R. D., Stafford, W. F. and Pollard, T. D., 1997. Structure, subunit topology, and actin-binding activity of the Arp2/3 complex from *Acanthamoeba*. *J. Cell Biol.* **136**: 331–343.

Mullins, R. D., Heuser, J. A. and Pollard, T. D., 1998. The interaction of Arp2/3 complex with actin: nucleation, high affinity pointed end capping, and formation of branching networks of filaments. *Proc. Natl. Acad. Sci. USA* **95**: 6181–6186.

Olazabal, I. M. and Machesky, L. M., 2001. Abp1p and cortactin, new 'hand-holds' for actin. *J. Cell Biol.* **154**: 679–682.

Pollard, T. D., Blanchoin, L. and Mullins, R. D., 2000. Molecular mechanisms controlling actin filament dynamics in nonmuscle cells. *Annu. Rev. Biophys. Biomol. Struct.* **29**: 545–576.

Robinson, R. C., Turbedsky, K., Kaiser, D. A., Marchand, J. B., *et al.*, 2001. Crystal structure of Arp2/3 complex. *Science* **294**: 1679–1684.

Rohatgi, R., Ma, L., Miki, H., Lopez, M., Kirchhausen, T., *et al.*, 1999. The interaction between N-WASP and the Arp2/3 complex links Cdc42-dependent signals to actin assembly. *Cell* **97**: 221–231.

Rohatgi, R., Ho, H. Y. and Kirschner, M. W., 2000. Mechanism of N-WASP activation by CDC42 and phosphatidylinositol 4, 5-bisphosphate. *J. Cell Biol.* **150**: 1299–1310.

Rohatgi, R., Nollau, P., Ho, H. Y., Kirschner, M. W. and Mayer, B. J., 2001. Nck and phosphatidylinositol 4,5-bisphosphate synergistically activate actin polymerization through the N-WASP-Arp2/3 pathway. *J. Biol. Chem.* **276**: 26448–26452.

Saller, E., Tom, E., Brunori, M., Otter, M., *et al.*, 1999. Increased apoptosis induction by 121F mutant p53. *EMBO J.* **18**: 4424–4437.

Seastone, D. J., Harris, E., Temesvari, L. A., Bear, J. E., *et al.*, 2001. The WASp-like protein scar regulates macropinocytosis, phagocytosis and endosomal membrane flow in *Dictyostelium*. *J. Cell Sci.* **114**: 2673–2683.

Soto, M. C., Qadota, H., Kasuya, K., Inoue, M., *et al.*, 2002. The GEX-2 and GEX-3 proteins are required for tissue morphogenesis and cell migrations in *C. elegans*. *Genes Dev.* **16**: 620–632.

Stradal, T., Courtney, K. D., Rottner, K., Hahne, P., *et al.*, 2001. The Abl interactor proteins localize to sites of actin polymerization at the tips of lamellipodia and filopodia. *Curr. Biol.* **11**: 891–895.

Suetsugu, S., Miki, H. and Takenawa, T., 1999. Identification of two human WAVE/SCAR homologues as general actin regulatory molecules which associate with the Arp2/3 complex. *Biochem. Biophys. Res. Commun.* **260**: 296–302.

Suzuki, T., Nishiyama, K., Yamamoto, A., Inazawa, J., *et al.*, 2000. Molecular cloning of a novel apoptosis-related gene, human Nap1 (NCKAP1), and its possible relation to Alzheimer disease. *Genomics* **63**: 246–254.

Symons, M., Derry, J. M., Karlak, B., Jiang, S., *et al.*, 1996. Wiskott–Aldrich syndrome protein, a novel effector for the GTPase CDC42Hs, is implicated in actin polymerization. *Cell* **84**: 723–734.

Tilney, L. G., Bonder, E. M., Coluccio, L. M. and Mooseker, M. S., 1983. Actin from Thyone sperm assembles on only one end of an actin filament: a behavior regulated by profilin. *J. Cell Biol.* **97**: 112–124.

Welch, M. D., Iwamatsu, A. and Mitchison, T. J., 1997. Actin polymerization is induced by Arp2/3 protein complex at the surface of *Listeria monocytogenes. Nature* **385**: 265–269.

Westphal, R. S., Soderling, S. H., Alto, N. M., Langeberg, L. K. and Scott, J. D., 2000. Scar/WAVE-1, a Wiskott–Aldrich syndrome protein, assembles an actin-associated multi-kinase scaffold. *EMBO J.* **19**: 4589–4600.

Wilkins, A. and Insall, R. H., 2001. Small GTPases in *Dictyostelium*: lessons from a social amoeba. *Trends Genet.* **17**: 41–48.

Witke, W., Podtelejnikov, A. V., Di Nardo, A., Sutherland, J. D., *et al.*, 1998. In mouse brain profilin I and profilin II associate with regulators of the endocytic pathway and actin assembly. *EMBO J.* **17**: 967–976.

Yamamoto, A., Suzuki, T. and Sakaki, Y., 2001. Isolation of hNap1BP which interacts with human Nap1 (NCKAP1) whose expression is down-regulated in Alzheimer's disease. *Gene* **271**: 159–169.

Zalevsky, J., Grigorova, I. and Mullins, R. D., 2001a. Activation of the Arp2/3 complex by the *Listeria* acta protein. Acta binds two actin monomers and three subunits of the Arp2/3 complex. *J. Biol. Chem.* **276**: 3468–3475.

Zalevsky, J., Lempert, L., Kranitz, H. and Mullins, R. D., 2001b. Different WASP family proteins stimulate different Arp2/3 complex-dependent actin-nucleating activities. *Curr. Biol.* **11**: 1903–1913.

Zallen, J. A., Cohen, Y., Hudson, A. M., Cooley, L., *et al.*, 2002. SCAR is a primary regulator of Arp2/3-dependent morphological events in *Drosophila. J. Cell Biol.* **156**: 689–701.

17

Directional Sensing: Subcellular Targeting of GPCR Downstream Effectors During Chemotaxis

Satoru Funamoto and **Richard A. Firtel**

Directional cell movement mediated by an extracellular substance is called chemotaxis, which is involved in various biological aspects such as wound healing, metastasis in cancer, embryogenesis, morphogenesis and axonal guidance. The formation of cell polarity is an initial event in chemotaxis. Cells are able to polarize and to move up a gradient of chemoattractant as low as 2%, indicating that cells must amplify this extracellular gradient to produce an asymmetrical distribution of regulatory information inside the cell. Phosphoinositides (D3-PI) such as phosphatidylinositol(3,4,5)-trisphosphate [$PI(3,4,5)P_3$] and phosphatidylinositol(3,4)-bisphosphate [$PI(3,4)P_2$], products of phosphatidylinositol 3-kinases (PI3Ks), mediate the directional sensing and the downstream establishment and maintenance of cell polarity. Cells in a chemoattractant gradient locally accumulate $PI(3,4,5)P_3/PI(3,4)P_2$ at the leading edge, which results in the binding of a subfamily of cytosolic pleckstrin homology (PH) domain-containing proteins to these lipid-enriched membrane domains and the localized assembly of the actin machinery. Recently, we and others defined the mechanisms controlling the spatial and temporal regulation of the PI3Ks and the negative regulator, the phosphatidylinositol 3′ phosphatase PTEN. In additional studies, we discovered that the *Dictyostelium* MAP kinase kinase MEK1, which plays a key role in the establishment of cell polarity and chemotaxis, also preferentially localizes to the leading edge. This

Cell Motility. Edited by Anne Ridley, Michelle Peckham and Peter Clark.
 ISBN 0 470 84872 3

pathway may be independent of the PI3K pathway and involves the chemoattractant-mediated SUMOylation of MEK1 and the downstream MAP kinase ERK1. Together, these findings shed light on the mechanisms regulating directional sensing and how these control the establishment of cell polarity and chemotaxis.

Introduction

Chemotaxis, or directional cell movement up a chemical gradient, is a basic property of eukaryotic cells, and is an essential component of a wide variety of cellular processes, including inflammation, wound healing, embryogenesis and morphogenesis, in diverse cell types such as neutrophils, macrophages and *Dictyostelium* cells (Devreotes and Zigmond, 1988; Firtel and Chung, 2000; Rickert *et al.*, 2000). A chemoattractant gradient causes asymmetric polymerization of filamentous actin (F-actin), leading to the protrusion of the plasma membrane towards the highest concentration of chemoattractant (leading edge), while myosin II-based contraction occurs in the posterior of the cell body (uropod; Chung *et al.*, 2001a). During these processes, cells move up the gradient. It is impressive that cells are able to sense and respond to gradients as shallow as 2%, which indicates there must be an asymmetric distribution of intracellular components that amplifies the external chemoattractant gradient. In *Dictyostelium*, as well as neutrophils, the chemoattractant receptors and the coupled heterotrimeric G proteins are uniformly localized along the cell surface or found in a very shallow gradient that cannot account for the signal amplification (Xiao *et al.*, 1997; Servant *et al.*, 1999; Jin *et al.*, 2000; Janetopoulos *et al.*, 2001). Recent studies that employ GFP fusions of a variety of signalling proteins, combined with genetic studies of genes required for directional sensing, have yielded new insights into how chemotaxis is regulated. Here, we review our current understanding of establishment of the internal signalling gradient in chemotaxis.

GPCR-mediated lipid signalling in chemotaxis of amoeboid cells

Chemotaxis requires multiple, well-organized steps (Chung *et al.*, 2001a; Katanaev, 2001; Parent and Devreotes, 1999; Rickert *et al.*, 2000): (1) binding of the chemoattractant to cell-surface G protein-coupled receptors (GPCRs) in the case of *Dictyostelium* cells and leukocytes; (2) amplifying an asymmetric intracellular signal and the establishment of cell polarity; (3) the formation of the leading edge and protrusion of the anterior part of cell (the leading edge is rich in F-actin controlled by small GTPases of the Rho family); and (4) myosin II-mediated contraction of the posterior part of the cell body. Chemoattractants, such as formyl-Met-Leu-Phe (f-MLP) in neutrophils and cyclic adenosine monophosphate (cAMP) in *Dictyostelium*, trigger signalling by activating specific GPCRs leading to the dissociation of the Gα and G$\beta\gamma$ subunits of the coupled heterotrimeric G protein.

The first insights into the establishment of an intracellular signalling gradient came from experiments in *Dictyostelium* in which a subfamily of PH domains that preferentially bind PI(3,4)P_2 and PI(3,4,5)P_3, the lipid products of Class I PI3Ks, rapidly and transiently translocate to the plasma membrane in response to a global stimulation of cAMP (a rapid non-spatial rise in chemoattractant so all receptors are uniformly activated) and to the leading edge in chemotaxing cells (Figure 17.1A,B; Parent *et al.*, 1998; Meili *et al.*, 1999; Funamoto *et al.*, 2001). These studies were subsequently extended to neutrophils (Servant *et al.*, 2000). This translocation is not dependent on the actin cytoskeleton, as it occurs in the presence of the actin polymerization inhibitor latrunculin A (Parent *et al.*, 1998).

Mechanisms controlling PH domain localization and the role of PI3K in chemotaxis

Binding of a chemoattractant to a GPCR results in recruitment of PH domain-containing proteins to the plasma membrane. These proteins include CRAC, a cytosolic regulator of adenylyl cyclase, the protein kinase Akt/PKB, and the protein PhdA (Parent *et al.*, 1998; Meili *et al.*, 1999; Funamoto *et al.*, 2001). As these PH domains bind to the D3-PIs (phosphatidyl inositols phosphorylated on the 3′ position of the inositol ring) PI(3,4)P_2 and PI(3,4,5)P_3, PH domain translocation in living cells suggests rapid and transient accumulation of PI(3,4)P_2/PI(3,4,5)P_3 on the membrane, while the asymmetric localization of the PH domains in chemotaxing cells reflects the preferential accumulation of D3-PIs at the leading edge.

These results suggested that PH domain localization is dependent on the activation of PI3Ks. This was confirmed by demonstrating that PH domain localization in *Dictyostelium* cells is abrogated in cells in which the genes encoding two of the three Class I PI3Ks, PI3K1 and PI3K2, are disrupted (*pi3k1/2* null cells) or by treating wild-type cells with the PI3K inhibitor LY294002 (Chung *et al.*, 2001b, Funamoto *et al.*, 2001). In mammalian cells, the Class 1B PI3Kγ (p110γ) is activated downstream of the fMLP and C5a chemoattractant receptors (Stephens *et al.*, 1994; Stoyanov *et al.*, 1995). In the case of PI3Kγ, the p110γ/p101 holoenzyme (p110γ catalytic subunit and the adaptor protein p101) exhibit an increased lipid kinase activity in response to binding by the G$\beta\gamma$ subunit of the coupled heterotrimeric G protein,

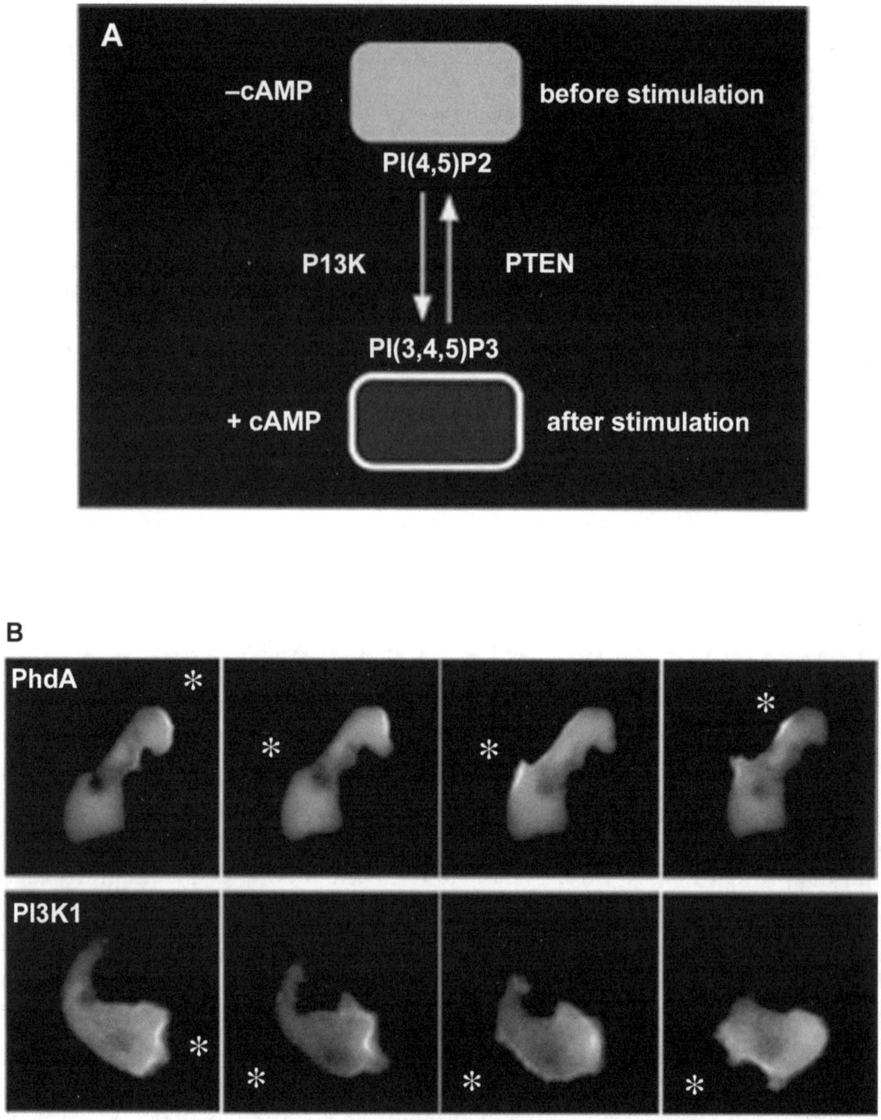

Figure 17.1 PH domain translocation regulated by PI3K and PTEN. (A) The PH domain distributes uniformly in the cytosol before stimulation. After global stimulation with the chemoattractant cAMP, PI3K phosphorylates the 3 position of the inositol ring of PI(4,5)P_2, producing PI(3,4,5)P_3 on the membrane. The PH domain binds to the PI(3,4,5)P_3 specifically and translocates to the membrane. PTEN dephosphorylates PI(3,4,5)P_3 and produces PI(4,5)P_2. (B) Cells were transformed with either the PhdA-GFP or the PI3K1-GFP fusion construct. Translocation of PhdA and PI3K1 is shown in upper and lower panels, respectively. Asterisks indicate the position of a micropipette filled with the chemoattractant cAMP. As the chemoattractant source moves, PhdA and PI3K1 localize to the new leading edge of the cell

suggesting that the $G\beta\gamma$ plays a central role in activating the response downstream from mammalian GPCR chemoattractant receptors (Stephens *et al.*, 1997; Krugmann *et al.*, 1999). Genetic disruption of p110γ and treatment of macrophages or neutrophils with PI3K inhibitors LY294002/Wortmannin have been used to demonstrate PH domain translocation is also PI3K-dependent in mammalian cells. Macrophages and neutrophils isolated from p110γ-null mice show no accumulation of $PI(3,4,5)P_3$ after chemoattractant stimulation (Hirsch *et al.*, 2000; Hannigan *et al.*, 2002; Li *et al.*, 2000; Rickert *et al.*, 2000; Sasaki *et al.*, 2000; Stephens *et al.*, 2002; Vanhaesebroeck *et al.*, 1999).

PH domain localization and chemotaxis efficiency were linked through the examination of chemotaxis of neutrophils, macrophages and *Dictyostelium* cells in which PI3K activity was abrogated by creation of PI3K gene knockouts, through the injection of PI3K isoform-specific antibodies and treatment of cells with PI3K inhibitors. *Dictyostelium pi3k1/2* null cells or cells treated with LY294002 exhibit a loss of cell polarity and defective cell migration (Chung *et al.*, 2001a,b; Meili *et al.*, 1999; Funamoto *et al.*, 2001). Activation of Akt/PKB is abolished, translocations of PH domains are abrogated in response to global stimulation, and PH domains do not localize to the leading edge. Mammalian neutrophils and macrophages experience a similar loss of chemotaxis efficiency, Akt activation, and PH domain localization (Hirsch *et al.*, 2000; Hannigan *et al.*, 2002; Li *et al.*, 2000; Rickert *et al.*, 2000; Sasaki *et al.*, 2000; Stephens *et al.*, 2002; Vanhaesebroeck *et al.*, 1999). Those observations indicate that localized activation of PI3K recruits the PH domain to the plasma membrane, leading to actin assembly, cell polarization and directional movement.

PI3K translocates upon stimulation with a chemoattractant

In order to investigate the basis of PH domain localization at the leading edge, we used GFP fusions of PI3K1 and PI3K2 in *Dictyostelium*. We showed that PI3K1-GFP or PI3K2-GFP rapidly and transiently translocates to the plasma membrane upon stimulation by a chemoattractant (Funamoto *et al.*, 2002). We discovered rapid accumulation of PI3K1-GFP and PI3K2-GFP to a new leading edge in response to movement of the chemoattractant source (Figure 17.1B). PI3K1 and PI3K2, like mammalian Class I PI3Ks, possess a Ras binding domain, a C2 domain, and a lipid kinase and associated domains. Surprisingly none of these domains are important for membrane translocation. Expression of truncated mutants of PI3K1 revealed that the N-terminal region (1–492) is necessary and sufficient for PI3K translocation to the membrane. This N-terminal region is unique in the protein database,

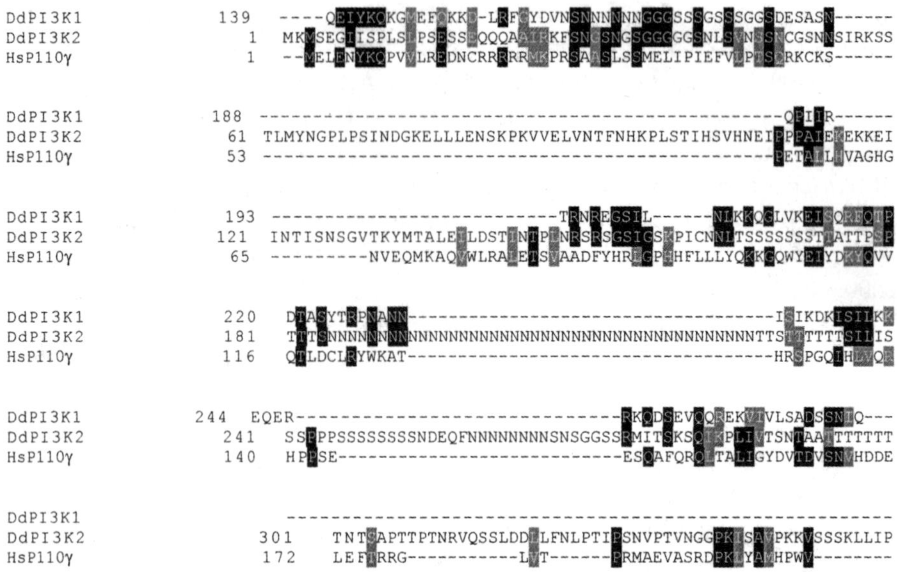

Figure 17.2 ClustalW sequence alignment of N-terminal domains of DdPI3K1, *Dictyostelium* PI3K1 (GenBank accession #P54673); DdPI3K2, *Dictyostelium* PI3K2 (accession #P54674); and HsPI3Kγ, human PI3Kγ (accession #P48736). The N-terminal domain of PI3K1 starts from 139 and ends at 271. This is the minimum region responsible for the translocation of PI3K1 (S. Funamoto and R. A. Firtel, unpublished observation)

but slightly conserved between PI3K1, PI3K2, and p110γ (Figure 17.2). The N-terminal region translocates in *pi3k1/2* knockout cells and wild-type cells treated with LY294002, suggesting that PI3K translocation is independent of its kinase activity. Expression of the N-terminal deleted mutant of PI3K1 (Δ2-492) failed to complement the phenotype of the *pi3k1/2* knockout mutant. By co-expression of PI3K1-CFP with PhdA-YFP as a D3-PI probe in living cells, it was found that PI3K1 translocates with kinetics similar to those of PH domain translocation (Funamoto *et al.*, 2002). The data suggest that targeting of PI3Ks to the membrane by the novel N-terminal domain is one of the important steps for its activation.

Constitutive localization of PI3K1 on the membrane through the use of an N-terminal myristoyl tag from Src did not cause constitutive activation of downstream PI3K1 effector pathways, suggesting that PI3K localization and activation are independent events. A point mutation in the PI3K1 Ras-binding domain blocked PI3K activation, suggesting that PI3K interaction with an active Ras protein through the Ras-binding domain is a key event to activate PI3K1 localized on the membrane (Pacold *et al.*, 2000).

The phenotype of *pi3k1/2* null cells expressing myr-PI3K1 provided new insights into the mechanisms controlling leading edge function. The analysis of *pi3k1/2* null cells demonstrated that PI3K is required for proper chemotaxis and cell polarity but did not provide direct evidence that PI3K plays an instructional role in controlling leading edge formation. Expression of myr-PI3K1 in *pi3k1/2* null cells leads to the formation of multiple lateral pseudopodia when the cells are placed in a chemoattractant gradient, indicating that activation of PI3K localized uniformly along the cell's membrane can induce F-actin polymerization. Evidence that PI3K is activated along the cell's periphery is provided by the localization of PH domain-containing proteins at the sites of the multiple pseudopodia. In contrast to other mutants that result in multiple pseudopodia such as *paka* null cells (Chung *et al.*, 2001b), PH domain-containing proteins also localize to the part of the cell opposite the chemoattractant source (cell's 'posterior'). Interestingly, these cells still move directionally towards the chemoattractant source, although less efficiently than wild-type cells, suggesting that these cells must have other mechanisms by which they sense the chemoattractant gradient. One mechanism could be PI3K3, which is still present in these cells and may localize to the leading edge, as do PI3K1 and PI3K2. MEK1 and ERK1, components of a MAP kinase pathway required for proper chemotaxis, also localize to the leading edge (Ma *et al.*, 1997; Sobko *et al.*, 2002).

PTEN as a negative regulator of the D3-PI signalling pathway in chemotaxis

PTEN, originally identified as a putative protein tyrosine phosphatase and a tumour suppressor gene from human brain, breast and prostate cancer cells (Li *et al.*, 1997; Steck *et al.*, 1997), is a 3′-specific phosphatidylinositol phosphatase (Maehama and Dixon, 1998). Overexpression of PTEN blocks the increase of PI(3,4,5)P_3 upon stimulation of insulin in glioma cells lacking PTEN, and Akt/PKB is hyperactivated and the D3-PI levels are elevated (Haas-Kogan *et al.*, 1998; Maehama and Dixon, 1998; Maehama *et al.*, 2001).

Identification of the mechanisms necessary to antagonize and down-regulate the D3-PI signalling pathway is very important to understand the regulatory control of chemotaxis. Recently, we and others demonstrated the involvement of PTEN in chemotaxis by controlling the levels of D3-PI in *Dictyostelium* (Iijima and Devreotes, 2002; Funamoto *et al.*, 2002; Comer and Parent, 2002). In the wild-type strain of *Dictyostelium*, the peak of PH domain accumulation to the membrane is $\sim$6 s after stimulation, whereupon the PH domain dissociates from the membrane within $\sim$20 s. In cells with decreased levels of PTEN function (*pten* null cells or PTEN hypomorphs), a substantial

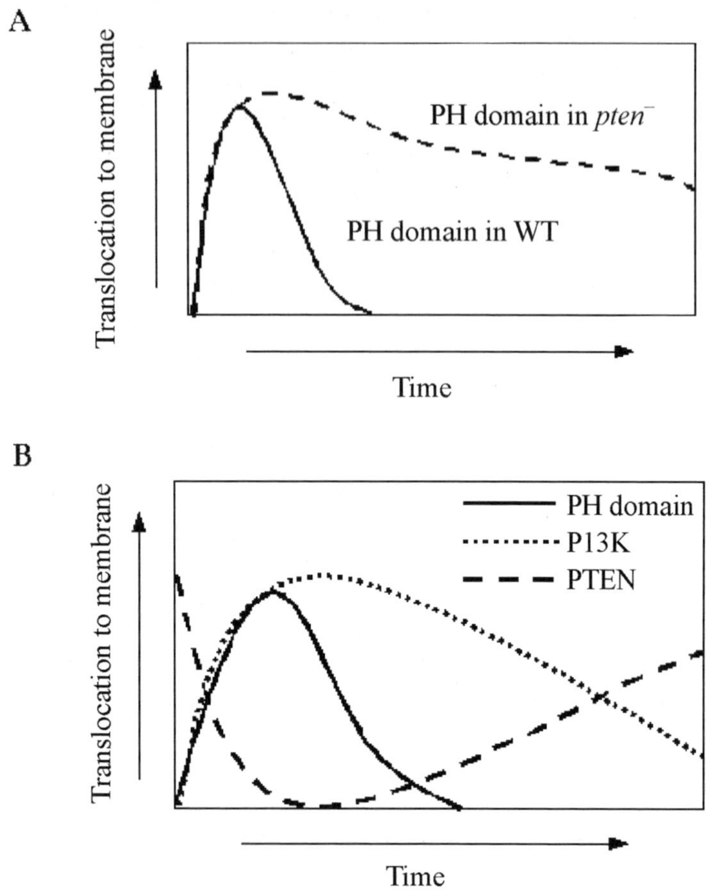

Figure 17.3 Translocation kinetics of the PH domain, PI3K, and PTEN. (A) The wild-type peak of translocation of the PH domain is 6 s, and the PH domain disassociates from the membrane within 20–30 s. In *pten* null cells, a prolonged PH domain is observed (> 3 min). (B) The translocations of PI3K and PTEN are mirror images of each other. After global stimulation, PI3K and the PH domain translocate to the membrane. It is notable that PI3K stays longer on the membrane than the PH domain does. For technical reasons, it is difficult to determine which goes to membrane first. On the other hand, PTEN delocalizes from the membrane as PI3K jumps to the membrane

amount of the PH domain remains associated with the membrane even after 3 min (Figure 17.3A). These cells also exhibit a prolonged and increased level of actin polymerization (Iijima and Devreotes, 2002) and increased activation of Akt/PKB (Funamoto *et al.*, 2002), as well as a chemotaxis defect. Interestingly, *pten* null cells are hypersensitive to stimulation by cAMP (10 nM cAMP compared with 1 nM cAMP for the wild-type; Iijima and Devreotes, 2002). This finding suggests that input from receptors is usually down-regulated by the D3-PI phosphatase activity of PTEN and this activity is essential for chemotaxis. Interestingly, *pten* null cells exhibit chemotaxis defects and properties of PH domain localization similar to those of *pi3k1/2* null cells expressing myr-PI3K1 (Iijima and Devreotes, 2002; Funamoto *et al.*, 2002).

PTEN-GFP is localized to the plasma membrane in resting cells. Surprisingly, PTEN delocalizes from the membrane upon chemoattractant stimulation within a few seconds and relocalizes in 30–60 s, kinetics similar to the localization and delocalization of PI3K (Iijima and Devreotes, 2002; Funamoto *et al.*, 2002). This dynamism of PTEN is a mirror image of the translocation of PH domains and PI3Ks (Figure 17.3B). In chemotaxing cells, PTEN localizes to the lateral and posterior parts of migrating cells but not at the leading edge. The localization of PI3K to the leading edge and the delocalization of PTEN from the leading edge provide a mechanism for sharpening the PI3K signal to the leading edge, thus restricting the localization of downstream effector pathways to the very front of the cells.

Chemotaxis regulated by MEK kinase signalling

MAP kinase cascades are activated by cell surface receptors and regulate various intracellular responses, such as cell growth, differentiation and cell viability (Bokemeyer *et al.*, 1996; Herskowitz, 1995; Levin and Errede, 1995; Marshall, 1995; Segall *et al.*, 1995). In *Dictyostelium*, two MAP kinase cascades have been identified, one of which is required for proper chemotaxis toward cAMP (Gaskins *et al.*, 1994; Ma *et al.*, 1997). Genetic deletion of the MAP kinase ERK1 or its upstream activator MEK1 results in chemotaxis and developmental defects (Gaskins *et al.*, 1994; Sobko *et al.*, 2002). ERK1 is not activated in *mek*1 null cells, indicating that ERK1 and MEK1 are in the same regulatory pathway (Ma *et al.*, 1997; Sobko *et al.*, 2002). Both *erk1* and *mek1* null strains produce very small aggregates. In single-cell chemotaxis assays using a micropipette to provide a directional chemoattractant gradient, the cells are very poorly polarized and produce multiple lateral pseudopodia simultaneously. The constitutively active form of MEK1 ($MEK1^{S444E,T448E}$) complements the null aggregation defect, although the aggregates arrest at the mound stage and are unable to undergo morphogenesis. $MEK1^{S444A,T448A}$, which cannot be phosphorylated and thus cannot be activated, fails to rescue the null phenotype and functions as a dominant negative form when expressed in the wild-type cells. Fibroblasts from $mek1^{-/-}$ or $mekk^{-/-}$ mice exhibit defects in cell migration, suggesting that the involvement of the MAP kinase kinase pathway in migration could be evolutionarily conserved from *Dictyostelium* to mammals (Giroux *et al.*, 1999; Yujiri *et al.*, 2000).

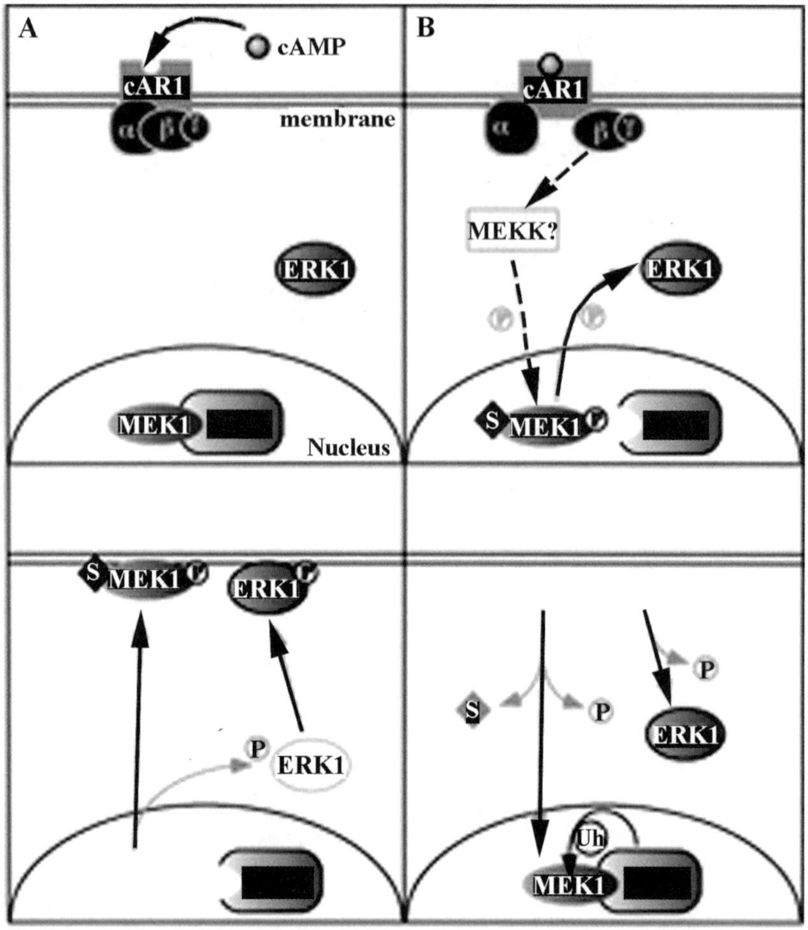

Figure 17.4 Translocation of MEK1 and ERK1 regulated by MEK1 SUMOylation. (A) MEK1 is concentrated in the nucleus, and ERK1 localizes in cytoplasm rather than the nucleus in resting cells. (B) Stimulation of GPCR by cAMP causes activation of putative MEKK and phosphorylation/SUMOylation of MEK1, which activates MEK1 and phosphorylates ERK1. (C) Recruitment of MEK1 and ERK1 to the membrane with the same kinetics as SUMOylation of MEK1. (D) MEK1 is released from the membrane by de-SUMOylation and ubiquitinated by MIP1

A small ubiquitin-related modifier, SUMO, is structurally similar to ubiquitin and covalently attached to a lysine residue on the substrate protein. Modification of the substrate by SUMO (SUMOylation) can control subcellular localization of the substrate protein and protect it from ubiquitination and proteasomal degradation by competitive attachment to the same lysine residues (Buschmann *et al.*, 2000; Muller *et al.*, 2001).

Recently, we demonstrated that the *Dictyostelium* MEK1 is transiently SUMOylated in response to stimulation by the chemoattractant cAMP (Sobko *et al.*, 2002), suggesting that SUMOylation may play a role in regulating MEK1.

We examined the subcellular localization of ERK1 and MEK1 by using epitope tags. In resting cells, MEK1 is concentrated in the nucleus, while ERK1 localizes in the cytoplasm (Figure 17.4A). Upon cAMP stimulation, MEK1 and ERK1 localize to the plasma membrane and are found at the leading edge of chemotaxing cells, with kinetics similar to those observed for PH domain-containing proteins and PI3K (Figure 17.4C). The kinetics of MEK1 translocation are indistinguishable from those of MEK1 SUMOylation of MEK1. Mutational analysis identified $MEK1^{K105}$ as the site of SUMOylation. $MEK1^{K105R}$, which carries a K-to-R substitution at the site of SUMOylation, remains in the nucleus. Constitutively active MEK1 ($MEK1^{S444E,T448E}$) is constitutively SUMOylated, constitutively cytosolic and partially membrane-associated. In contrast, $MEK1^{S444A,T448A}$ is not SUMOylated and remains nuclear after stimulation. These data suggest that MEK1 translocation requires SUMOylation and that MEK1 activation may be the direct signal that leads to MEK1 SUMOylation (Figure 17.4B). MEK1 is rapidly dephosphorylated and de-SUMOylated, whereupon it relocalizes to the nucleus and associates with MIP1, a MEK1-interacting protein identified via a yeast two-hybrid screen. MIP1 functions in nuclear sequestration of MEK1 in unstimulated cells. MEK1 is also ubiquitinated. MIP1 has a RING domain and possesses ubiquitin E3 ligase activity. Mutational analysis indicated that MIP1 mediates MEK1 ubiquitination. It is possible that MEK1 ubiquitination, which is highest in time points after MEK1 relocalizes to the nucleus, is a further mechanism of pathway down-regulation and may mediate MEK1 degradation, although this model has not been proven (Figure 17.4D). SUMOylation consensus sites are found in MEKs from a wide variety of organisms, suggesting that SUMOylation of MAP kinase kinases could be conserved and represent an ancient mechanism controlling subcellular localization of the cascade.

The initial asymmetric signal and downstream asymmetric signals

PH domain translocation to the leading edge, an initial event to establish cell polarity, is regulated by spatial and temporal activation of its upstream kinase, PI3K, and the phosphatase PTEN. PI3K jumps to the leading edge, whereas PTEN delocalizes from the edge after stimulation. What makes PI3K/PTEN localization polarized? Interestingly, the PH domain and PI3Ks do not translocate in the *gβ*-null cell line *in vivo* or *in vitro* (S. Funamoto and R. A. Firtel, unpublished observations). These observations suggest that the translocation is dependent on the Gβ subunit, possibly by directly binding to PI3K or an associated protein, as reported for PI3Kγ in mammalian cells. However, it is unlikely that binding to Gβ is responsible for the steep gradient of membrane localization of PI3K, as the anterior–posterior gradient of Gβ is relatively shallow. Moreover, fluorescence resonance energy transfer (FRET) technology revealed that G protein subunits remain disassociated as long as receptors are occupied, which removes the possibility of asymmetric localization of activated G proteins (Janetopoulos *et al.*, 2001). We are presently missing the polarized signal upstream of PI3K/PTEN.

References

Bokemeyer, D., Sorokin, A. and Du, M., 1996. Multiple intracelluar MAP kinase signaling cascades. *Kidney Int.* **49**: 1187–1198.

Buschmann, T., Fuchs, S., Lee, C., Pan, Z. and Ronai, Z., 2000. SUMO-1 modification of Mdm2 prevents its self-ubiquitination and increases Mdm2 ability to ubiquitinate p53. *Cell* **101**: 753–762.

Chung, C. Y., Funamoto, S. and Firtel, R. A., 2001a. Signaling pathways controlling cell polarity and chemotaxis. *Trends Biochem. Sci.* **26**: 557–566.

Chung, C. Y., Potikyan, G. and Firtel, R. A., 2001b. Control of cell polarity and chemotaxis by Akt/PKB and PI3 kinase through the regulation of PAKa. *Mol. Cell* **7**: 937–947.

Comer, F. I. and Parent, C. A., 2002. PI 2-kinases and PTEN: How opposites chemoattract. *Cell* **109**: 541–544.

Devreotes, P. N. and Zigmond, S. H., 1988. Chemotaxis in eukaryotic cells: a focus on leukocytes and *Dictyostelium*. *Annu. Rev. Cell Biol.* **4**: 649–686.

Dowler, S., Currie, R. A., Downes, C. P. and Alessi, D. R., 1999. DAPP1: a dual adaptor for phosphotyrosine and 3-phosphoinositides. *Biochem. J.* **342**: 7–12.

Dowler, S., Currie, R. A., Campbell, D. G., Deak, M., *et al.*, 2000. Identification of pleckstrin-homology-domain-containing proteins with novel phosphoinositide-binding specificities. *Biochem. J.* **351**: 19–31.

Escobedo, J. A., Navankasattusas, S., Kavanaugh, W. M., Milfay, D., *et al.*, 1991. cDNA cloning of a novel 85 kd protein that has SH2 domains and regulates binding of PI3-kinase to the PDGF beta-receptor. *Cell* **65**: 75–82.

Firtel, R. A. and Chung, C. Y., 2000. The molecular genetics of chemotaxis: sensing and responding to chemoattractant gradients. *BioEssays* **22**: 603–615.

Funamoto, S., Milan, K., Meili, R. and Firtel, R. A., 2001. Role of phosphatidylinositol 3′ kinase and a downstream pleckstrin homology domain-containing protein in controlling chemotaxis in *Dictyostelium*. *J. Cell Biol.* **153**: 795–810.

Funamoto, S., Meili, R., Lee, S., Parry. L. and Firtel, R. A., 2002. Spatial and temporal regulation of 3-phosphoinositides by PI 3-kinase and PTEN mediates chemotaxis. *Cell* **109**: 611–623.

Gaskins, C., Clark, A. M., Aubry, L., Segall, J. E. and Firtel, R. A., 1994. The *Dictyostelium* MAP kinase ERK2 regulates multiple, independent developmental pathways. *Genes Dev.* **10**: 118–128.

Giroux, S., Tremblay, M., Bernard, D., Cardin-Girard, J., *et al.*, 1999. Embryonic death of Mek1-deficient mice reveals a role for this kinase in angiogenesis in the labyrinthine region of the placenta. *Curr. Biol.* **9**: 369–372.

Hannigan, M., Zhan, L., Li, Z., Ai, Y., *et al.*, 2002. Neutrophils lacking phosphoinositide 3-kinase gamma show loss of directionality during N-formyl-Met-Leu-Phe-induced chemotaxis. *Proc. Natl. Acad. Sci. USA* **99**: 3603–3608.

Herskowitz, I., 1995. MAP kinase pathways in yeast: for mating and more. *Cell* **80**: 187–197.

Hiles, I. D., Otsu, M., Volinia, S., Fry, M. J., *et al.*, 1992. Phosphatidylinositol 3-kinase: structure and expression of the 110 kd catalytic subunit. *Cell* **70**: 419–429.

Hirsch, E., Katanaev, V. L., Garlanda, C., Azzolino, O., *et al.*, 2000. Central role for G protein-coupled phosphoinositide 3-kinase in inflammation. *Science* **287**: 1049–1053.

Hurley, J. H. and Meyer, T., 2001. Subcellular targeting by membrane lipids. *Curr. Opin. Cell Biol.* **13**: 146–152.

Iijima, M. and Devreotes, P., 2002. Tumor suppressor PTEN mediates sensing of chemoattractant gradients. *Cell* **109**: 599–610.

Isakoff, S. J., Cardozo, T., Andreev, J., Li, Z., *et al.*, 1998. Identification and analysis of PH domain-containing targets of phosphatidylinositol 3-kinase using a novel *in vivo* assay in yeast. *EMBO J.* **17**: 5374–5387.

Janetopoulos, C., Jin, T. and Devreotes, P. N., 2001. Receptor-mediated activation of heterotrimeric G proteins in living cells. *Science* **291**: 2408–2411.

Jin, T., Zhang, N., Long, Y., Parent, C. A. and Devreotes, P. N., 2000. Localization of the G protein $\beta\gamma$ complex in living cells during chemotaxis. *Science* **287**: 1034–1036.

Katanaev, V. L., 2001. Signal transduction in neutrophil chemotaxis. *Biochemistry* **66**: 351–368.

Klippel, A., Reinhard, C., Kavanaugh, W. M., Apell, G., *et al.*, 1996. Membrane localization of phosphatidylinositol 3-kinase is sufficient to activate multiple signal-transducing kinase pathways. *Mol. Cell Biol.* **16**: 4117–4127.

Krugmann, S., Hawkins, P. T., Pryer, N. and Braselmann, S., 1999. Characterizing the interactions between the two subunits of the p101/p110gamma phosphoinositide 3-kinase and their role in the activation of this enzyme by G beta gamma subunits. *J. Biol. Chem.* **274**: 17 152–17 158.

Levin, D. E. and Errede, B., 1995. The proliferation of MAP kinase signaling pathways in yeast. *Curr. Opin. Cell Biol.* **7**: 197–202.

Li, J., Yen, C., Liaw, D., Podsypanina, K., *et al.*, 1997. PTEN, a putative protein tyrosine phosphatase gene mutated in human brain, breast, and prostate cancer. *Science* **275**: 1943–1947.

Li, Z., Jiang, H., Xie, W., Zhang, Z., *et al.*, 2000. Roles of PLC-b2 and -b3 and PI3Kg in chemoattractant-mediated signal transduction. *Science* **287**: 1046–1049.

Ma, H., Gamper, M., Parent, C. and Firtel, R., 1997. The *Dictyostelium* MAP kinase kinase DdMEK1 regulates chemotaxis and is essential for chemoattractant-mediated activation of guanyly cyclase. *EMBO J.* **16**: 4317–4332.

Maehama, T. and Dixon, J. E., 1998. The tumor suppressor, PTEN/MMAC1, dephosphorylates the lipid second messenger, phosphatidylinositol 3,4,5-triphosphate. *J. Biol. Chem.* **273**: 13 375–13 378.

Maehama, T., Taylor, G. S. and Dixon, J. E., 2001. PTEN and myotubularin: novel phosphoinositide phosphatases. *Annu. Rev. Biochem.* **70**: 247–279.

Marshall, C. J., 1995. Specificity of receptor tyrosine kinse signaling: transient versus sustained extracellular signal-regulated kinase activation. *Cell* **80**: 179–185.

Meili, R., Ellsworth, C., Lee, S., Reddy, T. B., Ma, H. and Firtel, R. A., 1999. Chemoattractant-mediated transient activation and membrane localization of Akt/PKB is required for efficient chemotaxis to cAMP in *Dictyostelium*. *EMBO J.* **18**: 2092–2105.

Muller, S., Hoege, C., Pyrowolakis, G. and Jentsch, S., 2001. SUMO, ubiquitin's mysterious cousin. *Nat. Rev. Mol. Cell Biol.* **2**: 202–210.

Otsu, M., Hiles, I., Gout, I., Fry, M. J., *et al.*, 1991. Characterization of two 85 kd proteins that associate with receptor tyrosine kinases, middle-T/pp60c-src complexes, and PI3-kinase. *Cell* **65**: 91–104.

Pacold, M. E., Suire, S., Perisic, O., Lara-Gonzalez, S., *et al.*, 2000. Crystal structure and functional analysis of Ras binding to its effector phosphoinositide 3-kinase γ. *Cell* **103**: 931–943.

Parent, C. A. and Devreotes, P. N., 1999. A cell's sense of direction. *Science* **284**: 765–770.

Parent, C. A., Blacklock, B. J., Froehlich, W. M., Murphy, D. B. and Devreotes, P. N., 1998. G protein signaling events are activated at the leading edge of chemotactic cells. *Cell* **95**: 81–91.

Rickert, P., Weiner, O., Wang, F., Bourne, H. and Servant, G., 2000. Leukocytes navigate by compass: roles of PI3Kg and its lipid products. *Trends Cell Biol.* **10**: 466–473.

Sasaki, T., Irie-Sasaki, J., Horie, Y., Bachmaier, K., *et al.*, 2000. Colorectal carcinomas in mice lacking the catalytic subunit of PI(3)Kgamma. *Nature* **406**: 897–902.

Segall, J. E., Kuspa, A., Shaulsky, G., Ecke, M., *et al.*, 1995. A MAP kinase necessary for receptor-mediated acctivation of adenylyl cyclase in *Dictyostelium*. *J. Cell Biol.* **128**: 405–413.

Servant, G., Weiner, O. D., Neptune, E. R., Sedat, J. W. and Bourne, H. R., 1999. Dynamics of a chemoattractant receptor in living neutrophils during chemotaxis. *Mol. Biol. Cell* **10**: 1163–1178.

Servant, G., Weiner, O. D., Neptune, E. R., Sedat, J. W. and Bourne, H. R., 2000. Polarization of chemoattractant receptor signaling during neutrophil chemotaxis. *Science* **287**: 1037–1040.

Sobko, A., Ma, H. and Firtel, R. A., 2002. Regulated SUMOylation and ubiquitination of DdMEK1 is required for proper chemotaxis. *Dev. Cell* **2**: 745–756.

Steck, P. A., Pershouse, M. A., Jasser, S. A., Yung, W. K., *et al.*, 1997. Identification of a candidate tumour suppressor gene, MMAC1, at chromosome 10q23.3 that is mutated in multiple advanced cancers. *Nat. Genet.* **15**: 356–362.

Stephens, L., Smrcka, A., Cooke, F. T., Jackson, T. R., *et al.*, 1994. A novel phosphoinositide 3 kinase activity in myeloid-derived cells is activated by G protein beta gamma subunits. *Cell* **77**: 83–93.

Stephens, L. R., Eguinoa, A., Erdjument-Bromage, H., Lui, M., *et al.*, 1997. The G beta gamma sensitivity of a PI3K is dependent upon a tightly associated adaptor, p101. *Cell* **89**: 105–114.

Stephens, L., Ellson, C. and Hawkins, P., 2002. Roles of PI3Ks in leukocyte chemotaxis and phagocytosis. *Curr. Opin. Cell Biol.* **14**: 203–213.

Stoyanov, B., Volinia, S., Hanck, T., Rubio, I., 1995. Cloning and characterization of a G protein-activated human phosphoinositide-3 kinase. *Science* **269**: 690–693.

Vanhaesebroeck, B., Jones, G. E., Allen, W. E., Zicha, D., *et al.*, 1999. Distinct PI(3)Ks mediate mitogenic signalling and cell migration in macrophages. *Nat. Cell Biol.* **1**: 69–71.

Wymann, M. P. and Pirola, L., 1998. Structure and function of phosphoinositide 3-kinases. *Biochim. Biophys. Acta.* **1436**: 127–150.

Xiao, Z., Zhang, N., Murphy, D. B. and Devreotes, P. N., 1997. Dynamic distribution of chemoattractant receptors in living cells during chemotaxis and persistent stimulation. *J. Cell Biol.* **139**: 365–374.

Yujiri, T., Ware, M., Widmann, C., Oyer, R., *et al.*, 2000. MEK kinase 1 gene disruption alters cell migration and c-Jun NH2-terminal kinase regulation but does not cause a measurable defect in NF-κB activation. *Proc. Natl. Acad. Sci. USA* **97**: 7272–7277.

Zhou, K. M., Takegawa, K., Emr, S. D. and Firtel, R. A., 1995. A phosphatidylinositol (PI) kinase gene family in *Dictyostelium discoideum*: biological roles of putative mammalian p110 and yeast Vps3p PI 3-kinase homologs during growth and development. *Mol. Cell Biol.* **15**: 5654–5656.

18

Cell Crawling, Cell Behaviour and Biomechanics During Convergence and Extension

Ray Keller and **Lance Davidson**

The morphogenic movements of convergence and extension (narrowing and lengthening) of embryonic tissues play major roles in gastrulation, neurulation and body axis formation in vertebrates, and also function in many other morphogenic processes in both vertebrates and invertebrates. These movements are driven by polarized cell behaviours that appear to pull cells between one another along the mediolateral axis and thereby elongate or extend the tissue. Here we discuss the biomechanical relationship between these polarized cell behaviours and the patterns of cell intercalation, during the active, force-producing convergence and extension movements occurring in early *Xenopus* morphogenesis.

Introduction

A major unresolved issue in morphogenesis is how polarized cell motility produces the active, orientated intercalation of cells that drives the convergence and extension movements found throughout the morphogenesis of metazoans. Convergence and extension refer to the coordinated narrowing (convergence) and lengthening (extension) movements of tissues during morphogenesis. Convergence and extension elongate the anterior–posterior

Cell Motility. Edited by Anne Ridley, Michelle Peckham and Peter Clark.
 ISBN 0 470 84872 3

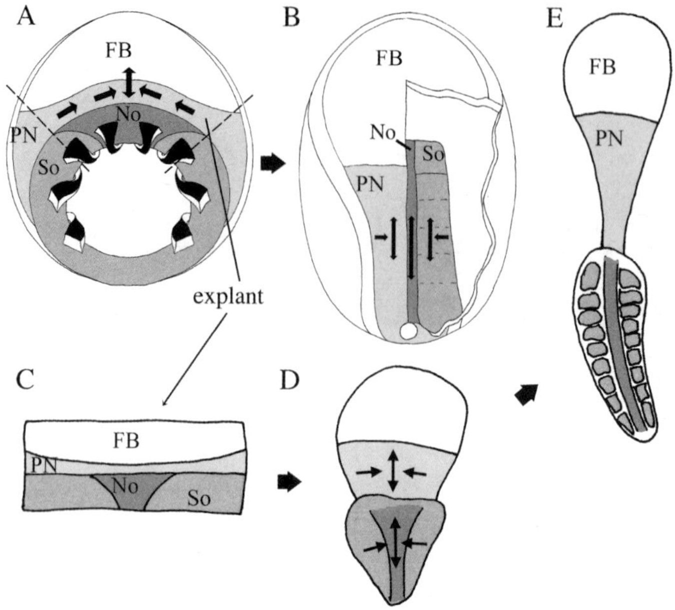

Figure 18.1 Diagrams show how the morphogenic movements of convergence (narrowing) and extension (lengthening) of the presumptive notochordal (No), somitic (So), and posterior neural (PN) tissues contribute to gastrulation and body axis elongation. As the mesodermal tissues involute, they converge and constrict the blastopore (black arrows, A). After involution they continue to converge in the mediolateral direction and extend in the anterior–posterior direction, and thereby elongate the body axis (B, arrows, No, So). Meanwhile, the posterior neural tissue (PN) converges and extends in a congruent manner on the outside (A, B, arrows). These movements are produced by internally generated forces as shown by the fact that explants excised from the early gastrula (A, dashed lines), and sandwiched with their inner surfaces together (C), will also show convergence and extension of both the mesodermal and neural components (C–E). Forebrain: FB

body axis during gastrulation and neurulation of many species of vertebrates, and occurs during many other morphogenic events in both vertebrates and invertebrates (Keller, 2002). One example of these movements occurs during gastrulation of the amphibian *Xenopus laevis*. As the presumptive notochordal and the somitic tissues involute, or turn inside the blastopore, they converge in their mediolateral dimension and extend in their anterior–posterior aspect, which elongates the body axis and at the same time constricts the circumference of the closing blastopore (Figure 18.1A–B). Simultaneously, the overlying prospective hindbrain and spinal cord tissue of the neural plate also converges and extends congruently with the underlying mesoderm but on the outside (Figure 18.1A–B). These movements elongate and narrow the neural plate and also aid the mesoderm in constricting the blastoporal lip (Keller *et al.*, 2000) (Figure 18.1A–B). Convergence and extension also occurs during gastrulation of other amphibians (Vogt, 1929; Shook *et al.*, 2002), teleost fish (Glickman *et al.*, 2003; Trinkaus *et al.*, 1992; Trinkaus, 1998), ascidians (Miyamoto and Crowther, 1985; Munro and Odell, 2002a,b), birds (Schoenwolf and Alvarez, 1989; Schoenwolf *et al.*, 1992; Sausedo and Schoenwolf, 1993; Lawson and Schoenwolf, 2001a,b), and mammals (Sausedo and Schoenwolf, 1994). However, the biomechanical consequences and the morphogenic function of these movements vary among these groups of organisms.

During gastrulation and neurulation of amphibians studied thus far, these movements are active processes, driven by internal forces rather than passive responses to forces developed elsewhere in the embryo. Convergence and extension occur in cultured explants of the anuran (tail-less) amphibians, *Xenopus laevis* (Keller and Danilchik, 1988; Elul *et al.*, 1997) and *Hyla regilla* (Schechtman, 1942), and in the urodele (tailed) amphibians, *Taricha torosus*, *Ambystoma mexicanum* and *Ambystoma maculatum* (Shook *et al.*, 2002) (Figure 18.1C–D). In *Xenopus*, both the mesodermal and neural regions actively converge and extend in explants (Keller and Danilchik, 1988; Elul *et al.*, 1997; Elul and Keller, 2000) and *in vivo* (Wallingford and Harland, 2001). It is not known whether the neural region actively extends in other species.

Convergence and extension in *Xenopus* occur by active movement of cells between one another, a process initially referred to as 'interdigitation' (Keller, 1984) but later changed to the more appropriate 'intercalation' (Keller *et al.*, 1985; Keller, 1986). Two types of cell intercalation occur. First, radial intercalation occurs, a process in which several layers of deep cells intercalate between one another along the radius of the embryo to produce fewer layers of greater length (Figure 18.2A). Then at the midgastrula stage, mediolateral intercalation begins. Mediolateral intercalation is a process in which cells intercalate mediolaterally to produce a narrower, longer array (Figure 18.2B).

Mediolateral intercalation is driven by polarized cell motility. Before convergence and extension begin, the protrusive activity of the cells is not orientated within the plane of the tissue (Figure 18.3A). At the onset of convergence and extension and cell intercalation, the protrusive activity of the deep mesodermal cells becomes polarized. Large lamelliform protrusions form at the medial and lateral ends of the cells, and numerous, small filiform protrusions form at their anterior and posterior surfaces (Figure 18.3D). The large medial and lateral protrusions appear to exert traction on adjacent cells, and the cells become mediolaterally elongated, and orientated parallel to one another (Figure 18.3B). They then move between one another along the mediolateral axis, producing a longer, narrower array (Figure 18.3C–D) (Shih and Keller, 1992a,b; Keller *et al.*, 1991, 1992a). The cells appear to use one another as movable substrates. As they exert traction on one another's surfaces (pointers, Figure 18.3B–D), they pull themselves between one another

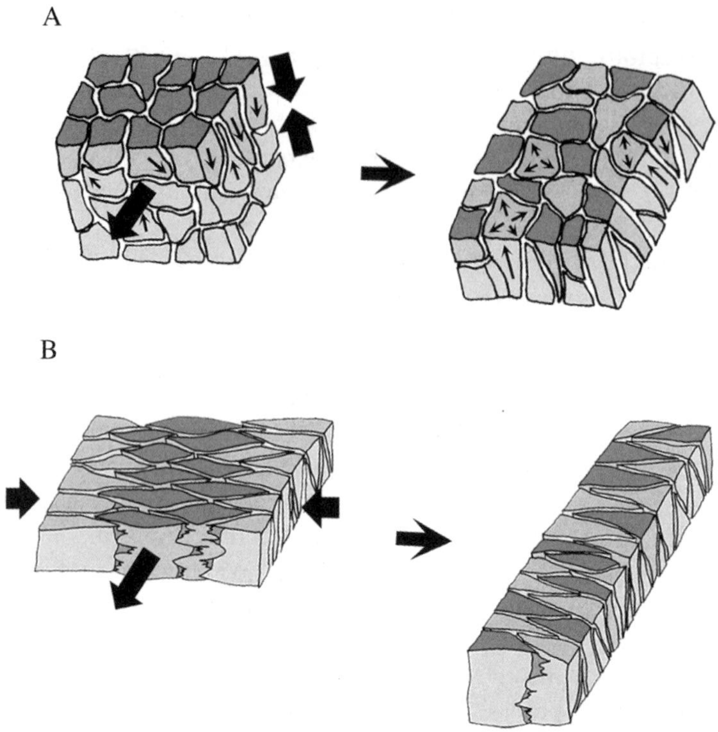

Figure 18.2 Convergence and extension of *Xenopus* mesodermal and neural tissues occurs by two processes of cell intercalation. During radial intercalation (A), several layers of deep mesenchymal cells intercalate between one another along the radius of the embryo to form a tissue that is thinner and longer (A). In this case, radial intercalation produces thinning and extension. During mediolateral intercalation (B), cells intercalate between one another along the mediolateral axis to form an array that is narrower (convergence) and longer (extension), and usually somewhat thicker as well. These intercalating deep cells are normally associated with an overlying epithelial layer that forms the outside of the embryo (not shown) and does not appear actively to converge and extend but does so passively by virtue of its attachment to the underlying deep cells. Modified from Keller *et al*., 1992a

(Figure 18.3C–D). Macroscopically, the expression of polarized protrusive activity is patterned in progressively developing arc-like patterns, which consist of a contiguous array of intercalating cells, anchored at both ends (Figure 18.3E). The arcs are shortened as the cells pull between one another, resulting in some cells being displaced to join other similarly displaced cells to form new arcs (Figure 18.3E). For example, cells 1–4 initially form a tensile unit that is shortening by virtue of mutual traction (Figure 18.3C). But as cells 2 and 4 intercalate between cells 1 and 3 (Figure 18.3C), and begin to exert traction on one another (Figure 18.3D), cells 1 and 3 are displaced anteriorly

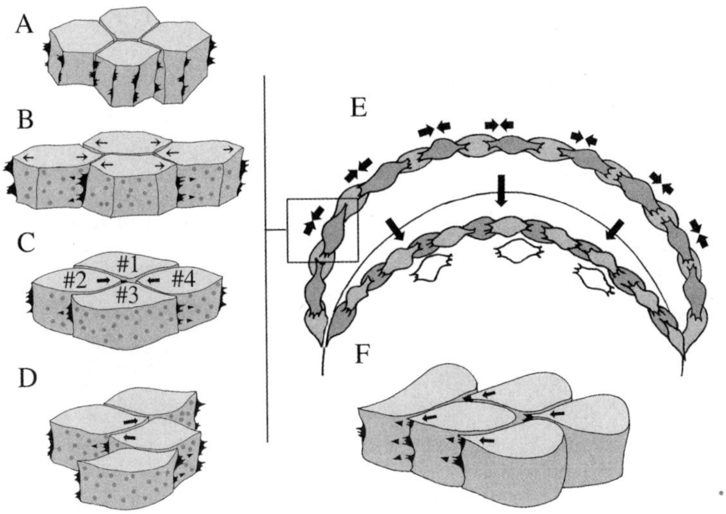

Figure 18.3 Mediolateral cell intercalation occurs as a result of polarized protrusive activity. Before cell intercalation begins, deep mesodermal cells are pleiomorphic and their lamelliform and filo-lamelliform protrusive activity (black protrusions in A–D) is unpolarized in the plane of the tissue (A). At the onset of cell intercalation, these lamelliform or filo-lamelliform protrusions become concentrated at the medial and lateral ends of the cells and are thought to exert traction on adjacent cells (pointers, B–D). The traction developed by these protrusions is thought to elongate the cells mediolaterally (small arrows, B) and pull them between one another along the mediolateral axis (arrows, C–D). These behaviours effectively form tensile arcs spanning the mediolateral aspect of the prospective mesoderm, which are anchored at both ends near the vegetal endoderm (E). It is the shortening of these arcs by cell intercalation (apposed arrows, E) that squeezes the blastopore shut and aids involution (arrows, E). Deep neural cells are thought to intercalate by using the same type of traction on one another, but in this case they appear to do so by heavily biasing their lamelliform protrusive activity toward the midline (F). The polarized cells are connected to one another at their anterior and posterior surfaces by numerous, small contacts (dark grey patches, B–D)

or posteriorly. There they begin to exert traction on similarly displaced, neighbouring cells located medial or lateral to them, and essentially participate in a new arc (unshaded cells, Figure 18.3E). An important element of this mechanism is that the cell body acts as a stable substrate on which the lamelliform medial and lateral protrusions can exert traction (Keller *et al.*, 2000).

Neural cells also become polarized during mediolateral intercalation with one major difference. Instead of being bipolar, with the lamelliform protrusive activity orientated mediolaterally, this protrusive activity is heavily biased toward the midline of the embryo (Elul and Keller, 2000) (Figure 18.3F). This medially-directed protrusive activity is thought to drive cell intercalation by a mutual cell-on-cell traction, similar to that seen in the bipolar mode (Keller *et al.*, 2000). Neural cells can intercalate using a bipolar mode, similar to that used by the mesoderm, in absence of midline tissue and underlying mesoderm (Elul *et al.*, 1997) but normally when these tissues are present, neural deep cells express the medially-biased (monopolar) mode of cell intercalation. This medial-directed protrusive activity is dependent on a signal from the midline notoplate/notochord tissues (Ezin *et al.*, 2003).

Similar protrusive activity may underlie cell intercalation in other systems. Cells near the dorsal midline in the fish embryo may intercalate using a mechanism similar to the bipolar one seen in frogs. This notion is based on the fact that the cells are similarly elongated and aligned mediolaterally (Sepich *et al.*, 2000; Glickman *et al.*, 2003; Marlow *et al.*, 2002; Topczewski *et al.*, 2001). However, the protrusive activity of these cells should be described in order to confirm that the same mediolateral polarization underlies their characteristic shape. Laterally, the cells of the germ ring show protrusive activity and movement directed dorsally toward the embryonic shield (Trinkaus *et al.*, 1992; Trinkaus, 1998; Marlow *et al.*, 2002). In this regard, these cells resemble the neural cells of the amphibian. In the extending notochord of ascidians, cells elongate, align and intercalate mediolaterally much like their amphibian counterparts (Miyamoto and Crowther, 1985). This cell intercalation involves transversely orientated protrusions at the leading edges of the intercalating cells, suggesting that similar polarized protrusive activity is a general mechanism underlying transversely orientated cell intercalation in the chordates (Munro and Odell, 2002a). Notochord cell intercalation in ascidians is dependent on interactions with surrounding tissues (Munro and Odell, 2002b), which is similar to the dependency of *Xenopus* notochord extension on the somitic mesoderm (Wilson *et al.*, 1989).

Recent evidence shows that the vertebrate homologues of the genes in the planar cell polarity pathway of *Drosophila* (Adler, 2002) are essential for convergence and extension in frogs and fish. This pathway includes the frizzled ligand, Wnt 11 (Tada and Smith, 2000; Heisenberg *et al.*, 2000), its serpentine receptor, Frizzled (Dijane *et al.*, 2000), the cytoplasmic signalling proteins Dishevelled (Sokol, 1996; Wallingford *et al.*, 2000; Tada and Smith, 2000). Dishevelled signals downstream through DAAM1 (Dishevelled Associated Activator of Morphogenesis) (Habas *et al.*, 2002), and the small GTPase and cytoskeletal regulator, Rho, and Rho kinase (Habas *et al.*, 2002; Marlow *et al.*, 2002). Another member of the Rho family of small GTPases, Cdc42, is also important in convergent extension (Choi and Han, 2002), as well as maintaining separation between the mesoderm and overlying neural tissue (Winklbauer *et al.*, 2001). Both of these functions of Cdc42 appear to be regulated not by the Dishevelled-dependent, planar cell polarity pathway, but by the Wnt/Ca^{++} pathway (Winklbauer *et al.*, 2001; Choi and Han, 2002; Sheldahl *et al.*, 1999). Finally, Strabismus (also known as van Gogh), a putative membrane protein regulating polarity in *Drosophila* (Adler *et al.*,

2000), is involved in some unknown way in convergence and extension (Darken *et al.*, 2002; Goto and Keller, 2002; Jessen *et al.*, 2002; Park and Moon, 2002) (Figure 18.2).

Convergent extension has been reviewed several times recently with a focus on various aspects, including the involvement of the planar cell polarity (PCP) pathway (Wallingford *et al.*, 2002; Mlodzik, 2002), the cell biological and biomechanical aspects (Keller *et al.*, 2000), a focus on these movements in teleosts (Myers *et al.*, 2002), and the relation of polarized cell behaviour to organismal polarity (Keller, 2002). The cell-traction/cell substrate model that we proposed for convergence and extension by cell intercalation in *Xenopus* some years ago (Keller *et al.*, 1991, 1992a), and revisited recently (Keller *et al.*, 2000; Keller, 2002) (Figure 18.1E,F), has been widely interpreted and reinterpreted in the context of many experimental results on amphibians and other organisms. Here we will revisit some of the key issues about convergence and extension by cell intercalation with a focus on the mechanism by which cell behaviour and cell motility generates active, force-producing cell intercalation.

The diversity and complexity of convergence and extension

It is a mistake to assume that convergence and extension are universal, uniform processes. Convergence and extension simply means narrowing and lengthening, without an implication of mechanism. Some examples of these movements occur passively in response to forces generated elsewhere, and others are active, and driven by forces generated within the tissues. Both the mesodermal and neural tissues in *Xenopus* actively extend when isolated as explants, and they are capable of exerting a pushing force (Keller and Danilchik, 1988; Moore, 1992; Moore *et al.*, 1995). However, the deep mesenchymal cells of these regions appear to be the active force-generating elements of these tissues, whereas the epithelial component seems to converge and extend passively by virtue of its attachment to the underlying deep cells (Keller and Danilchik, 1988). The superficial epithelial cells initially respond to being stretched in the axis of extension by elongating and narrowing, but then they slide by one another and intercalate to form a longer, narrower array and become less elongated (Keller, 1978). However, in most cases of convergence and extension that have been described, it is not known whether the movements observed are active or passive, nor has the relative importance of internally and externally generated forces been evaluated.

Secondly, although convergence and extension are often referred to collectively as 'convergent extension', this convenient somewhat misleading term oversimplifies the fundamental nature of an underlying machine, of which the paired movements of convergence and extension are only one manifestation. At the core of the mechanism of convergence and extension by cell intercalation is the idea that cells can forcibly intercalate between one another along one axis, thus reducing that dimension of the tissue and simultaneously forcing an increase in another, transverse dimension. The reduction in one dimension and increase in another is coupled biomechanically by conservation of volume. That is, as cells intercalate between one other along one axis, they force an increase in the dimension of a perpendicular axis. Ideally, halving the width of the tissue could produce a doubling of its length for maximum efficiency of translating convergence into extension (Figure 18.4A, left). But in fact, convergence generally results in increase in both length and thickness (Figure 18.4A, right). For example, convergence in *Xenopus* mesoderm and neural tissue produces both extension and thickening, and each tissue appears to have a specific conversion ratio of convergence-in-to-extension and convergence-in-to-thickening (Keller *et al.*, 1989; Wilson *et al.*, 1989; Shih and Keller, 1992a,b). Complete translation of convergence to extension seems never to occur, but instead, part of the reduction in width is absorbed by thickening. According to our model, as cells actively intercalate mediolaterally, they wedge between one another, and push one another apart along the anterior–posterior axis, thus producing the extension (Figure 18.3B–D). However, the compression forces that cells come under in the anterior–posterior axis during this wedging would also tend to force the cells to move either above or below the plane of convergence and extension; that is, it would tend to form a second layer and thicken the tissue (Figure 18.4B).

What is the process that resists thickening, or restores cells to their original layer? The fact that tissues vary in the efficiency of translation of convergence into extension versus thickening suggests that whatever the process, the tissues vary in its effectiveness. The obvious process is radial intercalation, the process that thinned the tissue in the first place (Figure 18.2A). The initial event in the overall process of convergence and extension is thinning and extension, driven by radial intercalation (Keller and Danilchik, 1988). If the thinning forces generated by radial intercalation continue during mediolateral intercalation, it would tend to reduce the tendency of the latter to produce thickening by reversing the process of radial intercalation (Figure 18.4C).

The converse question is why does radial intercalation give rise to extension rather than uniform spreading in all directions during the initial phases of convergence and extension (Figure 18.2A) (Wilson *et al.*, 1989; Wilson and Keller, 1991)? Radial intercalation seems to be an isotropic process in the plane of the tissue (Wilson and Keller, 1991), and therefore it should give rise to spreading in all directions. It does so during epiboly of the animal cap (Keller, 1978, 1980), but in the initial phases of convergence and extension of the marginal zone, it produces extension. Using the same argument applied to mediolateral intercalation above, radial intercalation, even if it were biased to occur only between anterior–posterior neighbours, would tend to also produce

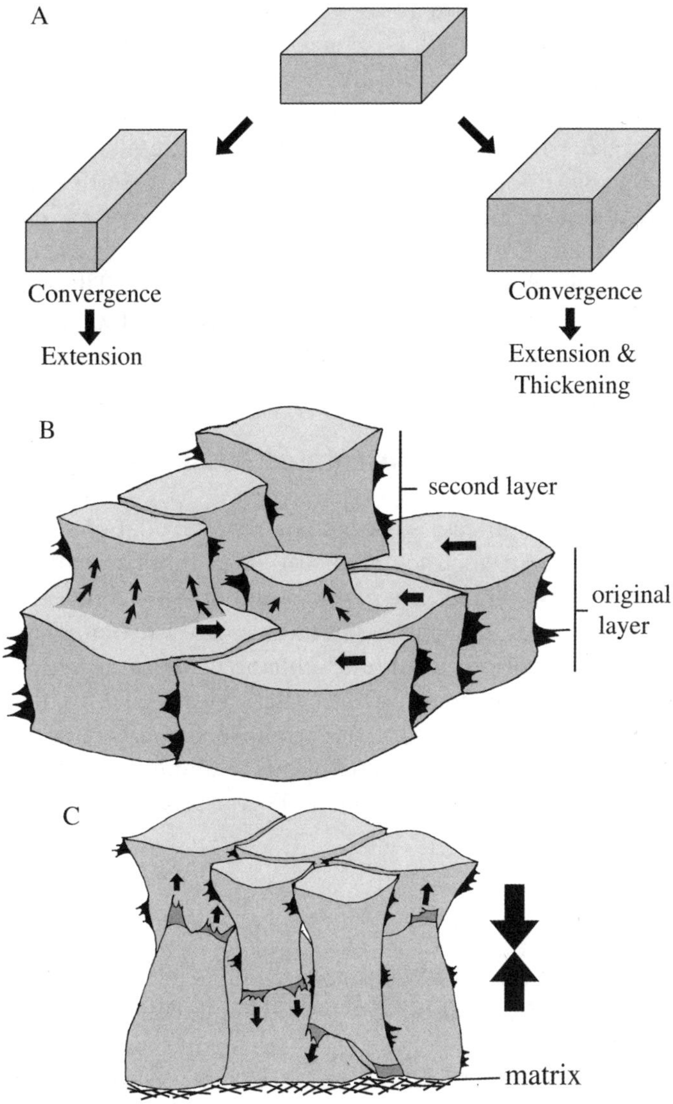

Figure 18.4 Active, force-producing convergence and extension in early *Xenopus* development is thought to involve convergence of the mediolateral aspect of the tissue by virtue of the mediolaterally (in the mesoderm) or medially (in the neural tissue) polarized protrusive activity (A, top). Volume is conserved, and therefore, if thickening does not occur, convergence would result in a proportional extension (A, left). In fact, convergence results in some extension and some thickening (A, right). Thickening can be explained as follows. Although the protrusive activity driving convergence is orientated in the mediolateral axis (or directed medially in the case of the neural tissue), as the cells wedge between one another, they come under compression in the anterior–posterior axis. Such compression would tend to squeeze the cells into a taller shape, or squeeze them out of the layer to form a second layer (B). This tendency toward multilayering may be resisted by continued radial intercalation (C)

some spreading in the mediolateral orientation as a result of the resistance to extension in the anterior–posterior orientation. As the cells intercalate and come under compressive force in the anterior–posterior direction, they would tend to be forced between mediolateral neighbours as well, unless there were a countering process. Could it be that cryptic mediolateral intercalation has already begun during radial intercalation? Or is it the case that some other process resists the tendency for uniform spreading? In any case, radial intercalation produces thinning and extension and provides an example of why extension is not necessarily linked to convergence. The cellular and mechanical basis of the channelling of the forces generated in one axis to the perpendicular axes deserves further investigation.

Extension by shape change, orientated division and growth

Extension may occur by processes other than cell intercalation. For example, if cuboidal cells divide in an orientated fashion without growth, the linear dimensions of the array will increase in the axis of division and decrease in the transverse axis. Moreover, if the cells grow (increased in volume) as well, the effect of orientated division would be enhanced. Such a process would produce extension but not convergence. There is evidence that orientated division plays a role in elongation of the primitive streak (Wei and Mikawa, 2000) as well as in extension of the axial tissue in amniotes (Schoenwolf and Alvarez, 1989; Sausedo and Schoenwolf, 1993, 1994; Sausedo *et al*., 1997).

Bipolar traction

In the bipolar mode of cell intercalation, each cell is presumed exert traction on adjacent, lateral and medial cells, using the large lamelliform protrusions found at the medial and lateral ends. This bipolar traction is thought first to stretch the cells mediolaterally, thus elongating them, and then, when the cell resistance to stretch balances the resistance to intercalation, the traction pulls the cells between one another along the mediolateral axis (Keller *et al*., 1991, 1992a, 2000). There is no direct evidence that the large medially and laterally orientated lamelliform protrusions actually exert traction. However, similar protrusions of cells cultured on elastic membranes pull wrinkles in these membranes, demonstrating that they are the primary tractive organelles of the cell (Harris *et al*., 1980). The regions of the cell between such large protrusions are under tension (Kolega, 1986). These facts support the idea that the large lamelliform protrusions generate the tension that elongates the cells mediolaterally, and that these protrusions pull between one another along this axis.

The elongation of the cells is correlated with the onset of the polarized protrusive activity. The cells initially show rapid filo-lamelliform protrusive activity in all directions, and they are rotund in the planar aspect of the tissue. As they restrict their protrusive activity to the medial and lateral ends of the cells, they also become elongated in this axis and begin to intercalate along this axis (Shih and Keller, 1992a) (Figure 18.3). Manipulation of components of the planar cell polarity pathway, such as Dishevelled (Wallingford *et al.*, 2000) and Strabismus (Goto and Keller, 2002), result in random protrusive activity, and manipulation of other components, such as Rho kinase and Strabismus in fish (Marlow *et al.*, 2002; Jessen *et al.*, 2002) show loss of the mediolaterally elongated phenotype. These results argue that the polarized protrusive activity is, in fact, the basis of the elongate, fusiform morphology and the intercalation parallel to the mediolateral axis.

Do the intercalating cells actually 'migrate' on one another?

One of the most intriguing aspects of convergence and extension is that these movements can occur in sandwich explants that are unattached to any external substratum, and in open-faced explants (ones exposing the actively intercalating deep cell cells) cultured on non-adhesive substrates, such as agarose (Keller and Danilchik, 1988; Shih and Keller, 1992a,b). However, there may be an internal substrate, other than the surfaces of adjacent cells, that might be used by the cells to move between one another. A web of matrix could surround all of the cells. The fibronectin-containing matrix separating the neural and mesodermal tissue could serve this function. These potential substrates could be carried along in cultured explants. Both the neural and the mesodermal tissues have protrusions in the vicinity of this matrix (Keller and Schoenwolf, 1977). However, the pattern of cell movements argues that the intercalating cells, in fact, use one another as substrates. If cells intercalate mediolaterally at a locally constant rate of overlap by using one another as substrates, regardless of whether it is by the bipolar (Figure 18.5A) or the monopolar mode (Figure 18.5B), the cumulative effect is progressively faster displacement of cells toward the centre of the array at distances farther from the centre (Figure 18.5A,B). In both mesodermal (Shih and Keller, 1992a) and neural tissues (Keller *et al.*, 1992b) this is what is observed. In contrast, if cells translocate at a constant rate toward the midline on an external substrate, no additive effect is realized and the cells farther from the midline approach it at an equal rate (Figure 18.5C), which is inconsistent with what is observed. In order to intercalate, the cells farther away on the axis of convergence would have to translocate progressively faster on an external substrate in order to overtake and get between their medial counterparts. There is no evidence of

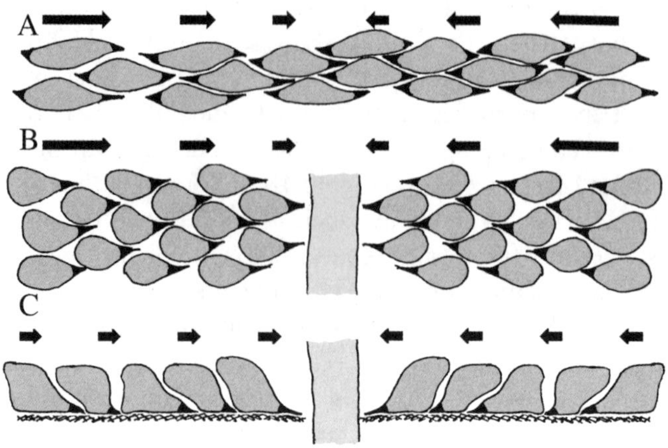

Figure 18.5 The relationship between the rates of local cell intercalation or migration and tissue-level displacement argues that the cells crawl on one another rather than on an external substratum. Assume that the cells exercise a constant rate of translocation on one another's surfaces in the mediolateral axis, either by the bipolar, mediolaterally orientated mode (A) or the monopolar, medially-directed mode (B). Under this assumption, the rates of approach to the midline will be progressively greater with distance from the midline of the array (arrows, A, B). Intercalation would occur throughout the array, because the cells, using one another as substrates, would move between one another uniformly. However, if the cells exercise a uniform, medially-directed rate of translocation on an external substrate, the rate of approach to the midline of the array will be the same with distance from the midline (C). Intercalation would occur only at the row of cells next to the midline, assuming that the second row would continue to move until it reached the midline and then stop

such progressively increasing rates of local activity with distance from the midline in converging mesodermal or neural tissues of *Xenopus*.

Interestingly, the convergence of the cells of the teleost germ ring toward the embryonic shield has the reverse pattern from what is expected by the cell-traction/cell substrate model of cell intercalation: they move faster as they approach the shield (Trinkaus *et al.*, 1992; Trinkaus, 1998; Marlow *et al.*, 2002). This fact suggests that they are migrating on an 'external' substrate, in this case, either the overlying enveloping layer or the underlying yolk syncytial layer, rather than one another, although they do interact with one another and even intercalate between one another (Trinkaus *et al.*, 1992). However, they do not undergo the organized intercalation associated with convergence-driven extension, and these cells do not produce extension of the germ ring as they converge. This movement has all the hallmarks of a directed migration, and the fact that the cells appear to be more persistent in their movement as they approach the midline suggests that they are responding to a midline-generated chemotactic signal. In contrast, in the dorsal sector of the blastoderm, convergence and extension occurs by cell intercalation and in this case the medial movement of the cells shows the progressive increase in rate with distance from the midline (Glickman *et al.*, 2003), characteristic of intercalation by cell-on-cell traction model.

These differences in patterns of movements raise the issue of the definition of convergence. On the one hand, convergence can be defined as the movement of cells toward a midline, but that kind of movement does not necessarily produce extension, for a number of reasons. If all move to the midline at the same rate, no extension will be produced, at least not by virtue of this movement. If there is a gradient in the rate of movement, those farther away moving faster, intercalation should occur, but whether extension or thickening will be produced, or how much of each, is dependent on additional constraints on the intercalation process. These constraints would determine whether intercalation occurred between radial neighbours to produce thickening, or between mediolateral neighbours to produce extension. The cell biological determinants of these alternatives of cell behaviour remain unresolved and at once warn against oversimplifying the processes of convergence and extension and assure that there is much more work to be done on the cellular biomechanics of these processes.

Dynamics and stiffness: how does a dynamic tissue push?

The dorsal mesodermal tissues increase in stiffness during convergence and extension and generate a pushing force of about a half micronewton (Moore *et al.*, 1995; Moore, 1992). How can cells shear between one another and yet maintain a cohesive tissue sufficiently stiff to exert a pushing force? Our mechanical model of convergent extension by cell intercalation holds that the cells actively wedge between one another by mediolaterally orientated traction, and as they do so, they push one another apart preferentially along the anterior–posterior axis (Keller *et al.*, 1992a, 2000). The wedging forces generated by all of the individual cells are summed and generate the total pushing force of the tissue. However, in order to push the cells must form an array stiff enough to bear whatever compressive load the tissue will come under as it pushes, without buckling. Mechanical measurements show that the converging and extending mesodermal tissues become stiffer by a factor of three or four, compared with pre-convergent extension tissues (Moore *et al.*, 1995), and that these tissues can push with about a half micronewton of force (Moore, 1992).

Cell intercalation, the rearrangement of the structural elements of the tissue, would seem to be at odds with forming a stiff structure. If a cell intercalates, it must detach from old neighbours and reattach to new ones. If too many cells did this at one time, the tissue would presumably become less stiff, because external forces would be met with weaker intercellular adhesion, on average.

Of course, at tissue level, not all cells rearrange at once, and at cell level, an individual cell does not exchange all of its adhesions at once, but loses attachments to old neighbours and makes new ones with prospective neighbours over some period of time.

We have postulated that the stiffness of the intercalating array of cells may be a function of the less obvious aspect of the polarization of these cells, the numerous small contacts along the anterior and posterior sides of these cells. In scanning electron micrographs these contacts appear as filopodia (Figure 9 in Keller *et al.*, 2000), but their length is probably partially due to shrinkage of the cells, which can be to 10% or more (Keller and Schoenwolf, 1977). SEM shows that these contacts are present in large numbers, but their number is underestimated in the original, standard low light fluorescent movies of cells filled with fluorescent dextrans, conditions under which few of them are visible (Shih and Keller, 1992a). A sectional view with scanning confocal fluorescence microscopy of cells labelled with a membrane targeted GFP shows that these protrusions are numerous and turning over rapidly, on average about every 2 min, and they appear as short, sharply tapered protrusions, most resembling short filopodia (Keller, 2002, movie S6). These small, dynamic contacts represent dynamic adhesions that allow the cells to shear past one another and thus intercalate, but also may lock the intercalating array into a rigid structure capable of bearing a compression load and exerting a pushing force (Keller *et al.*, 2002). These contacts could also contribute to the elongate cell shape and the formation of the parallel array of elongated cells by sticking the long sides of the cells together. However, we believe that it is unlikely to play a large role in the initial polarization and elongation of the cells, and that the larger role is played by the mediolateral, bipolar filo-lamelliform protrusive activity. However, selective adhesion of the elongate sides to one another could promote formation of a stable, parallel array (Elsdale and Wasoff, 1976).

A key to the mechanics of convergent extension by cell intercalation may be regulation of the dynamics of cell–cell adhesion. These contacts are constantly being made and broken but, at any one time, many of them attach cell to its neighbours. The local periodic breakdown of these adhesions would allow shearing of cells past one another and invasion of the intercalating, tractoring protrusions between cells. For example, as the lamelliform, tractoring protrusions advance, their way would be blocked for a time, but soon the filiform protrusions would be retracted, allowing advance of the lamelliform protrusion, and then new filiform protrusions would form contacts again on the sides of the intervening cell. The tissue can be self-deforming and stiff during cell intercalation if the adhesions are strong while they last, but have a regulated turnover. Turnover of adhesions, of course, is not an unusual property; it is essential for cell migration on a planar substrate. In the case of intercalation, it is simply a matter of migration on another cell. Adhesion is too often viewed as a static property, and it is obviously not. If it were, cells could not exchange neighbours or migrate.

Cell–cell and cell–matrix adhesion in convergent extension

The polarized deep mesodermal cells are obviously stuck to one another and to extracellular matrix as well, but little is known about the role of adhesion or its modulation, and whether adhesion is also a polarized property in parallel with the morphological cellular polarization. With the large filo-lamellipodia on the medial and lateral ends, and the small contacts along the anterior and posterior sides, the obvious question is whether there are particular types of cell adhesion molecules, or a specialized local activity of a cell adhesion molecule, associated with these polarized protrusions. The principal cadherin in the deep mesoderm is C-cadherin, and adhesion assays suggest that its activity decreases during convergence and extension (Brieher and Gumbiner, 1994; Zhong *et al.*, 1999). Whether C-cadherin mediates all cell–cell adhesions around the mesodermal cells, or just a subset, is not known. The decrease in C-cadherin-mediated adhesion, if it occurs *in vivo*, is very interesting in the context of stiffening of the mesoderm in compression–stress relaxation tests during convergent extension. Cells do not rearrange during the compression or the relaxation phase of the compression–stress relaxation test used to measure tissue stiffness (Moore *et al.*, 1995). This holds true before convergence and extension, and also after convergence and extension when the tissue is relatively stiffer. Therefore, increase in cell adhesion alone, and thus increased resistance to cell rearrangement under compression, does not account for the increase in tissue stiffness. A tissue could become stiffer as a result of cells being more difficult to rearrange under an external load, or as a result of the cells themselves becoming stiffer. Moreover, if C-cadherin does decrease in activity, it implies that adhesion can actually be weakened during convergence and extension without bringing passive cell rearrangement under a compression load into play as a limiting factor in tissue stiffness.

One explanation for this phenomenon is that cytoskeletal stiffness and adhesion are linked, due to coordinate organization of cytoskeleton in response to adhesion, but that the tissue stiffness component due to cytoplasmic stiffness is always limiting relative to the component due to cell adhesion. The fact that the cells are also actively rearranging, and therefore must be exchanging adhesions, suggests that the dynamics of adhesion is important, and raises the issue of what exactly it is that is important about cell–cell adhesion during convergent extension, and whether adhesion assays measure a parameter relevant to this function. The turnover of adhesions may be particularly important. Many rapidly turning over adhesions would perhaps be conducive to active rearrangement, which occurs over 20 or

30 min, and yet be unyielding in the face of a compression test done over 3 min.

Specific protocadherins, axial protocadherin (AXPC) and paraxial protocadherin (PAPC), are expressed at the RNA level in the notochord and somitic mesoderm, respectively, and they appear to have adhesion or signalling roles, and may function in recognition and separation of these tissues prior to or during convergence and extension (Kim *et al.*, 1998; Yamamoto *et al.*, 1998; Kuroda *et al.*, 2002). It is not known how these molecules function, and whether they are associated with specific aspects of cell polarity or with specific types of protrusions during convergent extension. The messages are expressed in the notochord and somitic mesoderm, respectively, but the expression pattern of the protein is not known. The message for AXPC appears quite late, during mid-neurulation, and much later than that for PAPC, which appears at mid-gastrula. When ectopically expressed in two cell populations, an AXPC-like construct and PAPC appear to mediate cell recognition and sorting out, and expression of PAPC appears to increase adhesion and limit cell mixing during epiboly (Kim *et al.*, 1998). Whether these molecules have a signalling function, an adhesive function, or both, in convergence and extension is not known.

Extracellular matrix and cell intercalation

It is also possible that adhesion to extracellular matrix may also function in mediolateral intercalation despite the fact it seems unlikely to be used as a substrate. A fibronectin-containing, fibrillar matrix forms on the blastocoele roof during gastrulation. This matrix lies on the ventral surface of the neural plate and between the prospective somitic mesoderm and the overlying prospective neural tissue (Nakatsuji and Johnson, 1982, 1983; Komazaki, 1988; Darribere *et al.*, 1990; Winklbauer and Stoltz, 1995). Inhibiting integrin-mediated cell interactions with matrix slows blastopore closure and embryo elongation (Ramos and DeSimone, 1996). This manipulation stops the migration of the leading edge mesendoderm of the embryo toward the animal pole (Davidson *et al.*, 2002), and this alone may slow convergence and extension by mechanical interference. However, inhibition of cell–matrix interactions also blocks radial intercalation and explant extension (Marsden and DeSimone, 2001, 2003), which drives the initial thinning and extension step of convergence and extension. The effect of blocking cell–matrix interactions could be solely due to blocking radial intercalation, or, in addition, it could be that mediolateral intercalation also depends on cell–matrix interactions. Also, radial intercalation may be necessary for the subsequent mediolateral intercalation, and blocking the first may indirectly inhibit the latter. It remains unknown whether integrin-mediated interactions with fibronectin, or other extracellular matrix, are directly involved in mediolateral cell intercalation.

The special function of bipolarity

A critical element of the intercalation of the deep mesodermal cells appears to be balanced, bipolar protrusive activity. It appears that cells can intercalate two ways: they can use the neural mode consisting of a strong bias of traction consistently in one direction, or they can use the mesodermal, bipolar mode, but if the latter is used, the bipolar protrusive activity, the cells must minimize any imbalance of protrusive activity and traction on the two ends of the cells at any given time in order for intercalation to occur. If the cell is biased first in the lateral direction, and then in the medial direction, while nearby cells similarly alternate from one polarity to the other, the cells would be expected to exchange places but produce little convergent extension. This type of exchange of places is seen during intercalation of neural deep cells in explants in which the midline notochord and notoplate have been removed and the monopolar mode lost. Under these conditions, the deep neural cells show bipolar protrusive activity, when activity is averaged over time, and in fact they intercalate, in the sense that a given cell moves first one way, either medially or laterally, and then back again, and other cells will do the same, resulting in a high degree of cell mixing. But this promiscuous mixing only involves exchange of mediolateral positions and results in only weak convergence and extension results (Elul *et al.*, 1997; Elul and Keller, 2000). In contrast, both the monopolar neural mode and the balanced bipolar mesodermal mode produce conservative intercalation that is productive in producing convergent extension (Elul and Keller, 2000; Shih and Keller, 1992a,b). Biomechanical modeling studies should be done in order to evaluate the tissue-level effects of balanced mediolaterally orientated traction, strongly biased traction directed toward the midline, and various patterns of alternating directionality of traction. The mechanism of establishing bipolarity also deserves more attention. Most cells translocate by polarizing their motility, and their traction on the substrate, in one direction such that one edge advances and the other retracts, forming a 'tail' (Trinkaus, 1976; Euteneuer and Schliwa, 1986; Dormann *et al.*, 2002). But a bipolar cell must maintain traction at opposite ends of the cell, and in addition, it appears that balancing this traction is important. It is not understood how cells maintain a balanced bipolarity.

References

Adler, P., 2002. Planar signaling and morphogenesis in *Drosophila*. *Dev. Cell* **2**: 525–535.

Adler, P., Taylor, J. and Charlton, J., 2000. The domineering non-autonomy of *frizzled* and *Van Gogh* clones in the *Drosophila* wing is a consequence of a disruption in local signaling. *Mech. Dev.* **96**: 197–207.

Brieher, W. M. and Gumbiner, B. M., 1994. Regulation of C-Cadherin function during activin induced morphogenesis of *Xenopus* animal caps. *J. Cell Biol.* **126**: 519–526.

Choi, S.-C. and Han, J.-K., 2002. *Xenopus* Cdc42 regulates convergent extension movements during gastrulation through Wnt/Ca^{2+} signalling pathway. *Develop. Biol.* **244**: 342–357.

Darribere, T., Guida, K., Larjava, H., Johnson, K. E., *et al.*, 1990. In vivo analyses of integrin β 1 subunit function in fibronectin matrix assembly. *J. Cell Biol.* **110**: 1813–1823.

Darken, R., Scola, A. M., Rakeman, S. S., Das, G., Mlodzik, M. and Wilson, P., 2002. The planar polarity gene strabismus regulates convergent extension movements in *Xenopus*. *EMBO J.* **21**: 976–985.

Davidson, L., Hoffstrom, B., Keller, R. and DeSimone, D., 2002. Mesendoderm extension and mantle closure in *Xenopus laevis* gastrulation: Combined roles for integrin a5b1, fibronectin, and tissue geometry. *Develop. Biol.* **242**: 109–129.

Djiane, A., Riou, J., Umbhauer, M., Boucaut, J. and Shi, D., 2000. Role of frizzled 7 in the regulation of convergent extension movements during gastrulation in *Xenopus laevis*. *Development* **127**: 3091–3100.

Dormann, D., Weijer, G., Parent, C. A., Devreotes, P. N. and Weijer, C. J., 2002. Visualizing PI3 kinase-mediated cell–cell signaling during *Dictyostelium* development. *Curr. Biol.* **12**: 1178–1188.

Elsdale, T. and Wasoff, F., 1976. Fibroblast cultures and dermatoglyphics: the topology of two planar patterns. *Wilhelm Roux's Arch. Develop. Biol.* **180**: 121–147.

Elul, T., Koehl, M. and Keller, R., 1997. Cellular mechanism underlying neural convergent extension in *Xenopus laevis* embryos. *Develop. Biol.* **191**: 243–258.

Elul, T. and Keller, R., 2000. Monopolar protrusive activity: a new morphogenic cell behavior in the neural plate dependent on vertical interactions with the mesoderm in *Xenopus*. *Develop. Biol.* **224**: 3–19.

Euteneuer, U. and Schliwa, M., 1986. The function of microtubules in directional cell movement. *Ann NY Acad. Sci.* **466**: 867–886.

Ezin, M., Skoglund, P. and Keller, R., 2003. The midline (notochord and notoplate) patterns cell motility underlying convergence and extension of the *Xenopus* neural plate. *Develop. Biol.* **256**: 100–113.

Glickman, N., Kimmel, C., Jones, M. and Adams, R., 2003. Shaping the zebrafish notochord. *Development*, **130**: 873–887.

Goto, T. and Keller, R., 2002. The planar cell polarity gene *Strabismus* regulates convergence and extension and neural fold closure in *Xenopus*. *Dev. Biol.* **247**: 165–181.

Habas, R., Kato, Y. and He, X., 2002. Wnt/Frizzled activation of Rho regulates vertebrate gastrulation and requires a novel Formin homology protein, Daam1. *Cell* **107**: 843–854.

Harris, A. K., Wild, P. and Stopak, D., 1980. Silicone rubber substrata: a new wrinkle in the study of cell locomotion. *Science* **208**: 177–179.

Heisenberg, C. P., Tada, M., Rauch, G.-J., Saude, L., *et al.*, 2000. Siberblick/Wnt11 activity mediates convergent extension movements during zebrafish gastrulation. *Nature* **405**: 76–81.

Jessen, J., Topczewski, J., Bingham, S., Sepich, D., *et al.*, 2002. Zebrafish trilobite identifies new roles for *Strabismus* in gastrulation and neuronal movements. *Nat. Cell Biol.* **4**: 610–615.

Keller, R. E., 1978. Time-lapse cinemicrographic analysis of superficial cell behavior during and prior to gastrulation in *Xenopus laevis*. *J. Morph.* **157**: 223–248.

Keller, R. E., 1980. The cellular basis of epiboly: An SEM study of deep-cell rearrangement during gastrulation in *Xenopus laevis*. *J. Embryol. Exp. Morph.* **60**: 201–234.

Keller, R., 1984. The cellular basis of gastrulation in *Xenopus laevis*: active, postinvolution convergence and extension by mediolateral interdigitation. *Am. Zool.* **24**: 589–603.

Keller, R. E., 1986. The cellular basis of amphibian gastrulation. In *Developmental Biology: A Comprehensive Synthesis, Vol. 2. The Cellular Basis of Morphogenesis* (Browder, L. Ed.). Plenum Press, N.Y., pp. 241–327.

Keller, R., 2002. Shaping the vertebrate body plan by polarized embryonic cell movements. *Science* **298**: 1950–1954.

Keller, R. E. and Schoenwolf, G. C., 1977. An SEM study of cellular morphology, contact, and arrangement, as related to gastrulation in *Xenopus laevis*. *Wilhelm Roux's Arch.* **182**: 165–186.

Keller, R. E. and Danilchik, M., 1988. Regional expression, pattern and timing of convergence and extension during gastrulation of *Xenopus laevis*. *Development* **103**: 193–209.

Keller, R. E., Danilchik, M., Gimlich, R. and Shih, J., 1985. The function and mechanism of convergent extension during gastrulation in *Xenopus laevis*. *J. Embyol. Exp. Morph.* **89**: 185–209.

Keller, R. E., Cooper, M., Danilchik, M., Tibbetts, P. and Wilson, P., 1989. Cell intercalation during notochord development in *Xenopus laevis*. *J. Exp. Zool.* **251**: 134–154.

Keller, R., Shih, J., Wilson, P. and Sater, A., 1991. Pattern and function of cell motility and cell interactions during convergence and extension in *Xenopus*. In *49th Symp. Soc. Develop. Biol. Cell–Cell Interactions in Early Development* (Gerhart, J. Ed.). Wiley and Sons, N. Y., pp. 93–107.

Keller, R., Shih, J. and Domingo, C., 1992a. The patterning and functioning of protrusive activity during convergence and extension of the *Xenopus* organizer. *Development* (Suppl 1992): pp. 81–91.

Keller, R., Shih, J. and Sater, A., 1992b. The cellular basis of the convergence and extension of the *Xenopus* neural plate. *Develop. Dynamics* **193**: 199–217.

Keller, R., Davidson, L., Edlund, A., Elul, T., *et al.*, 2000. Mechanisms of convergence and extension by cell intercalation. *Phil. Trans. R. Soc. Lond. B* **355**: 897–922.

Kim, S.-H., Yamamoto, A., Bouwmeester, T., Agius, E. and DeRobertis, E. M., 1998. The role of paraxial protocadherin in selective adhesion and cell movements of the mesoderm during *Xenopus* gastrulation. *Development* **125**: 4681–4691.

Kolega, J., 1986. Effects of mechanical tension on protrusive activity and microfilament and intermediate filament organization in an epidermal epithelium moving in culture. *J. Cell Biol.* **102**: 1400–1411.

Komazaki, S., 1988. Factors related to the initiation of cell migration along the inner surface of the blastocoelic wall during amphibian gastrulation. *Cell Differ.* **24**: 25–32.

Kuroda, H., Inui, M., Sugimoto, K., Hayata, T. and Asashima, M., 2002. Axial protocadherin is a mediator of prenotochord cell sorting in *Xenopus*. *Develop. Biol.* **244**: 267–277.

Lawson, A. and Schoenwolf, G. C., 2001a. Cell populations and morphogenetic movements underlying formation of the avian primitive streak and organizer. *Genesis* **29**: 188–195.

Lawson, A. and Schoenwolf, G. C., 2001b. New insights into critical events of avian gastrulation. *Anat. Rec.* **262**: 238–252.

Marlow, F., Topczewski, J., Sepich, D. and Solnica-Krezel, L., 2002. Zebrafish Rho kinase 2 acts downstream of Wnt11 to mediate cell polarity and effective convergence and extension movements. *Curr. Biol.* **12**: 876–884.

Marsden, M. and DeSimone, D., 2001. Regulation of cell polarity, radial intercalation and epiboly in *Xenopus*: novel roles for integrin and fibronectin. *Development* **128**: 3635–3647.

Marsden, M. and Desimone, D., 2003. Integrin–ECM interactions regulate cadherin-dependent cell adhesion and are required for convergent extension in *Xenopus*. *Curr. Biol.* **13**: 1182–1191.

Miyamoto, D. M. and Crowther, R. J., 1985. Formation of the notochord in living ascidian embryos. *J. Embryol. Exp. Morphol.* **86**: 1–17.

Mlodzik, M., 2002. Planar cell polarization: do the same mechanisms regulate *Drosophila* tissue polarity and vertebrate gastrulation? *Trends Genetics* **18**: 564–578.

Moore, S., 1992. Direct measurement of dynamic biomechanical properties of amphibian embryonic tissues. Ph.D. Thesis, University of California, Berkeley, Berkeley, CA, USA.

Moore, S., Keller, R. and Koehl, M., 1995. The dorsal involuting marginal zone stiffens anisotropically during its convergent extension in the gastrula of *Xenopus laevis*. *Development* **121**: 3131–3140.

Munro, E. M. and Odell, G. M., 2002a. Polarized basolateral cell motility underlies invagination and convergent extension of the ascidian notochord. *Development* **129**: 13–24.

Munro, E. M. and Odell, G., 2002b. Morphogenetic pattern formation during ascidian notochord formation is regulative and highly robust. *Development* **129**: 1–12.

Myers, D., Sepich, D. and Solnica-Krezel, L., 2002. Convergence and extension in vertebrate gastrulae: cell movements according to or in search of identity? *Trends Genetics* **18**: 447–455.

Nakatsuji, N. and Johnson, K., 1982. Movement and guidance of migrating mesodermal cells in *Ambystoma maculatum* gastrulae. *J. Cell Sci.* **56**: 207–222.

Nakatsuji, N. and Johnson, K., 1983. Conditioning of a culture substratum by the ectodermal layer promotes attachment and orientated locomotion by amphibian gastrula mesodermal cells. *J. Cell Sci.* **59**: 43–50.

Park, M. and Moon, R., 2002. The planar cell-polarity gene stbm regulates cell behaviour and cell fate in vertebrate embryos. *Nat. Cell Biol.* **4**: 20–25.

Ramos, J. W. and DeSimone, D. W., 1996. Xenopus embryonic cell adhesion to fibronectin: position-specific activation of RGD/Synergy site-dependent migratory behavior at gastrulation. *J. Cell Biol.* **134**: 1–14.

Sausedo, R. A. and Schoenwolf, G. C., 1993. Cell behaviors underlying notochord formation and extension in avian embryos: quantitative and immunocytochemical studies. *Anat. Rec.* **237**: 58–70.

Sausedo, R. A. and Schoenwolf, G. C., 1994. Quantitative analyses of cell behaviors underlying notochord formation and extension in mouse embryos. *Anat. Rec.* **239**: 103–112.

Sausedo, R. A., Smith, J. L. and Schoenwolf, G. C., 1997. Role of nonrandomly orientated cell division in shaping and bending of the neural plate. *J. Comp. Neurol.* **381**: 473–488.

Schechtman, A. M., 1942. The mechanics of amphibian gastrulation. *Univ. Calif. Publ. Zool.* **51**: 1–39.

Schoenwolf, G. C. and Alvarez, I., 1989. Roles of neuroepithelial cell rearrangement and division in shaping of the avian neural plate. *Development* **106**: 427–439.

Schoenwolf, G. C., Garcia-Martinez, V. and Dias, M., 1992. Mesoderm movement and fate during avian gastrulation and neurulation. *Develop. Dyn.* **193**: 235–248.

Sepich, D., Myers, D., Short, R., Topczewski, J., *et al.*, 2000. Role of the zebrafish trilobite locus in gastrulation movements of convergence and extension. *Genesis* **27**: 159–173.

Sheldahl, L, Park, M., Malbon, C. and Moon, R. 1999. Protein kinase C is differentially stimulated by Wnt and Frizzled homologs in a G-protein-dependent manner. *Curr. Biol.* **9**: 695–698.

Shih, J. and Keller, R., 1992a. Cell motility driving mediolateral intercalation in explants of *Xenopus laevis*. *Development* **116**: 901–914.

Shih, J. and Keller, R., 1992b. Patterns of cell motility in the organizer and dorsal mesoderm of *Xenopus*. *Development* **116**: 915–930.

Shook, D. R., Majer, C. and Keller, R., 2002. Urodeles remove mesoderm from the superficial layer by subduction through a bilateral primitive streak. *Develop. Biol.* **248**: 220–239.

Sokol, S. Y., 1996. Analysis of Dishevelled signalling pathways during *Xenopus* development. *Curr. Biol.* **6**: 1456–1467.

Tada, M. and Smith, J. C., 2000. Xwnt 11 is a target of *Xenopus* Brachyury: regulation of gastrulation movements via dishevelled, but not through the canonical Wnt pathway. *Development* **127**: 2227–2238.

Topczewski, J., Sepich, D., Myers, D., Walker, C., *et al.*, 2001. The zebrafish glypican knypek controls cell polarity during gastrulation movements of convergent extension. *Devel. Cell* **1**: 251–264.

Trinkaus, J. P., 1976. On the mechanism of cell movements. In *The Cell Surface in Animal Embryogenesis and Development* (eds Poste, G. and Nicolson, G. L.). New York: Elsevier/North-Holland Biomedical Press, pp. 225–329.

Trinkaus, J. P., 1998. Gradient in convergent cell movement during *Fundulus* gastrulation. *J. Exp. Zool.* **281**: 328–335.

Trinkaus, J. P., Trinkaus, M. and Fink, R., 1992. On the convergent cell movements of gastrulation in *Fundulus*. *J. Exp. Zool.* **261**: 40–61.

Vogt, W., 1929. Gestaltanalyse am Amphibienkein mit örtlicher Vitalfärbung. II.Teil. Gastrulation und Mesodermbildung bei Urodelen und Anuren. *Wilhelm Roux Arch. EntwMech. Org.* **120**: 384–706.

Wallingford, J. B. and Harland, R. M., 2001. *Xenopus* Dishevelled signaling regulates both neural and mesodermal convergent extension: parallel forces elongating the body axis. *Development* **128**: 2581–2592.

Wallingford, J. B., Rowning, B. A., Vogeli, K. M., Rothbacher, U., *et al.*, 2000. Dishevelled controls cell polarity during *Xenopus* gastrulation. *Nature* **405**: 81.

Wallingford, J. B., Fraser, S. E. and Harland, R., 2002. Convergent extension: the molecular control of polarized cell movement during embryonic development. *Develop. Cell* **2**: 695–706.

Wei, Y. and Mikawa, T., 2000. Formation of the avian primitive streak from spatially restricted blastoderm: evidence for polarized cell division in the elongating streak. *Development* **127**, 87–96.

Wilson, P. and Keller, R., 1991. Cell rearrangement during gastrulation of *Xenopus*: Direct observation of cultured explants. *Development* **112**: 289–300.

Wilson, P., Oster, G. and Keller, R. E. 1989. Cell rearrangement and segmentation in *Xenopus*: Direct observation of cultured explants. *Development* **105**: 155–166.

Winklbauer, R. and Stoltz, C., 1995. Fibronectin fibril growth in the extracellular matrix of the *Xenopus* embryo. *J. Cell Sci.* **108**: 1575–1586.

Winklbauer, R., Medina, A., Swain, R. and Steinbeisser, H. 2001., Frizzled-7 signalling controls tissue separation during *Xenopus* gastrulation. *Nature* **413**: 856–860.

Yamamoto, A., Amacher, S., Kim, S., Geissert, D., *et al.*, 1998. Zebrafish paraxial protocadherin is a downstream target of spadetail involved in morphogenesis of gastrula mesoderm. *Development* **125**: 3389–3397.

Zhong, Y., Brieher, W. and Gumbiner, B., 1999. Analysis of C-cadherin regulation during tissue morphogenesis with an activating antibody. *J. Cell Biol.* **144**: 351–359.

19

Regulation of Cell Migration *In Vitro* and *In Vivo*

Donna J. Webb, Karen Donais, Shin-ichi Murase, Hannelore Asmussen and **Alan F. Horwitz**

Understanding migration is challenging because it involves the integration of many different processes that occur in distinct locations in the cell. Considerable progress has been made in identifying molecules that contribute to migration, but the mechanisms that regulate its central processes, such as formation and disassembly of adhesions, remain largely unknown. As mechanisms that regulate migration emerge from *in vitro* studies, the next challenge is to determine whether these same mechanisms also regulate migration *in vivo*. We have developed assays to gain insight into these processes both *in vitro* and *in vivo*, which is the focus of this overview. To address the mechanisms of adhesion formation and disassembly, we visualize the dynamics of adhesion molecules fused to GFP in migrating cells. FAK, paxillin and zyxin localized in small clusters near the leading edge of the cell. The adhesions at the base of a protrusion disappear as new adhesions form near the leading edge, which we refer to as adhesion turnover. FAK and Src activity regulate turnover of these nascent adhesions at the cell front. In addition, FAK plays a role in the disassembly of adhesions at the rear of the cell. We then asked whether the same mechanisms regulate migration *in vitro* and *in vivo*. To address this, we are developing *in situ* migration systems that closely mimic the *in vivo* environment. To date, these include migration of muscle precursors from somites to the limb bud and migration of neuronal precursors in the rostral migratory stream (RMS). The basic migratory cycle

Cell Motility. Edited by Anne Ridley, Michelle Peckham and Peter Clark.
 ISBN 0 470 84872 3

observed *in vitro*, which includes extension of protrusions, formation of stable attachments at the leading edge, translocation of the cell body forward, and release at the cell rear, is recapitulated by the precursors migrating *in situ*. However, unlike cultured cells, both the muscle and neuronal precursors extended a single, long, persistent protrusion in the direction of migration, a process regulated by Rac activation. Thus, the present cell culture systems may not completely recapitulate the complex *in vivo* environment.

Introduction

Cell migration is central to many biological and pathological processes (Lauffenburger and Horwitz, 1996) and, thus, understanding the molecular basis of migration is of considerable importance. This is a challenging task since migration involves the integration of several different cellular processes, including the formation of protrusions and adhesions, translocation of the cell body and release of adhesions at the cell rear. While many proteins that contribute to migration have been identified, the mechanisms that regulate some of its central processes, such as the formation and disassembly of adhesions, are not well known. Understanding these processes should provide significant insight into the mechanisms that regulate migration.

Most of the studies examining the molecular basis of migration have utilized fibroblasts or tumour cells growing on tissue culture dishes. While this is an attractive model system because of the convenience, cost and ease with which gene expression can be manipulated, it is presently unclear how well migration in cell culture systems mimics that seen *in vivo*. As mechanisms that regulate migration emerge from *in vitro* studies, the next challenge is to determine if these same mechanisms also regulate migration *in vivo*. The technology to examine migration *in vivo* is rapidly emerging making this formidable task possible.

Cell migration begins with the extension of the membrane in the direction of movement. The assembly of actin filaments at the leading edge of the cell drives the initial extension of the plasma membrane known as a protrusion (Borisy and Svitkina, 2000; Carson *et al.*, 1986; Wang, 1985). For migration to occur, the protrusion must be stabilized by the formation of adhesion complexes, which usually consist of the integrin family of transmembrane receptors, signalling, adaptor and cytoskeletal proteins (Burridge and Chrzanowska-Wodnicka, 1996; Schoenwaelder and Burridge, 1999; Yamada and Miyamoto, 1995). The integrin serves as the link between the extracellular matrix and the cytoskeleton. Rac regulates the formation of the nascent adhesions at the cell front that drive rapid cell migration probably by serving as traction points for the propulsive forces that move the cell body forward (Beningo *et al.*, 2001; Galbraith and Sheetz, 1997; Lee *et al.*, 1994; Oliver *et al.*,

1999). The small adhesions can mature into larger, more organized structures, which tend to inhibit migration (Burridge and Chrzanowska-Wodnicka, 1996; Chrzanowska-Wodnicka and Burridge, 1996; Zamir and Geiger, 2001). Disassembly of adhesions at the rear results in the net translocation of the cell in the direction of movement.

This overview focuses on two issues that are of recent interest to us. The first is the mechanisms by which adhesions form and disassemble. To address this, we have initiated a ratio imaging approach using GFP variants fused to key adhesion, signalling, and cytoskeletal proteins. The second is a comparison between the mechanisms that regulate migration *in vitro* and *in vivo*. We have developed *in situ* migration systems that closely mimic the *in vivo* environment to answer this question. A major advantage of our *in situ* system is that we can express exogenous GFP-tagged proteins in the slices and observe the molecular and cellular dynamics of migration by time-lapse microscopy.

Adhesion dynamics in migrating cells

Adhesion formation

Most studies examining adhesion formation and disassembly have utilized quiescent fibroblasts, which are not highly motile and tend to form large adhesions. Since these processes are poorly understood in migrating cells, we are studying adhesion dynamics under conditions that promote prominent migration, such as plating cells on low concentrations of fibronectin for short periods of time (less than 1 h). Our approach is to visualize the dynamics of adhesion molecules fused to GFP, expressed either singly or in pairs, in these migrating cells (Laukaitis *et al.*, 2001). The molecules include $\alpha5$ integrin, focal adhesion kinase (FAK), zyxin, paxillin and α-actinin.

Since the leading edge of a cell is a site where new adhesions form, we examined the dynamics of adhesion formation in protrusive regions of the cell. At a 60 s time resolution, FAK, paxillin and zyxin enter the nascent adhesions at about the same time. Surprisingly, we were unable to detect visibly organized α-actinin or $\alpha5$ integrin in these nascent adhesions. However, after protrusive activity ceased, α-actinin entered adhesions, which was subsequently followed by detection of visible concentrations of $\alpha5$ integrin in these structures. The adhesions containing visibly organized α-actinin and $\alpha5$ integrin probably represent the more mature adhesions that are studied by most others (Burridge and Chrzanowska-Wodnicka, 1996; Chrzanowska-Wodnicka and Burridge, 1996).

The absence of organized $\alpha5$ integrin in the initial FAK/paxillin containing adhesions was surprising because it is generally assumed that integrin ligation and/or clustering serves to nucleate the formation of adhesions. Although it is

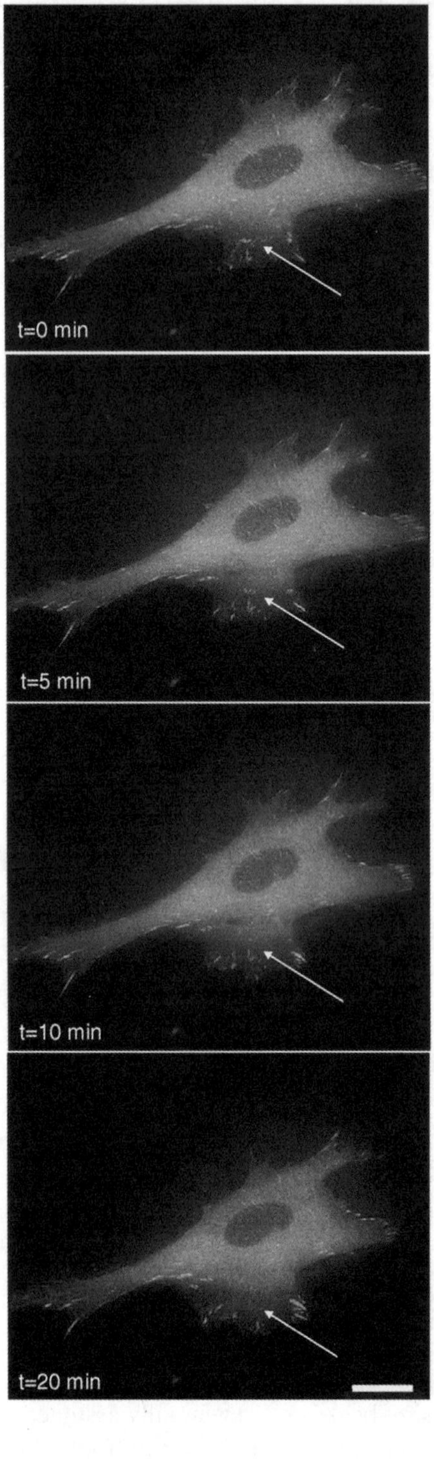
t=0 min
t=5 min
t=10 min
t=20 min

possible that other molecules may serve this function, it is clear that α5 is the major integrin in CHO cells and is required for adhesion and migration on fibronectin. In addition, antibody staining against various integrins and other putative adhesion molecules did not show readily detectable amounts of these proteins associated with the nascent adhesions. Recent data using time-lapse confocal microscopy suggests that the integrin serves as the nucleation site for the new adhesions (Claire Brown and Donna Webb, unpublished observation). We detect diffuse α5 integrin near the leading edge of the cell prior to the recruitment of paxillin, which overlays the integrin. Thus, the α5 integrin-containing nucleation sites are not organized as large visible complexes, but present in a concentration sufficient to stimulate the recruitment of other molecules, such as paxillin.

At this stage our studies support a regulated, sequential model for adhesion formation. We have clear evidence that some components enter the adhesions serially and with quite distinct kinetics, e.g., α-actinin and paxillin. The distinct kinetics suggest a role for key, uncharacterized regulatory events that mediate the entry of some components. It is also likely that some components enter adhesions simultaneously, either individually or as preformed complexes. For example, GIT1 appears to target to adhesions in a complex with signalling components including PIX and PAK (Manabe *et al.*, 2002). Higher resolution kinetics of other adhesion components, such as paxillin and zyxin, will help determine whether their entry is also coordinated. The next step in our studies of adhesion assembly is to use the ratio imaging approach with mutant molecules to determine what targets various molecules into nascent adhesions.

Adhesion turnover

In both fibroblasts and CHO cells, the paxillin clusters at the base of a protrusion disappear while adhesions form near the new leading edge of the cell (Figure 19.1). This turnover appears to differ from the breakdown of adhesions at the cell rear (see next section) and thus, to distinguish them, we refer to the former as turnover and the latter as disassembly. Since adhesion turnover has not been previously described, we examined it in more detail.

Figure 19.1 (*opposite*) Paxillin adhesions localize near the leading edge of a protrusion and turn over as new adhesions form. WI38 cells (human fibroblasts) expressing paxillin-GFP were plated on 1 μg/ml fibronectin for 1 h at 37°C and adhesion dynamics were visualized by time-lapse fluorescent microscopy. The paxillin clusters at the base of the protrusion disappear while adhesions form near the new leading edge of the cell. We refer to this as adhesion turnover. The arrow indicates a group of paxillin adhesions that turn over as the cell protrudes (compare t = 0 and t = 20 min). Scale bar = 20 μm

FAK and zyxin, but not visibly organized α-actinin, localize in these dynamic paxillin adhesions. The turnover is a property of the adhesion as a whole since other components turn over as well. For example, when paxillin-DsRed2 is co-expressed with either GFP-FAK or zyxin-GFP, all three components turn over at the base of the protruding lamellipodium. Thus, the adhesions appear to turn over in protruding regions of the cell. This disassembly of adhesions followed by formation of new adhesions appears to be important since we consistently see a coupling between the rate of adhesion turnover and protrusion frequency. The fate of the adhesion components after they turn over is presently unknown; however, one possibility is that some of this material is utilized in the formation of new adhesions.

Defects in migration of fibroblasts from mice lacking some tyrosine kinases and phosphatases have been speculated to arise from impaired turnover of adhesions (Angers-Loustau *et al.*, 1999; Ilic *et al.*, 1995; Klinghoffer *et al.*, 1999; Yu *et al.*, 1998). This points to the phosphorylation/dephosphorylation of their substrates as key regulators of adhesion dynamics. We investigated this by expressing paxillin-GFP and zyxin-GFP in fibroblasts from mice lacking FAK. We determined the rate constants for disassembly of adhesion components by measuring the total fluorescent intensity in specific adhesions as a function of time. The rate constant for paxillin disassembly decreased 14-fold in FAK null fibroblasts when compared with wild-type fibroblasts. A similar inhibition of disassembly was also observed with zyxin suggesting that it is a property of the entire adhesion. Src, another tyrosine kinase, showed a similar inhibition of adhesion turnover. Co-expression of kinase-deficient Src with paxillin-GFP decreased the rate constant for disassembly by 18-fold. Taken together, these results demonstrate that both FAK and Src regulate adhesion turnover.

Adhesion disassembly at the cell rear

Adhesion disassembly at the cell rear is not a simple reversal of the mechanisms of formation. Several studies have shown that a fraction of the integrin can be left behind on the substrate of migrating cells (Palecek *et al.*, 1996, 1998; Regen and Horwitz, 1992). In CHO cells and fibroblasts, we have made similar observations using α5 integrin-GFP (Figure 19.2). By contrast, neither paxillin nor α-actinin is detected in the tracks behind the cell indicating that the cleavage of the integrin–cytoskeletal linkage occurs at or very close to the integrin. Paxillin and α-actinin remain in clusters within the cell. These clusters move along the edge of the cell and are stable for over 30 min, but they eventually disperse. This contrasts the adhesion disassembly at the cell front in which paxillin, zyxin and FAK rapidly turn over with similar kinetics. Interestingly, in FAK null fibroblasts, paxillin, like integrin, is left behind on

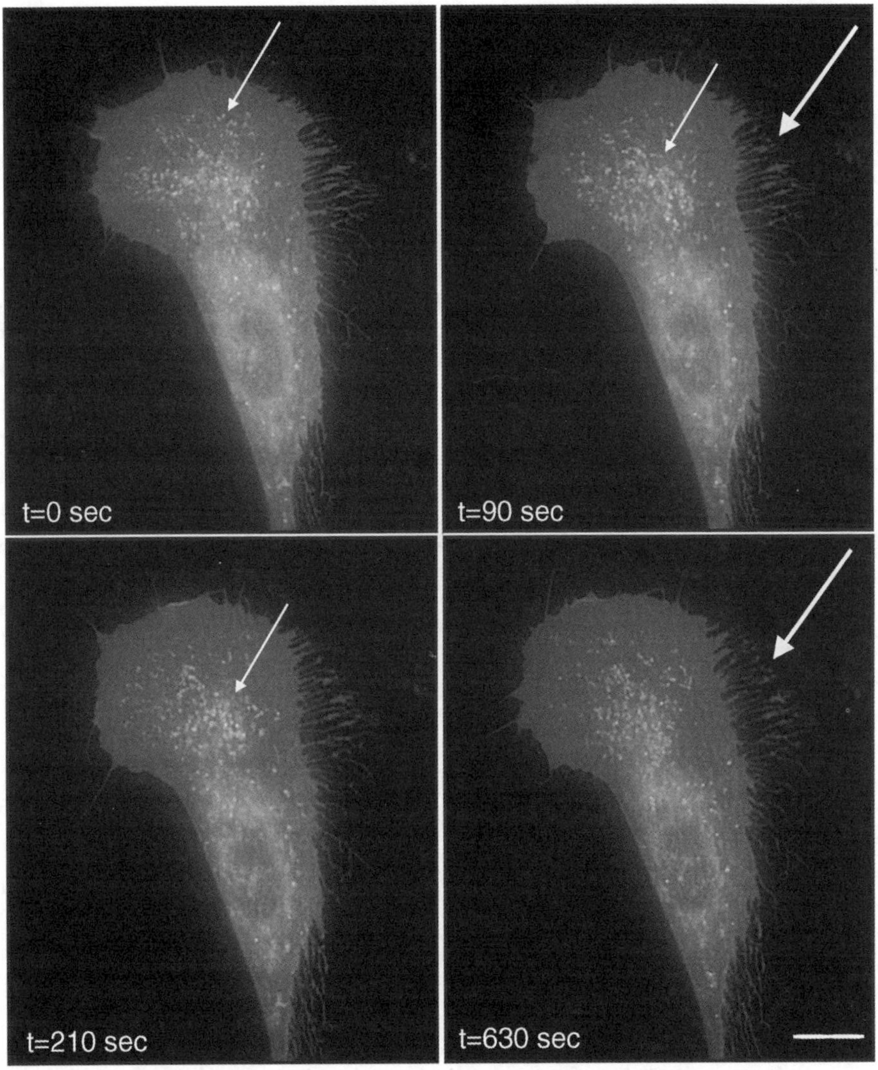

Figure 19.2 α5 Integrin resides in vesicles. In α5-GFP expressing WI38 cells, integrin vesicles emanated from the leading edge of a protrusion and moved toward a perinuclear region (small arrows). The large number of vesicular structures around the perinuclear region makes tracking individual vesicles difficult. Also note that in the migrating fibroblast, α5-GFP is detected in tracks behind the cell (large arrows). However, neither paxillin nor α-actinin is detected in these tracks indicating that the cleavage of the integrin–cytoskeletal linkage occurs at or very close to the integrin. Scale bar = 20 μm

the substrate as the rear of the cell retracts. These results point to a critical role for FAK in the disassembly of adhesions at the rear of the cell as well as at the front.

A working model

Our observations of adhesion formation, disassembly and turnover in migrating cells point to the following working model. Integrin ligation to the ECM initiates the recruitment of signalling and adaptor molecules to newly forming contact sites. These nascent adhesions, which contain FAK, paxillin and zyxin, are highly dynamic and in the presence of FAK and Src activity turn over at the base of the protrusion. The turnover is coupled to the formation of new protrusions and some components may be reutilized for the assembly of new adhesions. Although the phosphorylation targets for FAK and Src in these nascent adhesions are not known, it is interesting to note that paxillin is a substrate for both of these molecules (Turner, 2000). In the absence of FAK and Src activity, these developing adhesions grow in size and molecular complexity as structural molecules, like α-actinin, are recruited to this site. Once α-actinin is present in the adhesions, they no longer turn over, but tend to slide inward toward the cell body. Subsequently, visible concentrations of integrin enter the adhesions and function to stabilize them. At the cell rear, cleavage of the integrin–cytoskeletal linkage, at a site proximal to the integrin, initiates breakdown of adhesions. FAK also appears to be necessary for disassembly of adhesions at the rear of the cell. The remaining paxillin and α-actinin containing complexes move toward the cell body and then eventually disperse.

It is presently not known how FAK activity promotes adhesion turnover. Some possibilities include signalling to activate Rac (or inhibit Rho), inhibiting contractility, which leads to a loss of organization, and/or post-translational modifications of a key adaptor molecule resulting in weakened interactions with other adhesion components. In considering these hypotheses, the generally concerted nature of adhesion turnover points to an event that rapidly propagates through the adhesion.

Intracellular trafficking of adhesion molecules

$\alpha 5$ Integrin resides in vesicles in fibroblasts

In polarized, migrating cells, adhesion components from the front of the cell accumulate near the cell rear away from the leading edge. It is possible that mechanisms exist to shuttle these molecules from the cell rear toward the cell front where they can be reutilized for the formation of new adhesions. One can envision that adhesion molecules, such as integrins, may traffic in endocytic vesicles either directly from the rear to the front of the cell or through endocytic recycling pathways. We and others have observed integrin containing vesicles moving from the rear of the cell to a perinuclear region (Bretscher, 1984; Bretscher and Aguado-Velasco, 1998; Laukaitis *et al.*, 2001; Lawson and Maxfield, 1995; Palecek *et al.*, 1996; Pierini *et al.*, 2000; Regen and Horwitz, 1992). However, we have not yet detected $\alpha5$ integrin vesicles moving directly from the rear to the front of the cell. This may be due to our inability to track the integrin vesicles once they enter the perinuclear region where the volume of the cell and the large number of vesicular structures preclude detection. Recently, we observed a fraction of the cytoplasmic paxillin, like $\alpha5$ integrin, trafficking in vesicles that co-localized with an endocytic marker, FM 4-64. In migrating fibroblasts, in which we saw minimal membrane ruffling, $\alpha5$ containing vesicles moved from the perinuclear region to the base, but not into the lamellipodium. In CHO and WI38 cells, where membrane ruffling was apparent, we also observed $\alpha5$ integrin-GFP vesicles that emanated from the leading edge in protrusions and concentrated in a perinuclear region (Figure 19.2).

There is good evidence for the trafficking of adhesion molecules in vesicles from both the rear and leading edge of cells and from the perinuclear region to the base of the lamellipodia (Bretscher, 1984; Bretscher and Aguado-Velasco, 1998; Laukaitis *et al.*, 2001; Lawson and Maxfield, 1995; Palecek *et al.*, 1996; Pierini *et al.*, 2000; Regen and Horwitz, 1992). However, it is currently unknown whether trafficking represents a significant mechanism for supplying materials to the front of the cell. To clarify the role that trafficking plays in regulating cellular processes, additional questions need to be addressed. For example, what fraction of the integrins trafficking from the cell rear or front are recycled versus degraded in a lysosomal compartment and what is the role of de novo biosynthesis versus recycling in supplying material to new adhesions?

These trafficking questions may be more readily addressed using a highly polarized cell type, such as a neuron, since it would avoid the high concentration of vesicles in the perinuclear region and the trafficking can be easily seen and assayed. When $\alpha5$ integrin-GFP is expressed in hippocampal neurons, we observed trafficking of the integrin in vesicle-like structures, analogous to those seen in the fibroblasts. These vesicles moved from the soma along the neurites toward the growth cones (Figure 19.3). As the neurites extended, the integrin containing structures also moved from the growth cones toward the soma. Most importantly, tracking individual integrin-containing vesicles in these cells is quite feasible (Figure 19.3). Thus, this provides us with a system to address the mechanisms of intracellular trafficking of adhesion components.

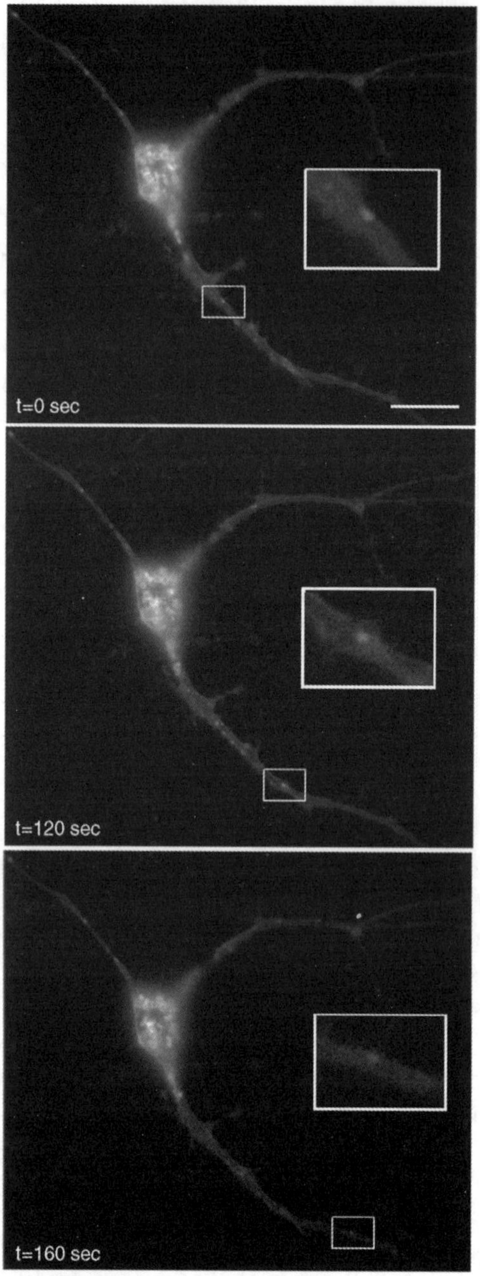

Figure 19.3 α5 Integrin traffics in vesicles in hippocampal neurons. In α5 integrin-GFP expressing hippocampal neurons, vesicles moved from the soma along the neurites toward the growth cones. An individual α5 vesicle, which is shown moving along the neurite (compare $t = 0$ and $t = 160$ s boxed regions), can be tracked over time. The inset shows a higher magnification of the α5 vesicle for each time point. Scale bar = 20 μm

Trafficking of adhesion components in complexes

It is becoming evident that signalling molecules may also traffic to and from functional regions of the cell via large, cytoplasmic complexes. GIT1, which is an adaptor protein that contains an ARF-GAP domain, cycles in large cytoplasmic complexes between the cytosol, adhesions and the leading edge (Manabe *et al.*, 2002). The GIT1 complexes are not vesicular, unlike what we observed with α5 integrin, since they do not co-localize with a number of different endocytic or Golgi markers. Instead they appear to represent large, motile, multi-molecular signalling complexes that contain molecules such as paxillin, PAK and PIX. By contrast, at least some of these adhesion components may also reside in vesicular structures (Di Cesare *et al.*, 2000; Paris *et al.*, 2002). These adhesion components can localize to distinct subcellular compartments, although their function is presently not known. Our studies indicate that the GIT1 complexes can target constitutively activated PAK to adhesions and the leading edge suggesting that these structures can regulate the intracellular distribution of signalling molecules.

Migration *in vivo*

Most studies examining mechanisms of cell migration have utilized cells growing in various culture systems. However, the ECM and growth factor environment *in vivo* are not generally known and differ from that *in vitro*. For example, cells migrating *in vivo*, unlike cells in culture, are exposed to a three-dimensional matrix of various ECM molecules. *In vivo*, cells encounter gradients of many different growth factors from the surrounding tissue while cells in culture are usually exposed to a homogeneous bolus of a single or small set of growth factors. In this context, it is unclear whether cells *in vitro* and *in vivo* use the same mechanisms to migrate. We have begun to address this issue by developing systems for studying cell migration *in situ*. The goal is to extend the measurements of migration, molecular localization and dynamics that are usually made in culture systems to cells migrating *in situ*.

Our approach is to image migrating cells in 200–300 μm slices using two different model systems. These include migration of muscle precursors from somites to the limb buds during development and migration of neuronal precursors from the subventricular zone to the olfactory bulb via the rostral migratory stream (RMS). We use time-lapse microscopy to observe the migration of individual cells that are either labelled with the fluorescent marker, DiI, or by expressing GFP-tagged fusion proteins in the slices. These studies have already provided unexpected differences in migration between the slice cultures and dissociated cell cultures *in vitro*. For example, the long, highly polarized, persistent protrusions observed in the slice cultures are not usually seen in migrating fibroblasts *in vitro*.

Somitic migration

Myogenic precursor cells migrate from the lateral part of the somites into the forelimb where they form muscles of the limb (Chevallier *et al.*, 1977; Christ *et al.*, 1977; Ordahl and Le Douarin, 1992; Williams and Ordahl, 1994). We prepared slice cultures from chick embryos and followed the dynamics of the migrating cells. Somitic cells, which were labelled with DiI, did not migrate along a highly constrained pathway. Rapid bursts of directed migration were followed by periods in which the cells wandered away from the restricted pathway. Unlike cultured cells, the somitic cells extended a single, long, persistent protrusion in the direction of migration. The protrusions moved laterally, but once stable attachments formed the cells moved forward. Some cells extended and retracted protrusions, often for long periods of time, without much accompanying movement suggesting that the formation of stable adhesions is not tightly coupled to the formation of a protrusion (Knight *et al.*, 2000). The exaggerated, persistent protrusions were not observed in cells that moved out of the slice. Instead, the cells extended random protrusions and migration was no longer directional indicating the influence of the surrounding environment on the defined migration.

The small GTPase Rac is a key regulator of protrusive activity in cultured cells (Ridley and Hall, 1992). We expressed wild-type Rac, dominant-negative Rac (N17Rac) and constitutively-active Rac (L61Rac) as GFP fusion proteins in the slices to determine whether this molecule contributed to the generation of the large protrusions in the embryo slices. The exaggerated, persistent protrusions are not observed in somitic cells expressing dominant negative Rac, and migration was inhibited. In cells expressing constitutively-active Rac, many small protrusions formed and retracted frequently, and migration was random. Thus, Rac regulates the polarization that characterizes the unusual protrusive activity observed in the somitic cells.

Migration of neuronal precursors in the rostral migratory stream

In rodents, olfactory interneuron precursors are generated in the subventricular zone (SVZ) of the forebrain in the embryo and throughout the adult life (Garcia-Verdugo *et al.*, 1998; Luskin, 1993). The neuronal precursors in the anterior part of the SVZ then migrate to their targets in the centre of the olfactory bulb and differentiate into mature interneurons. This restricted pathway composed of migrating neural precursor cells is known as the rostral migratory stream (RMS). Although the mechanisms that regulate migration along this pathway are poorly understood, studies from knockout mice suggest that the polysialic acid chains (PSA) on N-CAM molecules are involved in this process (Ono *et al.*, 1994; Tomasiewicz *et al.*,

1993). PSA may affect migration in the RMS by reducing the adhesiveness of N-CAM molecules; however, this effect of PSA cannot explain how cells migrate actively and unidirectionally toward the olfactory bulb. In addition, in mice deficient in N-CAM, migration in the RMS is only partially diminished (Chazal *et al.*, 2000) suggesting that other molecules contribute to the directed migration in this pathway.

To identify other molecules that regulate RMS migration, we developed a slice culture system that allows us to image the migrating cells at high temporal and spatial resolution in the presence and absence of function-blocking antibodies against integrins and other putative adhesion or guidance molecules (Murase and Horwitz, 2002; Webb *et al.*, 2002). We prepared 200 μm thick slices from the forebrains of postnatal mice and labelled the migrating neuronal precursors with DiI crystals. In migrating cells, we observed a single, long and persistent leading process orientated in the direction of migration (Figure 19.4). In contrast to the somitic cells, the neuronal precursors migrated along a highly constrained pathway; the average rate of migration was about 100 μm/h. When the slices were treated with function blocking integrin antibodies against $\alpha 1$, $\beta 1$ or αv at the stage of development in which the integrins were expressed, migration was inhibited. This suggests a role for integrins in the migration of neuronal precursors in the RMS.

Since the netrin family and their receptors are emerging as key molecules in guiding growing axons to their targets and directing cell migration during neural development, we asked whether the netrin-1 receptor, Deleted in Colorectal Carcinoma (DCC), played a role in RMS migration. When we incubated the brain slices with function blocking DCC antibodies, migration was no longer unidirectional. These cells continued to extend a single, long protrusion but it retracted frequently and was not directionally polarized. This suggests that DCC, possibly through an interaction with netrin-1, contributes to the formation of directed protrusions and migration of the neural precursors along a defined pathway. While both integrins and DCC appear to regulate migration in the RMS, these molecules have distinct functions. DCC is involved in directional, but not random, migration of the neural precursors in the RMS whereas integrin ligation is essential for any movement of these cells.

Conclusions

It is now apparent that the biological systems and imaging technology are available for studying cell migration *in situ*. While others have studied migration in similar systems, the focus has been on the migration pathways and guidance. The new focus is on using these systems to study the dynamics

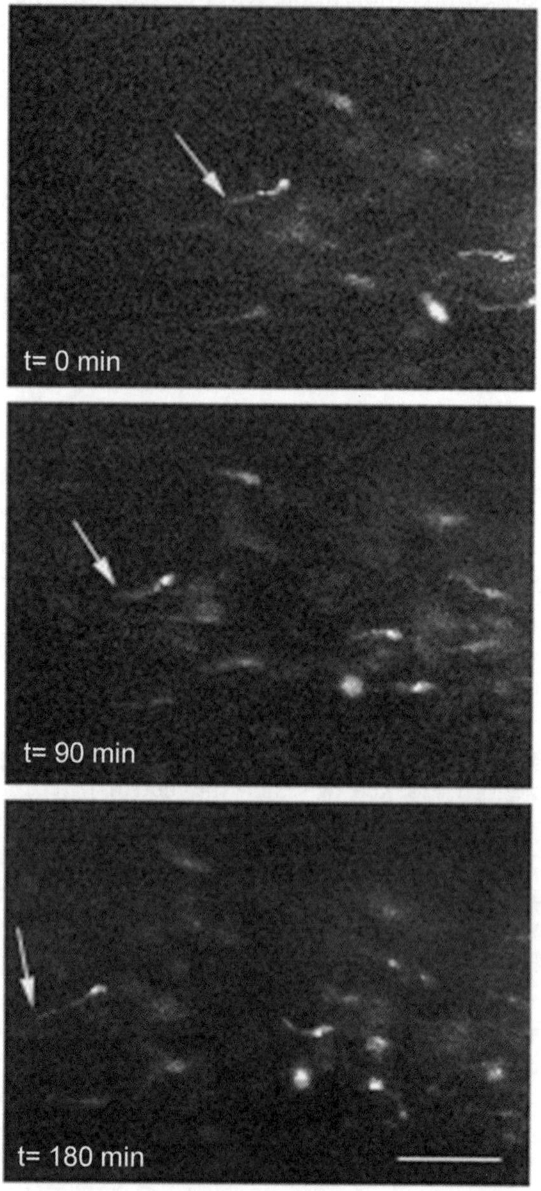

Figure 19.4 Migration of neuronal precursors in the rostral migratory stream (RMS). Brain slices from the forebrain, including the olfactory bulb, from a postnatal day 9 mouse were labelled with small DiI crystals and migration was visualized by time-lapse microscopy. Images are shown depicting unidirectional migration of neural precursors from the anterior part of the subventricular zone (right side of panel) to the centre of the olfactory bulb (left side of panel). The arrows indicate a single, long, leading process. Scale bar = 50 μm

of the migrating cells on a molecular level. The ability to introduce genes into the migrating cells in the slice cultures will allow us to determine whether the same processes that regulate migration *in vitro* also contribute to migration *in vivo*.

The slice technology can be extrapolated to other systems in both embryonic and adult tissues. Of particular interest will be studies observing the molecular dynamics of migrating cells from transgenic mice with selected genes deleted or mutated. Studies using knock-in mice with GFP-labelled wild-type and mutant proteins will also be especially informative. Finally, labelling the highly polarized cells that we observe *in situ* with antibodies against specific phosphorylation activation states should provide insight into which molecules are locally activated.

References

Angers-Loustau, A., Cote, J.-F., Charest, A., Dowbenko, D., *et al.*, 1999. Protein tyrosine phosphatase-PEST regulates focal adhesion disassembly, migration, and cytokinesis in fibroblasts. *J. Cell Biol.* **144**: 1019–1031.

Beningo, K. A., Dembo, M., Kaverina, I., Small, J. V. and Wang, Y.-L., 2001. Nascent focal adhesions are responsible for the generation of strong propulsive forces in migrating fibroblasts. *J. Cell Biol.* **153**: 881–888.

Borisy, G. G. and Svitkina, T. M., 2000. Actin machinery: pushing the envelope. *Curr. Opin. Cell Biol.* **12**: 104–112.

Bretscher, M. S., 1984. Endocytosis: relation to capping and cell locomotion. *Science* **224**: 681–686.

Bretscher, M. S. and Aguado-Velasco, C., 1998. Membrane traffic during cell locomotion. *Curr. Opin. Cell Biol.* **10**: 537–541.

Burridge, K. and Chrzanowska-Wodnicka, M., 1996. Focal adhesions, contractility, and signaling. *Annu. Rev. Cell. Dev. Biol.* **12**: 463–518.

Carson, M., Weber, A. and Zigmond, S. H., 1986. An actin-nucleating activity in polymorphonuclear leukocytes is modulated by chemotactic peptides. *J. Cell Biol.* **103**: 2707–2714.

Chazal, G., Durbec, P., Jankovski, A., Rougon, G. and Cremer, H., 2000. Consequences of neural cell adhesion molecule deficiency on cell migration in the rostral migratory stream of the mouse. *J. Neurosci.* **20**: 1446–1457.

Chevallier, A., Kieny, M. and Mauger, A., 1977. Limb–somite relationship:origin of the limb musculature. *J. Embryol. Exp. Morphol.* **41**: 245–258.

Christ, B., Jacob, H. J. and Jacob, M., 1977. Experimental analysis of the origin of the wing musculature in avian embryos. *Anat. Embryol.* **150**: 171–186.

Chrzanowska-Wodnicka, M. and Burridge, K., 1996. Rho-stimulated contractility drives the formation of stress fibers and focal adhesions. *J. Cell Biol.* **133**: 1403–1415.

Di Cesare, A., Paris, S., Albertinazzi, C., Dariozzi, S., *et al.*, 2000. p95-APP1 links membrane transport to Rac-mediated reorganization of actin. *Nat. Cell Biol.* **2**: 521–530.

Galbraith, C. G. and Sheetz, M. P., 1997. A micromachined device provides a new bend on fibroblast traction forces. *Proc. Natl. Acad. Sci. USA* **94**: 9114–9118.

Garcia-Verdugo, J., Doetsch, F., Wichterle, H., Lim, D. and Alvarez-Buylla, A., 1998. Architecture and cell types of the adult subventricular zone: in search of the stem cells. *J. Neurobiol.* **36**: 234–248.

Ilic, D., Furuta, Y., Kanazawa, S., Takeda, N., *et al.*, 1995. Reduced cell motility and enhanced focal adhesion contact formation in cells from FAK-deficient mice. *Nature* **377**: 539–544.

Klinghoffer, R. A., Sachsenmaier, C., Cooper, J. A. and Soriano, P., 1999. Src family kinases are required for integrin but not PDGFR signal transduction. *Embo J.* **18**: 2459–2471.

Knight, B., Laukaitis, C., Akhtar, N., Hotchin, N. A., *et al.*, 2000. Visualizing muscle cell migration in situ. *Curr. Biol.* **10**: 576–585.

Lauffenburger, D. A. and Horwitz, A. F., 1996. Cell migration: a physically integrated molecular process. *Cell* **84**: 359–369.

Laukaitis, C. M., Webb, D. J., Donais, K. and Horwitz, A. F., 2001. Differential dynamics of $\alpha5$ integrin, paxillin, and α-actinin during formation and disassembly of adhesions in migrating cells. *J. Cell Biol.* **153**: 1427–1440.

Lawson, M. A. and Maxfield, F. R., 1995. Ca^{2+}- and calcineurin-dependent recycling of an integrin to the front of migrating neutrophils. *Nature* **377**: 75–79.

Lee, J., Leonard, M., Oliver, T., Ishihara, A. and Jacobson, K., 1994. Traction forces generated by locomoting keratocytes. *J. Cell Biol.* **127**: 1957–1964.

Luskin, M. B., 1993. Restricted proliferation and migration of postnatally generated neurons derived from the forebrain subventricular zone. *Neuron* **11**: 173–189.

Manabe, R., Kovalenko, M., Webb, D. J. and Horwitz, A. R., 2002. GIT1 functions in a motile, multi-molecular signaling complex that regulates protrusive activity and cell migration. *J. Cell Sci.* **115**: 1497–1510.

Murase, S. and Horwitz, A. F., 2002. Deleted in colorectal carcinoma and differentially expressed integrins mediate the directional migration of neural precursors in the rostral migratory stream. *J. Neurosci.* **22**: 3568–3579.

Oliver, T., Dembo, M. and Jacobson, K., 1999. Separation of propulsive and adhesive traction stresses in locomoting keratocytes. *J. Cell Biol.* **145**: 589–604.

Ono, K., Tomasiewicz, H., Magnuson, T. and Rutishauser, U., 1994. N-CAM mutation inhibits tangential neuronal migration and is phenocopied by enzymatic removal of polysialic acid. *Neuron* **13**: 595–609.

Ordahl, C. P. and Le Douarin, N. M., 1992. Two myogenic lineages within the developing somite. *Development* **114**: 339–353.

Palecek, S. P., Schmidt, C. E., Lauffenburger, D. A. and Horwitz, A. F., 1996. Integrin dynamics on the tail region of migrating fibroblasts. *J. Cell Sci.* **109**: 941–952.

Palecek, S. P., Huttenlocher, A., Horwitz, A. F. and Lauffenburger, D. A., 1998. Physical and biochemical regulation of integrin release during rear detachment of migrating cells. *J. Cell Sci.* **111**: 929–940.

Paris, S., Za, L., Sporchia, B. and de Curtis, I., 2002. Analysis of the subcellular distribution of avian p95-APP2, and ARF-GAP orthologous to mammalian paxillin kinase linker. *Int. J. Biochem. Cell Biol.* **34**: 826–837.

Pierini, L. M., Lawson, M. A., Eddy, R. J., Hendey, B. and Maxfield, F. R., 2000. Oriented endocytic recycling of $\alpha5\beta1$ in motile neutrophils. *Blood* **95**: 2471–2480.

Regen, C. M. and Horwitz, A. F., 1992. Dynamics of $\beta1$ integrin-mediated adhesive contacts in motile fibroblasts. *J. Cell Biol.* **119**: 1347–1359.

Ridley, A. J. and Hall, A., 1992. The small GTP-binding protein rho regulates the assembly of focal adhesions and actin stress fibers in response to growth factors. *Cell* **70**: 389–399.

Schoenwaelder, S. M. and Burridge, K., 1999. Bidirectional signaling between the cytoskeleton and integrins. *Curr. Opin. Cell Biol.* **11**: 274–286.

Tomasiewicz, H., Ono, K., Yee, D., Thompson, C., *et al.*, 1993. Genetic deletion of a neural cell adhesion molecule variant (N-CAM-180) produces distinct defects in the central nervous system. *Neuron* **11**: 1163–1174.

Turner, C. E., 2000. Paxillin and focal adhesion signalling. *Nat. Cell Biol.* **2**: E231–E236.

Wang, Y.-L., 1985. Exchange of actin subunits at the leading edge of living fibroblasts: possible role of treadmilling. *J. Cell Biol.* **101**: 597–602.

Webb, D. J., Asmussen, H., Murase, S. and Horwitz, A. F., 2002. Cell migration in slice cultures. *Methods Cell Biol.* **69**: 341–358.

Williams, B. A. and Ordahl, C. P., 1994. Pax-3 expression in segmental mesoderm marks early stages in myogenic cell specification. *Development* **120**: 785–796.

Yamada, K. M. and Miyamoto, S., 1995. Integrin transmembrane signaling and cytoskeletal control. *Curr. Opin. Cell Biol.* **7**: 681–689.

Yu, D.-H., Qu, C.-K., Henegariu, O., Lu, X. and Feng, G.-S., 1998. Protein-tyrosine phosphatase Shp-2 regulates cell spreading, migration, and focal adhesion. *J. Biol. Chem.* **273**: 21 125–21 131.

Zamir, E. and Geiger, B., 2001. Molecular complexity and dynamics of cell–matrix adhesions. *J. Cell Sci.* **114**: 3583–3590.

20

Genes that Control Cell Migration during Mouse Development

Carmen Birchmeier

Cell migration is an important and frequent process in development, and the molecules that control migration events have received much attention. Of particular interest is the molecular nature of signals responsible for the release of cells from their origin, for the maintenance of cellular motility, or for target finding. Tyrosine kinase receptors play important roles in migration processes in development and metastasis. I have reviewed here evidence from my laboratory that demonstrated that tyrosine kinase receptors implicated in tumorigenesis, like the c-ErbB or c-Met, turn out to control decisive steps in migration of embryonic cells *in vivo*.

Introduction

During development, cell migration is an important process and it is frequently observed that cells are born at one position and subsequently move to their final locations. In development, migration events are tightly controlled. Cells are released at defined stages and positions, and they move along defined routes to their particular target sites. Molecules that control such migration events have received much attention. Of particular interest is the molecular nature of signals responsible for the release of the cells, for the

---

Cell Motility. Edited by Anne Ridley, Michelle Peckham and Peter Clark.
 ISBN 0 470 84872 3

maintenance of cellular motility, or for directed migration and target finding. I will summarize here data from my laboratory that contributed to an understanding of the molecular control of cell migration in the embryo. Over the years, we analysed genes that control migration of two developing cell types, neural crest cells and muscle progenitor cells. Both cell types are released from epithelial structures, and in order to detach and become motile they first undergo an epithelial–mesenchymal conversion. Mechanistically, the release of the cells and their subsequent motility resembles the process that is observed late during the progression of malignant carcinomas. There, cells detach from the primary tumour by epithelial–mesenchymal conversion and migrate in an uncontrolled manner to form metastases at sites distant of the primary tumour. Because of the mechanistic similarities, it is not so astonishing that genes implicated in tumour progression play also important roles in cell migration events during development. In particular, work from my laboratory demonstrated that tyrosine kinase receptors implicated in tumorigenesis, like the c-ErbB or c-Met tyrosine kinase receptors, turn out to control decisive steps in migration of embryonic cells.

Migration of neural crest cells

Neural crest cells are released from the dorsal portion of the neural tube during embryogenesis. The cells detach from the developing spinal cord and hindbrain, and migrate in a stereotypical manner to various sites in the embryo. Upon arrival at the targets, neural crest cells differentiate to form a wide variety of different cell types: neurons and glia of the peripheral nervous system, smooth muscle cells of the heart outflow tract, melanocytes, and also cartilage and bone of the head (Le Douarin and Kalcheim, 1999). Upon release from the neural tube, neural crest cells start to express a particular set of genes, for instance ErbB3 or Sox10, which can be used to identify the cells during migration.

ErbB receptors and their ligand, Nrg1

ErbB3 encodes a receptor of the type I family of tyrosine kinase receptors; the prototype member of this family is ErbB1 also known as the EGF receptor. ErbB receptors recognize specific ligands, which they bind with high affinity. Ligand binding leads to homo- or heterodimerization of ErbB receptors, inducing their tyrosine kinase activity and the tyrosine phosphorylation of the receptors and of other substrates. The phosphorylated tyrosine residues act as docking sites for various molecules, that subsequently transduce the signal in the cytoplasm (for a review of ErbB signalling see Yarden and Sliwkowski, 2001). Signals given by ErbB receptors can stimulate motility, proliferation, suppress apoptosis or induce differentiation. ErbB3, the receptor expressed at high levels in neural crest cells, recognizes neuregulins as ligands. The cytoplasmic domain of ErbB3 has homologies to tyrosine kinases, but the receptor appears to have little or no tyrosine kinase activity. Upon ligand binding, ErbB3 heterodimerizes with other ErbB receptors with activatable tyrosine kinase domains, which phosphorylate ErbB3 in trans and enable ErbB3 to signal. Neuregulins bind not only ErbB3, but also a different receptor of the family, ErbB4. However ErbB3 and ErbB4 are frequently expressed in distinct cell types and tissues and can therefore take over different and non-overlapping functions. In particular, ErbB3 but not ErbB4 is expressed in neural crest cells, and ErbB3 takes over decisive functions in development of this cell type.

Neuregulins, the ligands bound by the ErbB3 receptor, constitute a small family of EGF-like factors. Nrg1, the first identified member of the family, has been studied extensively. In cell culture, various biological activities of Nrg1 have been observed and were used to purify and characterize the factor: Nrg1 can induce growth of glial cells and was therefore also named GGF (Glial Growth Factor; Shah *et al.*, 1994). It can induce growth and differentiation of epithelial cells and was therefore named NDF (Neu Differentiation Factor; Marikovsky *et al.*, 1995) or Heregulin (Holmes *et al.*, 1992), and it can induce the expression of acetylcholine receptor in muscle cells and was therefore also named ARIA (Acetylcholine Receptor Inducing Activity; Falls *et al.*, 1993). My laboratory has investigated the functional role of Nrg1 and ErbB receptors and has demonstrated that the gene plays important roles in development of neural crest cells.

For this, we used mice as a model organism, and generated mutant alleles of ErbB3, ErbB2 and Nrg1 by the use of homologous recombination and the embryonal stem cell technology (Meyer and Birchmeier, 1995; Riethmacher *et al.*, 1997; Britsch *et al.*, 1998). Since the null-mutations interfere with survival of embryos, late developmental functions cannot be investigated in such mutant mice. We have therefore also used other, more sophisticated genetic techniques to overcome this limitation (Woldeyesus *et al.*, 1999; Garratt *et al.*, 2000). In general, we observe similar, if not identical, phenotypes in neural crest cells and their derivatives in ErbB2, ErbB3 and Nrg1 mutant mice. These results provide the genetic proof that Nrg1 signals via ErbB2/ErbB3 receptor heteromers in neural crest cells. It should be noted that Nrg1 can also signal via the ErbB2/ErbB4 receptor heteromers, for instance during heart development, but here only the roles of Nrg1/ErbB receptors in neural crest cells will be discussed.

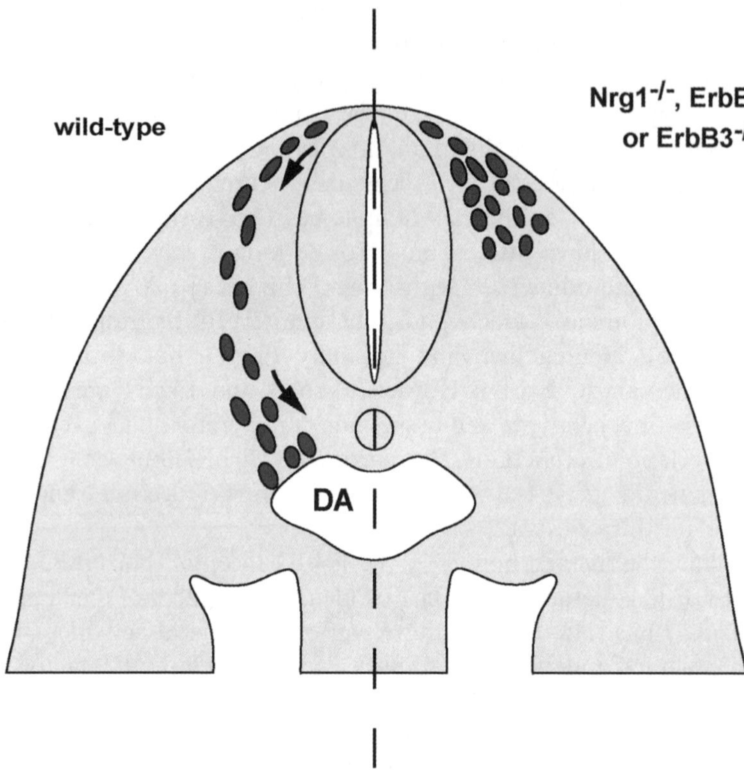

Figure 20.1 Migration of sympathogenic neural crest cells in wild-type and Nrg1, ErbB2 or ErbB3 mutant embryos. In control embryos, neural crest cells (red) are released from the dorsal neural tube and migrate in a ventral direction towards the dorsal aorta (DA), where they form the primary sympathetic ganglion chain (left). In Nrg1, ErbB2 or ErbB3 mutant embryos, sympathogenic neural crest cells stop their migration prematurely and are stuck at dorsal positions (right)

Nrg1, the ErbB2/ErbB3 receptors and migration of neural crest cells

The best evidence for a role of Nrg1 in migration of neural crest cells stems from a careful analysis of the neural crest cell population that forms the sympathetic nervous system (Britsch *et al.*, 1998). Sympathogenic neural crest cells of the trunk migrate along the medial path in the embryo to the mesenchyme lateral of the dorsal aorta (Figure 20.1), where they form the primary sympathetic ganglion chain; the differentiation into sympathetic neurons at this site requires signals provided by members of the BMP family (Shah *et al.*, 1996; Reismann *et al.*, 1996). In mice with mutations in Nrg1, ErbB2 or ErbB3, sympathogenic neural crest cells in the trunk migrate in an abnormal manner, i.e., the cells get stuck in dorsal positions, where dorsal root ganglia will form, and ventral migration is impaired. Consequently, the sympathetic nervous system of the trunk is severely hypoplastic. A particular Nrgl isoform, the type I isoform, provides the signal required for the migration of sympathogenic neural crest cells, and is expressed at the origin, along the migratory path and at the migration target of sympathogenic neural crest cells. The survival of neural crest cells during migration to the anlage of the sympathetic nervous system is not affected by the Nrgl, ErbB2 or ErbB3 mutations. The observed changes can therefore be attributed to an altered motility or directional migration of the neural crest cells (Britsch *et al.*, 1998).

Schwann cell precursors are also impaired in their development in Nrgl, ErbB2 or ErbB3 mutant mice. At E10 in the mouse, such glial progenitors can be observed to line peripheral axon bundles, but they do not yet express markers seen in Schwann cells, and instead express genes characteristic of neural crest cells. Thus, the cells are identifiable as Schwann cells progenitors only through their position, the association with the outgrowing axons. In Nrgl, ErbB2 or ErbB3 mutant mice, numbers of neural crest cells that line peripheral axons are already severely diminished at E10. The mechanism by which this arises has not been characterized in detail. Later during development, Nrgl plays a prominent role as a growth and survival factor for Schwann cell precursors. However, whether aberrant growth/survival is the mechanism that underlies the changed development of gliogenic neural crest cells in Nrgl, ErbB2 or ErbB3 mutant mice is less clear. Defective migration, i.e., an inability of neural crest cells to move out of dorsal root ganglia and to migrate along axon bundles, may be responsible or, at least, contribute. The isoform that is essential for the development of gliogenic neural crest cells corresponds to type III Nrgl, which is produced by neurons and inserted into the axonal membrane. Axonally presented Nrgl stimulates thus the development of gliogenic neural crest cells (Meyer *et al.*, 1997).

Sox10 controls the expression of ErbB3 during development of neural crest cells

In wildtype mice, Sox10 and ErbB3 are expressed in similar patterns in neural crest cells. We therefore tested whether the two genes interact genetically (Britsch *et al.*, 2001). This became possible since a spontaneous mutation of Sox10, the Sox10^{Dom} allele, was identified in mice (Southard-Smith *et al.*, 1998). The frameshift mutation present in the Sox10^{Dom} allele allows the production of a mutant mRNA that can be detected by *in situ* hybridization, which does however not allow the generation of a functional Sox10 protein. When ErbB3 is used as a probe to detect neural crest cells in Sox10 mutant mice, a marked downregulation of ErbB3 expression is apparent. Other genes expressed by the neural crest cells, like $p75^{NTR}$, Sox10 or Cadherin-6, are not downregulated. Thus Sox10 controls the expression of ErbB3 in neural crest cells. In accordance, all phenotypes present in ErbB3 mutants are also present in Sox10 mutant mice; e.g., the sympathetic nervous system is severely hypoplastic.

Other Sox10 target genes also exist, i.e., Sox10 controls not only ErbB3 expression, and phenotypes exist in Sox10 mutant mice that are not present in the ErbB3 mutants. We discovered one such phenotype: an impaired differentiation of neural crest cells into early Schwann cell precursors in the Sox10 mutant mice. Many genes expressed by neural crest cells are turned off in neurons, but remained expressed in the developing glial cells. Markers that distinguish neural crest cells from newly formed glia in the peripheral nervous system are scarce. However, brain-specific fatty acid binding protein (B-FABP) is not detectable in neural crest cells, but is easily detected in peripheral glial populations. For example, no B-FABP-positive cells along spinal nerves or in dorsal root ganglia are found at E10.5 in mice on lumbar axial levels by immunohistochemistry, but they are abundant at E11.5. Non-overlapping cell populations stain with B-FABP- and neuron-specific antibodies. Several days later, the S100-antigen appears in satellite cells of dorsal root ganglia or in Schwann cells and is co-expressed with B-FABP. B-FABP therefore distinguishes glia from neural crest cells and neurons during early stages of development of the peripheral nervous system. In homozygous Sox10 mutant mice, B-FABP-positive cells are missing along spinal nerves and in the dorsal root ganglia at E11.5, indicating that glial differentiation is affected. This does not reflect simply a delay of differentiation, since B-FABP-positive cells are also absent at E12.5 in the Sox10 mutant mice. However, remaining undifferentiated neural crest cells can be detected at these sites, indicating that the mutation does not simply affect the survival of gliogenic neural crest progenitors. In ErbB3 mutant embryos, B-FABP-positive cells are abundant in the ganglia but are rare along the nerves. Thus, glial cells that express B-FABP can be formed, but their numbers lining the nerves are reduced in accordance with the severe reduction in the numbers of neural crest cells that line developing axon tracts at earlier developmental stages in ErbB3 mutant mice (Britsch *et al*., 2001).

c-Ret and Eph tyrosine kinase receptors and the development of neural crest cells

In addition to the above described function of Nrg1 and the ErbB2/ErbB3 receptors in development of the neural crest analysed in my laboratory, further tyrosine kinase receptors and their ligands have been described by others to affect the development and, in particular, the migration of neural crest cells. c-Ret encodes a tyrosine kinase receptor that recognizes glial cell line-derived neurotrophic factor (GDNF) as a ligand. Mutations in c-Ret cause Hirschsprung disease, characterized by the absence of a portion of the enteric nervous system (for review see Taraviras and Pachnis, 1999). The majority of neurons and glia of the enteric nervous system are derived from the vagal neural crest. Shortly after emigration from the neural tube, these neural crest cells invade the anterior foregut and migrate in a rostro-caudal direction to colonize the remainder of the gut. c-Ret is expressed in the vagal neural crest and in their derivatives, the neurons of the enteric nervous system. Activation of the receptor tyrosine kinase c-Ret is required for development of the enteric nervous system (Schuchardt *et al.*, 1994). Several mechanisms were suggested to account for the observed absence of the enteric nervous system in c-Ret mutant mice. Since apoptotic cell death is enhanced in the foregut of embryos lacking c-Ret, it was originally suggested that c-Ret is required for survival of enteric neural crest cells and their derivatives (Taraviras *et al.*, 1999). However, a recent report demonstrates that also an impaired migration might contribute to the impaired development of the enteric nervous system (Natarajan *et al.*, 2002). Neural crest cells in explants of embryonic intestine migrate towards an exogenous source of GDNF in a c-Ret-dependent fashion. In addition, the rate of migration of neural crest cells in c-Ret mutant mice appears to be impaired. Consistent with a role of GDNF in the migration of enteric nervous system progenitors, GDNF is expressed at high levels in the gut of mouse embryos in a spatially and temporally regulated manner. During invasion of the foregut by vagal-derived neural crest cells, expression of GDNF was restricted to the mesenchyme of the stomach, ahead of the invading neural crest cells. Twenty-four hours later, when neural crest cells colonize the midgut, GDNF expression was upregulated in a more posterior region of the gut. Thus, c-Ret might be required for migration and survival of neural crest cells in the developing gut.

Eph tyrosine kinase receptors and their ligands recognize and provide repulsive cues during axonal pathfinding. Eph-mediated directional cues do not only guide axons, but also neural crest cells. Neural crest cells migrate in a segmental manner, for instance in the trunk when they pass on the ventral pathway through the rostral but not caudal somite. The addition of soluble ephrin-B1 to the developing chick embryo results in a loss of the metameric migratory pattern and in a disorganization of neural crest cell movement. In mice, the mutation of EphB2 results in an abnormal migration pattern of cranial neural crest. Thus, Eph receptor tyrosine kinases and their ligands can guide neural crest cells during their migration and mediate inhibitory activities necessary to constrain the migration to specific territories (reviewed by Drescher, 1997; Wilkinson, 2001).

Development of migrating muscle precursor cells

Muscles of the vertebrate trunk are generated from the dermomyotome, a derivative of the somite (Christ and Ordahl, 1995). A major subclass of hypaxial trunk muscles derives from long-range migrating progenitors. These migrating cells delaminate from the ventro-lateral dermomyotome on particular axial levels, and move as undifferentiated cells to specific targets where they form skeletal muscle. In mammals, the cells migrate to the limbs, the hypoglossal cord and the septum transversum, where they form the muscles of the extremities, tongue and diaphragm. Critical steps in the development of the migrating myogenic lineage are (1) the formation and specification of the precursor pool in the dermomyotome, (2) the delamination of the precursor cells and their migration to the correct target sites, (3) proliferation of precursors at their targets, and, finally (4) the activation of the myogenic programme at the targets by the myogenic regulatory factors. Genes that control general dermomyotome development or the myogenic programme have obviously broad roles in muscle development. However, additional steps are needed to specify development of the migrating lineage. Indeed, mutations that profoundly affect muscle groups that derive from migrating precursors, but not other muscles, have been described.

c-Met, its ligand SF/HGF and the Gab1 adaptor

The c-Met receptor tyrosine kinase was identified because of its oncogenic potential when mutated, and its biochemical properties and signalling capacity have been extensively studied (for a recent review see Furge *et al.*, 2000). Its ligand, scatter factor/hepatocyte growth factor (SF/HGF), was characterized as a factor that dissociates (scatters) cultured epithelial cells, and that stimulates growth of hepatocytes (Birchmeier and Gherardi, 1998). Biologically active SF/HGF is a two-chain heterodimer produced by proteolytic cleavage of an inactive precursor. The larger alpha chain contains four copies of the kringle domain. The smaller beta chain resembles a typical serine protease domain, but lacks enzymatic activity. All known biological functions of SF/HGF are mediated by its interaction with the c-Met tyrosine kinase receptor. Upon SF/HGF binding, the cytoplasmic tyrosine kinase activity of the c-Met receptor is increased, and the receptor is phosphorylated on tyrosine residues. Activation of the signal transduction pathways downstream of c-Met occurs mainly through the multidocking site, a short sequence motif located near the C-terminus of the receptor. Two tyrosine residues (Tyr_{1349} and Tyr_{1356}) located within the multidocking site are phosphorylated upon ligand binding, which play a critical role in the recruitment of several signal transducers and adaptor molecules including Gab1, Grb-2, SHC, PI-3 kinase, PLC-gamma and c-Src (Ponzetto *et al.*, 1994).

Gab proteins comprise a growing family of scaffolding/docking molecules involved in receptor tyrosine kinase signalling (for a recent review see Liu and Rohrschneider, 2002). They contain multiple functional motifs that allow them to mediate interactions with other signalling molecules. Gab1 (Grb2-associated binder 1), the first mammalian Gab cloned, was originally isolated as a Grb2-binding protein and is tyrosine phosphorylated in response to various stimuli (Holgado-Madruga *et al.*, 1996). It was also identified independently as a Met-receptor interacting protein in a yeast two-hybrid screen and as the major tyrosine phosphorylated protein in cells transformed by the Tpr-Met oncogene (Weidner *et al.*, 1996; Fixman *et al.*, 1997). Further biochemical studies demonstrated that Gab1 is also involved in a number of other signalling events mediated by interleukin, interferon, erythropoietin and thrombopoietin receptors. Gab1 contains an N-terminal Pleckstrin homology domain, a central proline-rich domain and multiple tyrosines within binding motifs favoured by various SH2-domain-containing proteins. Phosphorylated Gab1 can bind Grb2, the tyrosine phosphatase SHP2, the p85 subunit of PI-3K, PLC-gamma and Crk (Liu and Rohrschneider, 2002).

c-Met, SF/HGF and Gab1 control delamination of migrating muscle precursors from the dermomyotome

The important role that SF/HGF and c-Met play in the development of migratory muscle precursors was established by a genetic analysis in the mouse (Bladt *et al.*, 1995) (Figure 20.2). Mutation of the SF/HGF or c-Met gene results in complete absence of the muscle groups in the mouse embryo that derive from migrating cells, whereas other muscle groups form. In control embryos, migrating myogenic progenitors are generated from the dermomyotome by an epithelial–mesenchymal transition. The migrating cells cannot be observed in SF/HGF or c-Met null mutant mouse embryos. SF/HGF and c-Met regulate the detachment and emigration of myogenic precursor cells from the dermomyotome *in vivo*, a process that resembles the cellular response that led to the identification of SF/HGF as 'scatter factor' in cell culture. The c-Met and SF/HGF mutations do not interfere with development of the dermomyotome prior to delamination, nor do they impair the establishment of the muscle precursor pool in the dermomyotome (Dietrich *et al.*, 1999). The mutations preclude only the dispersal of the migratory progenitors, which are retained instead in the dermomyotome compartment. Interestingly, during normal development migratory progenitors are generated from the dermomyotome of specific somites only. The tight spatio-temporal control of the emigration

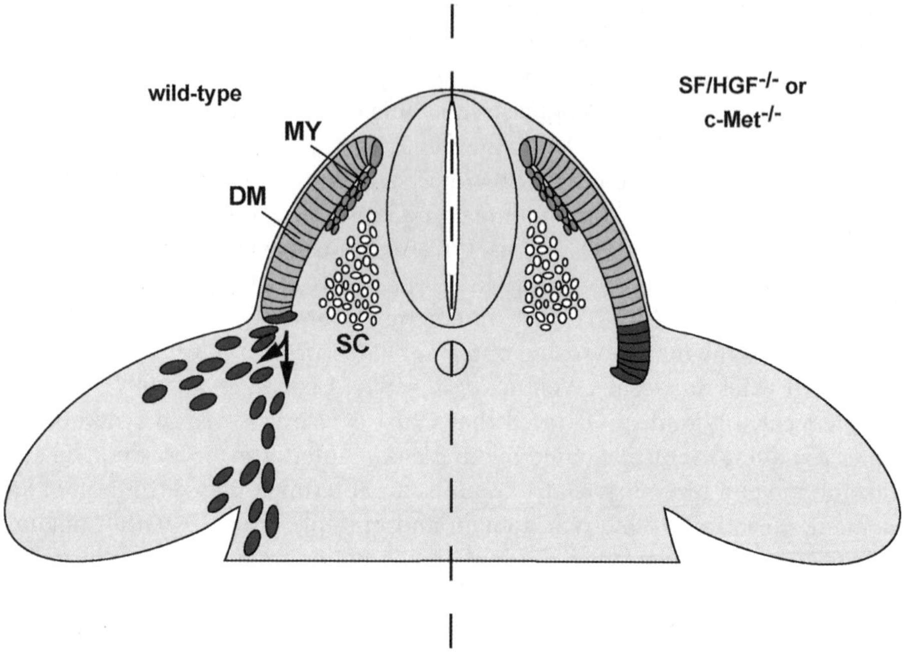

Figure 20.2 Myogenic precursor cells derive from the dermomyotome and are not released in c-Met or SF/HGF mutant embryos. Myogenic precursor cells are generated from the dermomyotome (DM). From the dorsal edge, cells are delaminated that form the myotome (MY). From the ventral edge, cells are released and migrate laterally into the limb, and ventrally towards the anlage of the diaphragm (left). In c-Met or SF/HGF mutant embryos, these cells retain their epithelial appearance and do not delaminate from the ventral dermomyotome. SC: sclerotome

appears to be regulated by restricted expression of SF/HGF, although additional signals could participate. Ectopic application of SF/HGF in the chick embryo induces ectopic epithelial–mesenchymal transitions and emigration of dermomyotomal cells (Brand-Saberi *et al.*, 1996; Heymann *et al.*, 1996).

During migration, hypaxial muscle precursors continue to express c-Met, but SF/HGF expression domains are highly dynamic. After migratory cells have delaminated, SF/HGF transcripts can be observed along the routes and at the targets of migrating muscle precursors. The characteristic expression domains are maintained in the absence of the migrating cells, demonstrating that the factor is produced by the mesenchyme and that it acts in a paracrine manner on the migrating progenitors. In limb mesenchyme, SF/HGF expression is controlled by signals that pattern the limb, which emanate from the apical ectodermal ridge and the zone of polarizing activity. During migration, the c-Met receptor expressing precursors receive a constant SF/HGF signal. The controlled spatial localization of SF/HGF expression indicates that c-Met signalling is of importance during the migration process (Dietrich *et al.*, 1999; Scaal *et al.*, 1999).

A genetic analysis in mice revealed the importance of the adaptor molecule Gab1 for signal transduction of c-Met. Gab1-mutant mice display phenotypes that closely resemble those present in c-Met or SF/HGF mutants. In particular, the development of muscle groups that derive from migrating progenitors is impaired; some of these muscles, like those present in the distal limbs, are completely absent, whereas others like the flexor muscles in the proximal forelimb are formed but are reduced in size. Delamination of muscle progenitor cells from the dermomyotome occurs, but is strongly reduced in efficiency. Compared with control mice, less cells leave the somites in Gab1 mutants, and the precursor stream headed towards the targets contains a reduced number of cells (Sachs *et al.*, 2000).

The homeobox gene Lbx1 is essential for correct target finding of migrating muscle precursor cells

Lbx1 encodes a mammalian homeodomain factor that was discovered because of its homology to the ladybird genes of *Drosophila*. During muscle development, expression of Lbx1 is restricted to the migratory lineage (Jagla *et al.*, 1995). Lbx1 is induced prior to delamination, maintained during migration and downregulated when the cells differentiate. This restricted expression pattern implied a role of the gene in the migrating lineage.

Migratory muscle precursors form and delaminate from the dermomyotome, but migrate in an abnormal manner in Lbx1 mutants (Schäfer and Braun, 1999; Brohmann *et al.*, 2000; Gross *et al.*, 2000). Most strongly affected are those cells destined to move to the limbs. They fail to move laterally towards the limbs and migrate ventrally instead. The misrouted cells accumulate in the mesoderm of the ventral body wall, and their migratory path is similar to that taken by cells that move towards the anlage of the diaphragm. A few progenitors reach the proximo-ventral field of the forelimb, but arrive delayed. As a consequence, most muscles in the hindlimbs and the extensor muscles in the forelimbs are absent, while flexor muscles in the forelimbs are reduced in size. Premature myogenic differentiation can interfere with migration of hypaxial muscle precursor cells, but is not responsible for the misrouting in Lbx1 mutants, since precursor cells are observed at aberrant positions prior to their differentiation. Rather, migrating cells appear to be unable to recognize or to interpret cues that direct them, implying a defective guidance mechanism in the mutants. It was therefore suggested that Lbx1 controls the expression of gene(s) that are essential to recognize or to interpret guidance cues that direct migrating muscle precursor cells to the limb buds.

Eph receptor signals during migration of muscle precursor cells to the limbs

In addition to the above described genes characterized in my laboratory, other researchers have reported that Eph receptors and their ligands affect migration of muscle progenitor cells to the limb. In particular, when muscle precursors delaminate from the dermomyotome and migrate into the forelimb in the chick embryo, the EphA4 receptor tyrosine kinase and its ligand, ephrin-A5, are expressed by muscle precursor cells and forelimb mesenchyme, respectively. Ectopic expression of ephrin-A5 in the presumptive limb mesoderm reduces the number of muscle precursor cells that reach the limb, and the cells congregate abnormally near the lateral dermomyotome. In stripe assays, isolated muscle precursor cells avoid substrate-bound ephrin-A5 and this avoidance is abolished by addition of soluble ephrin-A5. Thus, ephrin-A5 appears to restrict the migration of muscle precursor cells to their appropriate territories in the forelimb (Swartz *et al.*, 2001).

Conclusions

In cell culture experiments, ligands of tyrosine kinase receptors with oncogenic potential frequently affect not only growth but enhance cellular motility and act as chemoattractants. The genetic analysis of such receptors and their ligands in mice has demonstrated that they also control migration processes *in vivo*. However, these studies cannot distinguish whether the receptors are required for cell motility, or whether they provide directional cues and act as chemoattractants. So far, the best evidence that ligands of tyrosine kinase receptors can provide chemoattractive cues comes from an genetic analysis in *Drosophila*, where migrating border cells in the developing egg are guided by signals provided by the *Drosophila* PDGF/VEGF and EGF receptor (Rorth, 2002).

References

Birchmeier, C. and Gherardi, E., 1998. Developmental roles of HGF/SF and its receptor, the c-Met tyrosine kinase. *Trends Cell Biol.* **8**(10): 404–410.

Bladt, F., Riethmacher, D., *et al.*, 1995. Essential role for the c-met receptor in the migration of myogenic precursor cells into the limb bud. *Nature* **376**(6543): 768–771.

Brand-Saberi, B., Muller, T. S., *et al.*, 1996. Scatter factor/hepatocyte growth factor (SF/HGF) induces emigration of myogenic cells at interlimb level in vivo. *Dev. Biol.* **179**(1): 303–308.

Britsch, S., Li, L., *et al.*, 1998. The ErbB2 and ErbB3 receptors and their ligand, neuregulin-1, are essential for development of the sympathetic nervous system. *Genes Dev.* **12**(12): 1825–1836.

Britsch, S., Goerich, D. E., *et al.*, 2001. The transcription factor Sox10 is a key regulator of peripheral glial development. *Genes Dev.* **15**(1): 66–78.

Brohmann, H., Jagla, K., *et al.*, 2000. The role of Lbx1 in migration of muscle precursor cells. *Development* **127**(2): 437–445.

Christ, B. and Ordahl, C. P., 1995. Early stages of chick somite development. *Anat. Embryol. (Berl)* **191**(5): 381–396.

Dietrich, S., Abou-Rebyeh, F., *et al.*, 1999. The role of SF/HGF and c-Met in the development of skeletal muscle. *Development* **126**(8): 1621–1629.

Drescher, U., 1997. The Eph family in the patterning of neural development. *Curr. Biol.* **7**(12): R799–R807.

Falls, D. L., Rosen, K. M., *et al.*, 1993. ARIA, a protein that stimulates acetylcholine receptor synthesis, is a member of the neu ligand family. *Cell* **72**(5): 801–815.

Fixman, E. D., Holgado-Madruga, M., *et al.*, 1997. Efficient cellular transformation by the Met oncoprotein requires a functional Grb2 binding site and correlates with phosphorylation of the Grb2-associated proteins, Cbl and Gab1. *J. Biol. Chem.* **272**(32): 20 167–20 172.

Furge, K. A., Zhang, Y. W., *et al.*, 2000. Met receptor tyrosine kinase: enhanced signaling through adapter proteins. *Oncogene* **19**(49): 5582–5589.

Garratt, A. N., Voiculescu, O., *et al.*, 2000. A dual role of erbB2 in myelination and in expansion of the schwann cell precursor pool. *J. Cell Biol.* **148**(5): 1035–1046.

Gross, M. K., Moran-Rivard, L., *et al.*, 2000. Lbx1 is required for muscle precursor migration along a lateral pathway into the limb. *Development* **127**(2): 413–424.

Heymann, S., Koudrova, M., *et al.*, 1996. Regulation and function of SF/HGF during migration of limb muscle precursor cells in chicken. *Dev. Biol.* **180**(2): 566–578.

Holgado-Madruga, M., Emlet, D. R., *et al.*, 1996. A Grb2-associated docking protein in EGF- and insulin-receptor signalling. *Nature* **379**(6565): 560–564.

Holmes, W. E., Sliwkowski, M. X., Akita, R. W., Henzel, W. J., *et al.*, 1992. Identification of heregulin, a specific activator of p185erbB2. *Science* **256**: 1205–1210.

Jagla, K., Dolle, P., *et al.*, 1995. Mouse Lbx1 and human LBX1 define a novel mammalian homeobox gene family related to the *Drosophila* lady bird genes. *Mech. Dev.* **53**(3): 345–356.

Le Douarin, N. M. and Kalcheim, C., 1999. *The Neural Crest*. Cambridge University Press.

Liu, Y. and Rohrschneider, L. R., 2002. The gift of Gab. *FEBS Lett.* **515**(1–3): 1–7.

Marikovsky, M., Lavi, S., *et al.*, 1995. ErbB-3 mediates differential mitogenic effects of NDF/heregulin isoforms on mouse keratinocytes. *Oncogene* **10**(7): 1403–1411.

Meyer, D. and Birchmeier, C., 1995. Multiple essential functions of neuregulin in development. *Nature* **378**: 386–390.

Meyer, D., Yamaai, T., *et al.*, 1997. Isoform-specific expression and function of neuregulin. *Development* **124**(18): 3575–3586.

Natarajan, D., Marcos-Gutierrez, C., *et al.*, 2002. Requirement of signalling by receptor tyrosine kinase RET for the directed migration of enteric nervous system progenitor cells during mammalian embryogenesis. *Development* **129**(22): 5151–5160.

Ponzetto, C., Bardelli, A., *et al.*, 1994. A multifunctional docking site mediates signaling and transformation by the hepatocyte growth factor/scatter factor receptor family. *Cell* **77**(2): 261–271.

Reismann, E., Ernsberger, U., Francis-West, P. H., Rueger, D., Brickell, P. M. and Rohrer, H., 1996. Involvement of bone morphogenetic protein-4 and bone morphogenetic protein-7 in the differentiation of the adrenergic phenotype in developing sympathetic neurons. *Development* **122**(7): 2079–2088.

Riethmacher, D., Sonnenberg-Riethmacher, E., *et al.*, 1997. Severe neuropathies in mice with targeted mutations in the ErbB3 receptor. *Nature* **389**: 725–730.

Rorth, P., 2002. Initiating and guiding migration: lessons from border cells. *Trends Cell Biol.* **12**(7): 325–331.

Sachs, M., Brohmann, H., *et al.*, 2000. Essential role of Gab1 for signaling by the c-Met receptor in vivo. *J. Cell Biol.* **150**(6): 1375–1384.

Scaal, M., Bonafede, A., *et al.*, 1999. SF/HGF is a mediator between limb patterning and muscle development. *Development* **126**(21): 4885–4893.

Schäfer, K. and Braun, T., 1999. Early specification of limb muscle precursor cells by the homeobox gene Lbx1h. *Nat. Genet.* **23**(2): 213–216.

Schuchardt, A., D'Agati, V., *et al.*, 1994. Defects in the kidney and enteric nervous system of mice lacking the tyrosine kinase receptor Ret. *Nature* **367**: 380–383.

Shah, N. M., Marchionni, M. A., *et al.*, 1994. Glial growth factor restricts mammalian neural crest stem cells to a glial fate. *Cell* **77**(3): 349–360.

Shah, N. M., Groves, A. K., *et al.*, 1996. Alternative neural crest cell fates are instructively promoted by TGFbeta superfamily members. *Cell* **85**(3): 331–343.

Southard-Smith, E. M., Kos, L., *et al.*, 1998. Sox10 mutation disrupts neural crest development in Dom Hirschsprung mouse model. *Nat. Genet.* **18**(1): 60–64.

Swartz, M. E., Eberhart, J., *et al.*, 2001. EphA4/ephrin-A5 interactions in muscle precursor cell migration in the avian forelimb. *Development* **128**(23): 4669–4680.

Taraviras, S. and Pachnis, V., 1999. Development of the mammalian enteric nervous system. *Curr. Opin. Genet. Dev.* **9**: 321–327.

Taraviras, S., Marcos-Gutierrez, C. V., *et al.* 1999. Signalling by the RET receptor tyrosine kinase and its role in the development of the mammalian enteric nervous system. *Development* **126**(12): 2785–2797.

Weidner, K. M., Di Cesare, S., *et al.*, 1996. Interaction between Gab1 and the c-Met receptor tyrosine kinase is responsible for epithelial morphogenesis. *Nature* **384**: 173–176.

Wilkinson, D. G., 2001. Multiple roles of EPH receptors and ephrins in neural development. *Nat. Rev. Neurosci.* **2**(3): 155–164.

Woldeyesus, M. T., Britsch, S., *et al.*, 1999. Peripheral nervous system defects in erbB2 mutants following genetic rescue of heart development. *Genes Dev.* **13**(19): 2538–2548.

Yarden, Y. and Sliwkowski, M. X., 2001. Untangling the ErbB signalling network. *Nat. Rev. Mol. Cell Biol.* **2**(2): 127–137.

Index

Index compiled by Christine Boylan